EMPIRE OF LIBERTY

Portrait of Thomas Jefferson by Rembrandt Peale, 1805.
Courtesy of the New-York Historical Society.

EMPIRE OF LIBERTY

The Statecraft of Thomas Jefferson

Robert W. Tucker
David C. Hendrickson

NEW YORK   OXFORD
OXFORD UNIVERSITY PRESS
1990

Oxford University Press

Oxford   New York   Toronto
Delhi   Bombay   Calcutta   Madras   Karachi
Petaling Jaya   Singapore   Hong Kong   Tokyo
Nairobi   Dar es Salaam   Cape Town
Melbourne   Auckland

and associated companies in
Berlin   Ibadan

Copyright © 1990 by Oxford University Press, Inc.

Published by Oxford University Press, Inc.,
200 Madison Avenue, New York, New York 10016

Oxford is a registered trademark of Oxford University Press

Library of Congress Cataloging-in-Publication Data
Tucker, Robert W.
Empire of liberty: the statecraft of Thomas Jefferson/
Robert W. Tucker and David C. Hendrickson.
p. cm.   Includes bibliographical references.
ISBN 0-19-506207-8
1. Jefferson, Thomas, 1743–1826—Views on foreign relations.
2. Jefferson, Thomas, 1743–1826—Influence.
3. United States—Foreign relations—1783–1815.
I. Hendrickson, David C.   II. Title.
E332.45.T83   1990
327.73'009'034—dc20   89-25513   CIP

2 4 6 8 9 7 5 3 1
Printed in the United States of America
on acid-free paper

For Judith, Clelia, and Katherine

Preface

THE LATE EIGHTEENTH and early nineteenth centuries were years
of extraordinary upheaval in the Western world. The Wars of
the French Revolution and Napoleon, which with but one brief
interruption lasted from 1792 to 1815, posed formidable challenges to the security and well-being of the American republic.
The great debate over the writing and ratification of the United
States Constitution had scarcely concluded before the country faced
a host of external challenges that threatened to break its still precarious unity and undermine its fledgling institutions. That the nation might steer clear of foreign entanglements and preserve the
advantages of its peculiar geographical position was the fondest
hope of every American; but reality daily thrust problems of foreign policy to the top of the political agenda, where they remained
until the upheaval came to an end and peace was restored to the
Atlantic world.

Throughout this period, American statesmen faced predicaments similar in many respects to those provoked by the violent
upheavals of the twentieth century. The relative importance of
neutrality and the balance of power as objectives of American policy figured prominently in the early national period, just as they
would later do in the debates leading up to the military interven-
tions of the twentieth century. Then, as later, the prospect of war
was seen as the means by which the domestic experiment at home
might be deeply compromised, an idea that pointed toward a pol-

icy of isolation and noninvolvement in world affairs. Pointing in
a different direction were not only external challenges to the physical security of the country, but also certain deeply held economic
and political beliefs. One was that American prosperity (and, for
some, even the integrity of the nation's republican institutions) depended on an open trading system; another was the ideological
sympathy that Americans often felt for the combatants in Europe's
distant battles; and a third was the proclivity to see the challenges
of foreign policy against the backdrop of a moral order of universal significance. For all the momentous changes that have taken
place in the internal life and relative power of the United States
since Jefferson became its first Secretary of State in 1790, the debate over foreign policy continues to revolve around ideas and
alternatives that were given a full hearing in America's youth.

The widespread consciousness among the Founding Fathers that
they were the architects of a great empire may account in part for
the continuity of the debate over foreign policy. The United States,
said Alexander Hamilton, was a "Hercules in the cradle." It was
rapidly advancing, said Jefferson, to "destinies beyond the reach
of mortal eye." Intimations of future greatness and a special mission formed part of the American mind even in the colonial period
and were given a classic expression by Adam Smith at the outset
of the War of American Independence. "From shopkeepers,
tradesmen, and attorneys," he wrote in 1776, the American colonists "are become statesmen and legislators, and are employed in
contriving a new form of government for an extensive empire,
which, they flatter themselves, will become, and which, indeed,
seems very likely to become, one of the greatest and most formidable that ever was in the world."[1] Animated by the thirst for
fame, "the ruling passion of the noblest minds," the Founding Fathers self-consciously spoke to the ages, even if they often cannot
be heard through the harsh din of modern life. In foreign policy,
as elsewhere, their sayings still retain the capacity to illuminate
the predicaments of our own times; and they have a freshness and
a power that will doubtless continue to make them a recurring
source of political wisdom and insight for future generations.

The central figure of early American diplomacy was Thomas Jefferson. During the epoch that followed the achievment of American independence, Jefferson occupied positions of critical impor-

tance in the day-to-day conduct of foreign policy—as Minister to France (1784–89), Secretary of State (1790–93), and President (1801–9). The United States, he believed, was the bearer of a new diplomacy, founded on the confidence of a free and virtuous people, that would secure ends based on the natural and universal rights of man, by means that escaped war and its corruptions. This new diplomacy broke radically, Jefferson thought, from the practices and principles of the old European tradition of reason of state, with its settled belief in the primacy of foreign over domestic policy. That the security and even aggrandizement of the state ought to have priority over domestic welfare, and that the actions of the state ought to be judged according to a different moral calculus than the conduct of individuals, were ideas that Jefferson found utterly antithetical to human progress and enlightenment.

In constructing his "empire of liberty," Jefferson pursued ambitious ends—preeminently, territorial expansion and commercial reformation. In securing them, however, he was determined to dispense, so far as was possible, with the armies, navies, and diplomatic establishments that had badly compromised the prospects for political liberty and economic prosperity abroad and would do so at home if ever they became firmly entrenched. The conflict between an ambitious program in foreign policy and the rejection of the military and financial system characteristic of the Great Powers of Europe went to the core of Jefferson's dilemma in foreign policy. He resolved it by placing an inordinate degree of faith in his ability to "conquer without war"—to secure the objectives of the United States by economic and peaceable means of coercion.

Jefferson's road to the new diplomacy was not a direct one. In the 1780s, his outlook had been hardened by several years of war and was in many respects quite traditional. Only in the course of the subsequent decade and in the context of his bitter confrontation with Alexander Hamilton did Jefferson's new diplomatic outlook reach maturity. The rivalry between these two remarkable figures became the focal point not only of the emerging party system at home but also of two sharply different approaches to foriegn policy. The conflict between them has echoed throughout American history. Early in the twentieth century, the argument between Woodrow Wilson and the "neo-Hamiltonians"—Theodore Roosevelt, Henry Cabot Lodge, and Elihu Root—revived in signal respects the primordial dispute between Jefferson and Hamilton, while the debate among the Founding Fathers over the nature and operation of the balance of power is an almost perfect forerunner of the post-World War II dispute on the same subject. Differing systems of political economy, rival conceptions of the rights and duties imposed by the law of nations and the "dictates of national morality," varying ideas of the nature and necessity of military establishments—all these and more fueled the ferocious quarrel between the two men. The very purpose and meaning of the country's existence were thrown into the contest, which has never yielded a clear victor. Jefferson, of course, emerged triumphant from the immediate struggle, as he swept into office amid the ruins of a divided Federalist party, and he certainly evokes today a deeper strain of appreciation in the popular mind. Among historians, however, the contest has been more equal, as one or the other has seemed to successive generations to embody the distinctive merit or debility of the American approach to politics.

Jefferson's rejection of the diplomacy of the old regime pushed him in two very different directions. One part of him wished genuinely to reform the world, another feared contamination from it. Indeed, his own career as a statesman may plausibly stand as a reminder of the oft-remarked cyclical character of American foreign policy—that is, its alternation between interventionist and isolationist moods and policies. In either guise, Jefferson linked American foreign policy and the cause of human freedom in ways that would persist well beyond his death.

In the historic debate over American foreign policy, a consensus seldom existed over the relationship between liberty and the nation's diplomatic stance. Two broad views emerged. One side of the debate, well rooted in the sayings of the great Virginian, has sought to advance the cause of liberty by remaining separate from the world. Only by avoiding the enormous economic and constitutional threats produced by entangling alliances and foreign wars would America fulfill its mission as an example or asylum for oppressed peoples everywhere. The other side, which appears to have achieved an ironic fulfillment in our own time, has taken for granted that free political and economic institutions would flourish in America only if they took root elsewhere, an idea that has in turn underlain much of the crusading impulse in this century. That impulse, of which there are today both right- and left-wing variants, might also find support in Jefferson's thought, for alongside his isolationism lay the conviction that despotism meant war. On this view, the indispensable condition of a lasting peace was the replacement of autocratic regimes by governments that rested on consent. Neither perspective in this great debate is without its compelling arguments and irrefutable examples, though both have more than once been associated with dangerous excesses of wasteful intervention and precipitous withdrawal. Both for their inherent attractiveness and innate hazards, neither perspective is likely to ever gain a final victory over the other in American political thought.

Washington, D.C.   R. W. T.
Colorado Springs, Colo.   D. C. H.
June 1989

Acknowledgments

W E IMAGINE THAT nearly all others who have labored in the diplomatic history of the early republic have felt a keen awareness in working in the long shadow cast by Henry Adams and his masterpiece, *History of the United States during the Administrations of Jefferson and Madison*. We are much indebted to Adams, as the following pages will attest. We have also profited greatly from the works of subsequent generations of American historians—too numerous to mention here—and we hope that the notes and bibliography adequately convey our debt to them.

Support for this project was generously provided by the Bradley and Olin foundations; the book could not have been completed without the assistance they gave. To the librarians in Washington and Colorado who helped us, and to the educational institutions to which we are respectively attached, we also extend our thanks.

The readers and editors for Oxford University Press made a number of valuable suggestions in their review of the manuscript, as did Nicholas X. Rizopoulos. Mrs. Maria Valle typed part of the manuscript and organized the computer files for it; Marj Billings helped with the preparation of the bibliography and in giving order to piles of research. We thank all these individuals for their respective contributions. Among the students who worked with us on the project, we would like to single out the efforts of Theodore

J. Craig, whose indefatigable industry and helpful counsel made this a better book than it would otherwise have been.

All this help notwithstanding, we suspect that many errors of interpretation remain. We need hardly add that they are the responsibility of the co-authors.

Contents

I

AN AMERICAN
STATESMAN

1

The Man and the Nation

AMONG THE FOUNDING GENERATION of American statesmen, none
seems more elusive than Thomas Jefferson. That his was a
life of paradox is notorious. A Virginian nationalist, a slave-holding
philosophe, an aristocratic democrat, a provincial cosmopolitan, a
pacific imperialist—the paradoxes, it seems clear, are of no ordinary variety, reaching beyond the life of one man. They are as
large in meaning and as portentous in significance as America itself, and it is no surprise that his biographers, almost despite
themselves, have had no choice but to travel the formidable road
of the general interpretation of American history; or that Henry
Adams, in his monumental history of early America, should have
made Jefferson the central figure in his tale, the key to unlocking
the riddles of our national existence. "Almost every other American statesman," Adams held, "might be described in a parenthesis.
A few broad strokes of the brush would paint the portraits of all
the early Presidents with this exception, and a few more strokes
would answer for any member of their many cabinets; but Jefferson could be painted only touch by touch, with a fine pencil, and
the perfection of the likeness depended upon the shifting and uncertain flicker of its semi-transparent shadows."[1]

Jefferson's "peculiar felicity of expression," his propensity to
make of every particular controversy the raw matter for a general
theme transcending the immediate occasion, his genius at reducing
a tangled problem of foreign diplomacy or domestic policy to an

3

expression of such simplicity and elegance that it immediately fixed itself on the popular mind and became the "tocsin of party"—all this gave his writings an importance in subsequent political conflicts comparable to no other American statesman. That his writings might be invoked on every side of a given controversy added to the uses of the Jeffersonian past, and all the great conflicts of nineteenth-century American history—over slavery, union, and democracy—found partisans on either side appealing to the "sagacious aphorisms and oracular sayings" of the great Virginian.[2] That the real man was lost in the process became the ritual complaint of his twentieth-century biographers, who have done much to restore the original contours of his life and thought; yet still the mystery remains.

In lifting the veil from Jefferson's life, his modern biographers have been attracted to the theme of practical idealism, though it, too, in its marrying of apparent opposites, is suggestive of ambivalence and contradiction. Noting that Jefferson's opponents in his own day were often puzzled to decide whether he was a hopeless visionary or a shallow Machiavellian—a "misguided child of light or a demonic child of darkness"—Merrill Peterson argues that Jefferson, in truth, was neither. On this view, he was the consummate pragmatic statesman, "actively seeking realizable goals within the limits of principle."[3] Jefferson, notes Dumas Malone, not only placed the diplomacy of the young republic on an unusually high plane, but also displayed "high technical competence," "diplomatic skill," and a "basic realism in international matters."[4] Practical idealism was also the main theme of Gilbert Chinard's earlier biography of the Virginian. Chinard's Jefferson had "neither the pure and exalted morality of the political philosopher" nor the "cynical attitude of the political 'boss.' " He was not "a mere idealist, nor simply a practical politician." Rather, he made "persistent efforts to propagate that gospel of practical idealism that remains to this day one of the fundamental tenets of Americanism. In that respect," said Chinard, "party lines count little, and Lincoln was quite as much a disciple and a continuator of Jefferson as Woodrow Wilson." This combination of pragmatic and idealistic qualities was thought by Chinard to be the essence of the American character; in this light Jefferson stood as "the apostle of Americanism."[5]

It is not his practical idealism alone that makes Jefferson appear as the prototypical American among the founding generation. Jefferson was closely attuned to popular sentiment; and whereas most other leading men of his generation saw in democratic government an experiment to be limited by precautions and artificial barriers of every description, Jefferson himself imagined the popular mind to be in conformity with his own, and saw no danger in trusting its "natural integrity and discretion." The thought of his political rivals among the Federalists, by contrast, seems vaguely un-American. John Adams, with his profound debt to the traditions of English constitutionalism, and Alexander Hamilton, with his instinctive grasp of the classical tradition of European statecraft, may excite our admiration, or even be judged superior to Jefferson in, respectively, theoretical insight and diplomatic acumen. But of both it might be said, what Hamilton said of himself, that they were not made for this American world.

Among the leading men of his own party—James Madison, Albert Gallatin, and James Monroe—only Madison can stand in historical significance alongside Jefferson himself, and none approaches the third president in popular imagination. All three played an important role in Jefferson's presidency: Madison as Secretary of State, Gallatin as Secretary of the Treasury, and Monroe as the roving ambassador who, save for the triumph of Louisiana, was fated to suffer manifold defeats in London, Paris, and Madrid. Monroe seems clearly to be the lesser figure: he shifted too much with the political wind and tended to take on the coloration of his environment. Madison, by contrast, was made of sterner stuff. He played a critical role in the conduct of Jefferson's diplomacy, often supplying a clear direction when the President himself was racked by indecision. After Madison's break with Hamilton, which followed soon upon the formation of the new national government created at Philadelphia, the thinking of the two Virginians—lifelong friends and neighbors—ran along almost identical paths. Yet the identity was one of substance, not of style. Madison was a systematizer, Jefferson a phrase-maker; Madison, once he had fixed upon a position, was loathe to give it up, whereas Jefferson, whose thought was more languid, would flirt with alternatives and often seemed close to abandoning the whole of the Republican system. Madison normally succeeded in bringing him back, though the story is given the element of drama by virtue of the fact that Gallatin was also there, frequently pulling in a different direction.

Indispensable for his comprehension of financial matters, Gallatin functioned both as the conscience of the administration and as the man who brought the President down to the solid grounding of reality when he was most caught up in illusion. Of this triumvirate in Washington—as brilliant a collection of talents, it has been said, as ever graced an American administration—the Geneva-born Pennsylvanian was in many respects the most impressive. It would be impossible to find, in the annals of American history, a purer or more tragic devotion to public duty than Gallatin displayed in 1808, when he threw himself into enforcing the embargo he had come to detest. Yet in popular attachment, then as later, Jefferson stands clearly above his subordinates. Madison cannot be made the subject of myth, and Gallatin is forgotten.

The main source of Jefferson's continuing appeal lies in the facility with which he evoked the meaning of the American experiment in self-government. He thought of America the way we like to think of ourselves, and saw its significance, as we still tend to do, in terms larger than itself. Whether in relation to the domestic experiment at home or the conduct of the republic toward foreign powers, his most characteristic utterance was the contrast he drew between the high moral purpose that animated our own national life and action, and the low motives of power and expediency that drove others. Even the labor of the farm had a meaning beyond the hard drudgeries of existence—of clearing the forest, tending to crops, and providing shelter. To the meanest dirt farmer Jefferson gave the conviction that he—the American—was part of a form of civilization higher than the polished societies of Europe, with their artificial distinctions between social classes, their oppressive restrictions on human freedom, and their crushing burden of debt and taxes. That a republic so constituted should be guided in its foreign policy by the same calculations of power and expediency as animated the states of Europe was unthinkable. Those who felt power and forgot right could teach only negative lessons, by showing the path by which America might become corrupted.

Jefferson's vision was not without its critics, then or later. Europeans found laughable the notion that American civilization, with no art or literature worth mention, and animated in its daily life by the drives of sordid avarice, might be held to represent a higher form of communal life; and European diplomats gave no credence to the view that the external policy of the young republic was inspired by a noble moral purpose. The denunciations did not matter, and barely disturbed the American in his illusions, if illusions they were.[6] America—the "solitary republic of the world, the only monument of human rights, . . . the sole depository of the sacred fire of freedom and self-government, from hence it is to be lighted up in other regions of the earth, if other regions of the earth shall ever become susceptible to its benign influence"—this "hallowed ark of human hope and happiness" involved for Jefferson "everything dear to man."[7]

It is in Jefferson's sense of value that the deepest association exists between his own outlook and the American mind. The institutions that characterize American life today—the standing military establishments, the ballooning debt and high taxes, the whole complex of banks, corporations, and financial markets, the subordinate position of the state governments in relation to national power, the exalted status of federal judicature—all this he would have beheld with a kind of sacred horror, as constituting the victory of Hamilton's vision of American life. Jefferson himself contributed, to be sure, no small measure to the result. Both in the incorporation of Louisiana and in the prosecution of the embargo he abandoned constitutional principle, and thus provided a critical precedent for the consolidation of national power; but he detested his own creation and refused to the end to acknowledge that the government of extended powers that arose from the Republican ascendancy was in any way his own offspring. So in his old age he denounced these encroachments and consolidations, and considered a dissolution of the Union preferable, if the choice be necessary, to "submission to a government without limitation of powers."[8] Uncannily, however, the ideals of American life remain Jeffersonian, even in the midst of all these powerful and corrupting institutions; and we cannot help but turn to Jefferson even with the knowledge that the values he championed can be made a subject of reproach against himself.[9]

Among Jefferson's friendly biographers, few would adopt unreservedly the judgment that Jefferson's career was "guided by the moral principles and purpose set forth in the second paragraph of the Declaration of Independence."[10] The "darker side," in Leonard Levy's phrase, is too notorious.[11] Though Jefferson swore "upon the altar of god, eternal hostility against every form of tyranny over the mind of man," he did not renounce the doctrine of seditious libel, and counseled during his presidency that a few "wholesome" state prosecutions be undertaken against his Federalist enemies.[12] In regard to slavery, Jefferson steadily retreated from the implications of the "self-evident" truths he had proclaimed in the Declaration. Though he trembled for his country when he reflected that God is just, he drew back from the cause of emancipation in his declining years. At death he was able to free only five of his slaves, so enveloped was his estate in debt. His policy toward the Indian contemplated the steady purchase of tribal lands, and he aimed, he said, to teach the red man the civilized arts of cultivation; but when the Indian refused the proffered exchange and clung to the ancient manner of living of his forefathers, all that remained of this policy was expropriation. Jefferson's dark side is also the dark side of American history, and in this, too, the affinity between the man and the nation is complete.

Contemporaries often noted the resemblance between Jefferson and the popular mind, and ascribed it to the Virginian's consuming love of popularity. That Jefferson desired popular approbation is not to be doubted. "It is a charming thing to be loved by everybody," he told his grandchildren in 1802, when he stood high in popular favor.[13] He derived deep satisfaction from his sense that he enjoyed the confidence of the people. It fortified his inner strength; without it the burdens of public life would have been utterly insupportable. He was crushed and mortified when, as Governor of Virginia, resolutions of censure were introduced in the legislature over his conduct of the war; he bristled at the rank ingratitude displayed toward him by some of the people's representatives, and he withdrew into isolation, threatening to abandon public life altogether. The memory of the experience lay behind the reluctance with which he accepted the post of Secretary of State in Washington's cabinet in 1790. He entered "on it with gloomy forebodings from the criticisms and censure of a public, just indeed in their intentions, but sometimes misinformed and misled, and always too respectable to be neglected. I cannot but foresee," he said, "the possibility that this may end disagreeably for me, who, having no motive to public service but the public satisfaction, would certainly retire the moment that satisfaction should appear to languish."[14] Jefferson's forebodings proved to be all too accurate. By 1793, the burdens of his public duties had become too onerous to afford recompense, and he sought for happiness, as he had done previously (and as he would do again in 1809), "in the lap and love of my family, in the society of my neighbors and my books, in the wholesome occupations of my farm and my affairs." There had been a time, he told James Madison, "when perhaps the esteem of the world was of higher value in my eye than everything in it," but now the bargain had become unbearably unequal. "Giving everything I love, in exchange for everything I hate," Jefferson resolved to withdraw, and only in the shade of Monticello did the temptations of public life begin again to exert their hold.[15]

The intensity of the love of approbation is also a national trait distinguishing the American character from that of other peoples, as are the repeated disclaimers of ambition, whether individual or collective. Perhaps this oft-repeated rejection of the lure of power was the merest self-deception. So thought John Adams, who had a ready eye for human frailty in everybody but himself. It is probably closer to the truth, however, to find in the love of approbation, more than the love of power as such, the wellspring of both Jefferson's attraction to and repulsion from the political life, the motor force of that "cyclical rhythm of withdrawal and return" described by Merrill Peterson—a rhythm which, "beyond the limits of consciousness, expressed the deep ambivalence of his nature."[16] For in this motive we can discern much of Jefferson's political method: the public emphasis on conciliation in his First Inaugural, which covered his inextinguishable hatred of the Federalists; his refusal to become involved in partisan Republic quarrels within the states; his desire to stand above party, much as Washington had done. In this motive, perhaps unconscious but no less real, might also be found the reason for the scrupulous care Jefferson exercised as President to avoid displaying before the public the calculations of power and interest that normally fall to the statesman.

The high value that he placed on the esteem of the world does not explain, however, the affinity that existed—and continues to exist—between Jefferson and the American mind. This was not an artificial thing, the product, as his enemies charged, of a calculating and scheming nature, all the more disingenious for being hidden from public view. However much Jefferson's skill as a politician owed to the deceptions that normally accompany the political arts, the correspondence between his own outlook and the popular mind was authentic. That correspondence is revealed not only in Jefferson's social vision, his apotheosis of the agrarian way of life, his skeptical attitude toward government, and his dual commitment to liberty and equality. It is to be found as well in his articulation of an attitude toward the external world and the principles of foreign policy that would persist well beyond his death. Whether as Secretary of State or as President, he responded to nearly every diplomatic issue, great or small, by the invocation of principle that transcended the particular event. Standing over and above these principles, and partly determining their content, was Jefferson's conviction that relations among states ought to be based on the same spirit and principles that govern relations among individuals. "I know but one code of morality for men whether acting singly or collectively." [17] Given repeated expression over the course of his long career in high office, Jefferson never wavered in his verbal commitment to a principle that rejected the diplomacy of the *ancien régime* at the core of which was reason of state.

2

Jefferson and the Diplomacy
of the Old Regime

JEFFERSON IS THE GREAT EXEMPLAR, along with Woodrow Wilson, of the national conviction—so persistent and profound—that we have rejected an ancient reason of state, that we stand for something new under the sun, and that our destiny as a nation is to lead the world from the old to the new. No one gave more fervent and eloquent expression to this conviction than Wilson, and at a time when the nation's star, though already very high, was still rising. But the same conviction was apparent from the outset, as Jefferson testifies, though the circumstances attending it were quite modest. At the beginning of his first term as President, Jefferson wrote to Joseph Priestley: "We feel that we are acting under obligation not confined to the limits of our own society. It is impossible not to be sensible that we are acting for all mankind; that circumstances denied to others, but indulged to us, have imposed on us the duty of proving what is the degree of freedom and self-government in which a society may venture to leave its individual members."[18] America was destined to set an example to the world both in the principles of society it entertained at home and in the policies it followed abroad. "We are firmly convinced, and we act on that conviction," Jefferson declared in his Second Inaugural, "that with nations, as with individuals, our interests soundly calculated, will ever be found inseparable from our moral duties."[19] The foreign policy that faithfully reflected this conviction was bound to be radically different from the foreign policies of the European states.

The nineteenth-century historian of the diplomacy of the French Revolution, Albert Sorel, provided the classic description of the diplomacy that was the object of Jefferson's condemnation. "The final cause in all things was 'reason of state,' " Sorel wrote of eighteenth-century diplomacy,

> which meant the old doctrine of public safety, such as Rome had practiced and had taught to the world. . . . Reason of State directed all policy, and state interest was the only guarantee of any engagements. In other words, no guarantee existed at all. . . . War was the great instrument of rule, the supreme argument of reason of state. It was considered right as soon as it was judged necessary. It was waged to conquer or to preserve, in self defense against an attack or to forestall one.[20]

Sorel's harsh description of the diplomacy marking the *ancien régime* might just as well have been made by Jefferson. If they differed, it was only in the deadly caprice that Jefferson found in the behavior of the European monarchies. "Blessed effect of a kingly government," he once wrote from his ministerial post in Paris, "where a pretended insult to the sister of a king, is to produce the wanton sacrifice of a hundred or two thousand of the people who have entrusted themselves to his government, and as many of his enemies."[21] By contrast, Sorel had concluded that the "implacable realism of reason of state prevailed over all emotions based on honor, religion, and established law."[22]

Neither of these judgments, separated by nearly a century, was without basis. Reason of state had indeed been employed to justify the personal ambitions, lusts, and even whims of monarchs, just as it had been employed to justify measures that, though taken in pursuit of the state's vital interests, went beyond the limits sanctioned by law and ordinary morality. The former justification, however, represented the abuse of reason of state. Although it was true that in the age of absolutism the interest of the ruler became the interest of the state, a distinction was nevertheless drawn between the interests of the body politic and the merely personal interests of the prince. The attempt to equate the latter interests with reason of state was seen as illegitimate and a sign of tyranny.[23] By contrast, extreme measures constituted no abuse of reason of state so long as they were deemed necessary in pursuit of the state's vital interests. For reason of state taught that in conditions of necessity, the statesman was not bound by the constraints of positive law and morality. Instead, his highest duty was to protect and preserve the state's security, independence, and continuity. To carry out this task he might, when necessity decreed, violate legal and moral constraints otherwise applicable not only to individuals but to collectives as well.[24]

The idea of reason of state thus advanced the claim of the state's autonomy. Statecraft had its own rules and their character reflected the means distinctive to political life and, above all, to foreign policy. Statecraft also had its distinctive ends and their nature was determined by the law of state interest. To the idea of the autonomy of the state was added that of the supremacy of the state's vital interests. The political community's security, independence, and continuity took precedence over all other interests, private or public. The supremacy of foreign over domestic policy, of the state and its necessities over civil life, was the inevitable consequence of this ordering.[25]

The origins of the belief that this nation had rejected an ancient reason of state must in the first instance be found in the differences seen to separate republics from monarchies. If America rejected the reason of state that governed European diplomacy, this was initially seen to follow from the fact that this nation was a republic. The logic of reason of state was the logic of monarchies, not of republics. It was the logic of those who found in war the principal outlet for their passions and energies, who made of the "military system" the first principle of government. "Why are not republics plunged into war," Thomas Paine had asked, "but because the nature of their government does not admit of an interest distinct from that of the nation?"[26] Paine's answer might just as well have been given by Jefferson. Only when the decision for war rested on the will of the community rather than on the will of an unrepresentative government would this ever-present specter of the old diplomacy begin to recede and a great step toward permanent peace be taken. Hamilton had inveighed against the view that the "genius of republics is pacific," just as he had objected to the equation of peace with the spirit of commerce. "The causes of hostility among nations are innumerable," he had argued, and they operate independently of forms of government.[27]

The argument found no favor with Republicans. Indeed, it was generally at odds with the progressive thought of the time, and certainly at odds with the thought of the *philosophes,* which Jefferson found so congenial. And if Jefferson, and Madison as well, did not share the extreme view that saw in a world of republics the guarantee of universal and perpetual peace, they, and Republicans generally, did find in the advent of republican governments not only the prospect of a radical decline in the role played by war but the prospect as well of a virtual revolution in the conduct of diplomacy.

The distinctiveness of America's diplomacy was never considered to stem simply from the nation's republican character. That character provided the starting point for the claim to represent something new. From the outset, support for the claim was also found in the diplomatic record of the United States. No chapter in that record has appeared more instructive than Louisiana. For Louisiana seemed a startling vindication of the belief that America was not bound by the necessity that characterized the diplomacy of other states. Almost half a continent had been acquired by peaceful means. Not only at the time but even today the undoubted triumph of Louisiana is seen to represent something greater than either diplomatic skill or luck. Instead, this acquisition stands out as the justification of a diplomacy dedicated to the rejection of the necessity allegedly imposed by a traditional reason of state. The purchase of Louisiana, Jefferson's foremost living biographer has declared, was a "startling idea," one that "could only have arisen in a nation and with an administration determined to settle international disputes without resort to force."[28] Though determined to have Louisiana, Jefferson presumably was also determined to obtain it by means other than those characteristic of a conventional statecraft. Louisiana thus represents the triumph not only of interest but of ideals as well. It is held up as a striking example of a successful diplomacy that nevertheless abandoned the time-honored means of reason of state.

The belief that America had rejected an ancient reason of state was deepened still further by the identification of the nation's fate with the fate of freedom in the world, by the sense that the security and well-being of the United States were inseparable from the prospects of free government everywhere. This outlook, it seems almost redundant to observe, became deeply embedded in the nation's psyche and is today more triumphant than ever. Our purpose is not only seen to distinguish us from others but to give our interests a special character. The vital interests of other states, even of great states, are bounded ultimately by the state itself. But the same cannot be said of the state that stands for the freedom of people everywhere and on whose continued strength and well-being the hopes and future of freedom rest. The equation of America's security and survival with that of freedom in the world has not only given to our statecraft a dimension above and beyond a conventional reason of state but has made the two seem somehow qualitatively different. Jefferson was not the only early American statesman who articulated this credo. Yet none gave it more eloquent and enduring expression than he did.

Did Jefferson in fact reject a traditional reason of state? Certainly he appeared to reject the three claims that formed the basis of the doctrine: that statecraft constituted an autonomous realm governed by its own rules, that the vital interests of the state were supreme over the interests of civil society, and that the restraints of legality must give way before necessity. In place of the dual standard of reason of state, he had "but one system of ethics for men and for nations—to be grateful, to be faithful to all engagements under all circumstances, to be open and generous, promoting in the long run even the interests of both."[29] Against the assertion of the primacy of foreign over domestic policy, he insisted that the objectives of foreign policy were but a means to the end of protecting and promoting goals of domestic society, that is, the individual's freedom and society's well-being. "However . . . we may have been reproached for pursuing our Quaker system," Jefferson once explained to a European correspondent, "time will affix the stamp of wisdom on it, and the happiness and prosperity of our citizens will attest its merit. And this, I believe, is the only legitimate object of government, and the first duty of governors, and not the slaughter of men and devastation of the countries placed under their care, in pursuit of a fantastic honor, unallied to virtue or happiness."[30] And as against the view that necessity might override legal obligation, Jefferson held that the chief security of liberty lay in the iron constraints of a written constitution. The public confidence in the rulers prescribed by reason of state, he held, was the parent of despotism: "Free government is founded in jealousy, and not in confidence; it is jealousy and not confidence which prescribes limited constitutions, to bind down those whom we are obliged to trust with power. . . . In questions of power, then, let no more be heard of confidence in man, but bind him down from mischief by the chains of the Constitution."[31]

There are moments in Jefferson's life when his rejection of reason of state appears final and irrevocable, when the whole soul of his political existence seems directed against state power and its insidious auxiliaries. That rejection went deeper than an insistence on a strict construction of the Constitution and a belief that to take a single step beyond those limited grants of authority was to open wide the field to unbounded power and ambition. He rejected, in fact, the whole apparatus of the modern state that had emerged in Europe in the eighteenth century. The combination of funded debt, executive power, burdensome taxation, government-supported manufactures, and standing military establishments which characterized the Great Powers in the eighteenth century (and all of which characterize the modern state today, only more so) were thought by Jefferson to constitute the very essence of tyrannical government. In the Jeffersonian scheme of things, America was to be different. It was meant to escape the corruptions of Europe, and could not do so if it succumbed to the blandishments of the power state. When Hamilton threw his support behind measures designed to bolster the state in the approved European manner at the outset of the Washington administration, Jefferson detected a design to subvert the Constitution and abort the American experiment in liberty. To this supposed project, and to Hamilton's associated plans to establish a national bank, secure bounties for domestic manufactures, and expand the army and navy beyond the merest of constabulatory functions, Jefferson was passionately and systematically opposed.

These considerations go far toward explaining Jefferson's attitude toward war, that great instrument of reason of state. War, and the necessities that were regularly alleged to attend its conduct, constantly threatened the very institutions and values in defense of which it might be waged, and above all for republics. Hence, for Jefferson, and for Republicans generally, war was the great nemesis. "Of all the enemies to public liberty," James Madison wrote in *Political Observations*, an essay that stands out as a classic expression of the Republican credo, "war is, perhaps, the most to be dreaded, because it comprises and develops the germ of every other. War is the parent of armies; from these proceed debts and taxes; and armies, and debts, and taxes are the known instruments for bringing the many under the domination of the few."[32] This outlook did not prevent the Jeffersonians from embracing what would now be called police actions against either Indians or Algerines; it did, however, make the prospect of war with the Great Powers—preeminently, Britain and France—something to be avoided save in the worst extremity. War would introduce into the republic all the elements of its corruption: debt, taxes, standing armies, artificial privileges of all kinds, ultimately an enlargement of executive power that would lead to the reintroduction of monarchy. This was Jefferson's deepest instinct, and though he drew close to war on many occasions throughout his life and indeed sometimes wished dearly that it might come to purge his rage against foreign transgression, in the end he nearly always drew back.

If doubt must nevertheless persist that Jefferson rejected the central contentions of a traditional reason of state, it is not only because he employed most of the means characteristic of the old statecraft but also because of the ambitious objectives that he had for the United States. The means that he adopted often conflicted in fact with the principles he championed and seem on occasion scarcely reconcilable with gratitude, honest dealing, and a strict observance of legal obligations. The ends that he embraced, moreover, were scarcely modest. The belief in the necessity of territorial expansion, facilitated by the conviction that no constitution was as well suited as America's for "extensive empire and self-government"; the principle of freedom of commerce, which sought to free trade from the shackles of mercantilism; the attachment to neutral rights, or freedom of commerce in war, which aimed at mitigating the hardships that war imposed on neutral states while extending the benefits war brought to them; finally, the idea of the two hemispheres, which found its way into the Monroe Doctrine and declared the American continents beyond the reach of European war and despotism—all these objectives run throughout Jefferson's career as an American statesman. They constituted an imposing edifice; they described a structure of power, which, if realized, would place few obstacles in the way of the territorial and commercial expansion of the United States. What Montesquieu and Frederick the Great said of the diplomacy of the *ancien régime* might also be said of Jefferson's. In both cases, "the fundamental rule of governments" was "the principle of extending their territories."

3

"Conquering Without War"

THE GREAT DILEMMA of Jefferson's statecraft, then, lay in his apparent renunciation of the means on which states had always ultimately relied to ensure their security and to satisfy their ambitions, and in his simultaneous unwillingness to renounce the ambitions that normally led to the use of these means. He wished, in other words, that America could have it both ways—that it could enjoy the fruits of power without falling victim to the normal consequences of its exercise. He had good reasons for wanting both of these things, because both were indispensable to the realization of his vision of the American future. Both sprang from his vision of American society and reflect a classic instance of the primacy of domestic policy. But to pursue them together created a dilemma with which he wrestled throughout his presidency, one that forced him to articulate and ultimately to employ a new diplomatic method, sharply opposed to the classical statecraft of a Hamilton. "To conquer without war"—that objective, the French diplomat Turreau observed in 1805, was "the first fact" of Jeffersonian politics, and so it was.[33] Jefferson was not the first statesman to think of conquering without war, but he was the first to take the thought seriously and to seek to put it in practice. How he might do so constituted the essence of his problem in dealing with the external world.

The threats of war and alliance represented one way of conquering without war, and Jefferson resorted frequently to these

18

altogether traditional means of reason of state. He was never averse
to securing his aims by conjuring up before his adversary a diplomatic combination on the enemy's opposite flank, and he often
threatened war against Spain. But this was playing at the devil's
game, and he knew it. If forced to make good on either threat, he
understood that his enemies might have the "consolation of Satan
in removing our first parents from Paradise."[34] The core of his
diplomatic method lay elsewhere, in the instrument of "peaceable
coercion." By this he meant an ordering of American economic
relations that would leave the mercantilist states of Europe no choice
but to succumb to American demands. Premised on the great advantages the European powers derived from their commerce with
America, he thought that by mere domestic legislation he could
work his will upon other powers, despite their hostility. Discuss-
ing in May 1797 the prospect that a territorial exchange would
"give us new neighbors in Louisiana (probably the present French
armies when disbanded)," which in turn would open up "combinations of enemies on that side where we are most vulnerable,"
Jefferson reiterated his faith in the commercial propositions that
Madison had made earlier in the decade. Had they been adopted,
he told Thomas Pinckney, "we should at this moment have been
standing on such an eminence of safety & respect as ages can
never recover. But having wandered from that, our object should
now be to get back, with as little loss as possible, & when peace
shall be restored to the world, endeavor so to form our *commercial* regulations as that justice from other nations shall be their
mechanical result." That war was "not the best engine for us to
resort to," nature having "given us one *in our commerce*," became
his great article of faith. And it was of such potency that it could
contend as well with French armies on the Mississippi as with
British depredations on the high seas.[35]

The promise of Jefferson's statecraft was thus of a new diplomacy, based on the confidence of a free and virtuous people, that
would secure ends founded on the natural and universal rights of
man, by means that escaped war and its corruptions. He himself
never employed the term "new diplomacy," perhaps because the
French had already made use of it. But that it was in substance
new, that it constituted a radical break from the traditions of Old
World diplomacy—of this he was utterly persuaded. America stood
against force in two respects. Unlike other states, it did not make

the fatal confusion between might and right. In its stance toward the external world, no less than in its constitutional order at home, it stood on principles that were freely accessible to all men through reason, and which did not depend for their validity on local circumstances and particular interests, still less on raw power. These principles were to be validated, in turn, by a new diplomatic method that was coercive yet eschewed force. The traditional view, shared by Hamilton, had been stated most pithily by Frederick the Great: diplomacy without armaments, he had held, was like music without instruments. Jefferson meant to show the falsity of this doctrine, and he would ultimately come to stake nearly all of his statecraft on the demonstration.

To the old question raised by reason of state—whether domestic or foreign policy was primary—Jefferson thus gave a new resolution, one that differed significantly from those that had been proffered in the debates over the ratification of the Constitution. Patrick Henry, in the Virginia ratifying convention, had stated the case of the anti-Federalists most bluntly: "If we admit this consolidated government," he warned, "it will be because we like a great, splendid one. Some way or other we must be a great and mighty empire; we must have an army, and a navy, and a number of things. When the American spirit was in its youth, the language of America was different; liberty, sir, was then the primary object." [36] Faced with a similar objection in the Federal Convention, Hamilton had condemned its lack of realism: "It has been said that respectability in the eyes of foreign nations was not the object at which we aimed; that the proper object of republican Government was domestic tranquillity & happiness. This was an ideal distinction. No Governmt. could give us tranquillity & happiness at home, which did not possess sufficient stability and strength to make us respectable abroad." [37]

Jefferson was not alone among the Founding Fathers in wanting to have both empire and liberty, and in thinking that he could have the one without sacrificing the other. But the terms on which he proposed to resolve this ancient problem were indeed novel, and therein lay the distinctiveness and originality of his statecraft. His imperial aspirations, as reflected in the objectives he embraced on both sea and land, were ambitious—more expansive, indeed, than Hamilton's—yet his distrust of power and its auxiliaries rivaled that of the most confirmed anti-Federalist. To realize his objects, to be sure, required some degree of national consolidation. Even more, it meant rejecting the old teaching that republics had to remain small if they were to retain their virtuous character. But his empire was one of liberty and not of power; it did not require "an army, and a navy, and a number of things." It would be strong because it had the confidence of a virtuous citizenry, because in such a government every man "would fly to the standard of the law, and would meet invasions of the public order as his own personal concern."[38] It could rely in its foreign policy on the great need that other powers had of its friendship and commerce. But it would have few if any of the institutions and establishments that distinguish the modern state today, which Jefferson identified not with modernity but with the decay and corruption of Europe, and which he often thought and hoped were tottering on the brink of collapse.

In the presidency of Thomas Jefferson, a diplomatic pattern thus emerged that was something new under the sun. It constituted a novel experiment in statecraft; it was distinctly American. Its basic elements came together in Jefferson's mind well before he reached the presidency, during the formative years he served as Minister to France (1784–89) and as Secretary of State (1790–93). In those years, however, it had no decisive consequences for policy. While he was Minister to France, the national government (such as it was) lacked effective power, and there was a distinct limit to what Jefferson could accomplish at Versailles, however vigorous and artful his diplomacy. As Secretary of State, Jefferson gradually lost influence within the councils of the administration, ultimately losing the battle with Hamilton for the control of Washington's mind. He was in the government but not of it, an increasingly untenable position that left him no alternative but to resign. Still, it was during these years that his outlook hardened and took on the character of a system—a system just as comprehensive as Hamilton's but radically different in its practical consequences and, even more, in its philosophical foundations.

II

THE DEVELOPMENT OF
REPUBLICAN STATECRAFT
(1783–1801)

4

Commerce, Manufactures, and the West

WHEN THE AMERICAN STATES achieved their independence in 1783, they secured a peace treaty from Great Britain much more generous than seemed reasonable to expect only a few years before. Hoping to reconcile the disaffected colonies to their mother country, the Shelburne ministry decided not to contest the American claim to the great interior, bounded on the east by the Appalachian Mountains, on the north by the Great Lakes, on the south by the 31st parallel, and on the west by the Mississippi River. Shelburne, hoping that the settlement of the interior would allow the United States to accept British manufactures in ever greater numbers, believed that it might be possible to enjoy again the substance of empire even while sacrificing the form.[1] There were general expectations that a commercial accord might follow quickly on the heels of a peace settlement. The Americans, hoping for a general relaxation (if not indeed abolition) of trading barriers among nations, expected that they would enjoy most of the commercial advantages that belonged to them as members of the empire while being freed of the restraints imposed by the Acts of Trade and Navigation. The most important advantage they expected to retain was access to the British West Indies, with which they had established an intimate commercial connection under the empire. At the same time, the former colonists would be free from the British practice of confining their trans-Atlantic commerce to the

British entrepôt, allowing the United States to explore the advantages of a direct commerce with continental Europe.[2]

Shelburne's policies, however, were soon reversed, American expectations betrayed. Under the influence of Lord Sheffield, the British moved to reconstitute the Acts of Trade and Navigation and to recreate in some measure the self-sufficient empire that Englishmen had long regarded as the basis of their wealth and power. Canada would begin to supply lumber and foodstuffs to the West Indies, displacing the exports of the United States; those American goods for which there was no adequate Canadian replacement could come to the West Indies only on British vessels. The direct trade between Britain and America—which sent tobacco, rice, iron, and other products to the former mother country in return for manufactures of all descriptions—was also to be reestablished on a basis favorable to Great Britain, and American ships were duly placed under onerous restrictions in British ports. British merchants and Scottish factors returned to the colonies in force after the war, and the old pattern of debt and dependence—particularly in the southern colonies—reasserted itself. Moreover, Spain had always looked with hostility on American pretensions, seeing in them a threat to its own empire in the Americas. Its West Indian ports remained closed to American shipping in the aftermath of war. Even France, with which the United States was bound in perpetual alliance and from which the most in the way of commercial concessions had been expected, sharply restricted American trade with the French West Indies in 1784 in both flour and sugar, and direct trade between the United States and France was exceedingly slow to develop.

These commercial injuries were seen by a great number of Americans to constitute a betrayal of the promise of independence. They were compounded by a series of ominous developments in the interior. Great Britain refused to evacuate its military position along the lakes in the Old Northwest, arguing that the unwillingness or inability of the United States to execute the treaty of peace (particularly with respect to the substantial debts owed to British merchants from before the war) relieved it of any obligations on this score. It flirted with the design of erecting an Indian barrier in the Ohio country and entertained relations with the tribes that, from the American point of view, seemed positively menacing. Spain was also active in the Southwest. It claimed territory far above the 31st parallel (the line agreed to in the Anglo-American peace treaty, to which Spain was not a party); it intrigued with Indian tribes and disaffected westerners in the disputed territory; and it closed the navigation of the Mississippi to American citizens in 1784. Without an outlet to the sea, the settlement of the Ohio and Mississippi valleys would be arrested. Though settlement had progressed rapidly during and after the war, it was concentrated in a narrow salient in the Kentucky country (then still a part of Virginia), and the allegiance of the settlers to the weak government created by the Articles of Confederation could not be taken for granted. Such was the general sense of impotence that John Jay, in 1786, negotiated an agreement with Spain that surrendered the navigation of the Mississippi for twenty-five years in exchange for commercial concessions by Madrid which were of particular benefit to the northern states. The Jay–Gardoqui negotiation badly split the country on sectional lines. Though never ratified, it was an ominous portent. It reflected how far American expectations had been deflated in the few years since the war; it showed how a foreign power might pit one section against another; and it was a depressing reminder of the weakness of the Confederation and the lack of regard shown to it in foreign capitals.

All these developments had special meaning for Jefferson. As a young man, he had been deeply affected by the experience of Virginia before the Revolution; by his belief that the economic nexus between British merchants and Virginia planters was inherently exploitative; and by his determination to free the commerce of his native state—and of America generally—from the pattern that had prevailed before the Revolution and made it center in England. In his harsh indictment of King and Parliament before the war, he had placed much greater emphasis than most colonists on the terrible costs exacted by the British mercantile system. The British, he claimed, "have indulged themselves in every exorbitance which their avarice could dictate or our necessity extort: have raised their commodities called for in America, to the double and treble of what they sold for, before such exclusive privileges were given them, and of what better commodities of the same kind would cost us elsewhere; and at the same time, give us much less for what we carry thither, than might be had at more convenient ports."[3] The redirection of American trade away from Great

Britain he expected to follow naturally from the achievement of American independence, and when the direct trade returned to its prewar channels and American commerce with the West Indies was shut off by manifold restrictions, he believed that it badly compromised the independent status of America.[4]

Jefferson's efforts to reverse this trend while Minister to France were not particularly successful. French merchants and shippers faced numerous difficulties in establishing themselves in America. The linguistic problem alone was a major obstacle, and it was compounded by the customs and tastes of the former colonists, which the English knew well and the French hardly at all—an obstacle that even Jefferson occasionally acknowledged. French merchants were badly hurt in their first foray into the colonies at the end of the war, sustaining major losses in their accounts. France derived substantial revenues from the trade in American tobacco, and freeing it up threatened to disturb long-established relationships within France. Nor were French domestic suppliers of the French West Indies happy with American demands that would deprive them of their protected market. Jefferson knew of all these obstacles but nevertheless believed that, with time and patient diplomacy, they might be overcome. The main obstacle to doing so lay not so much in France as in America. It consisted of the multiple and inconsistent regulations made by each of the states in retaliation against the British restrictions, all of which had little impact in Great Britain. To establish the French trade, Jefferson came to believe, first required depriving the English of their unfair advantages in American ports; this in turn required a national power over commerce. Although there were many roads to union in the critical period before the drafting of the Constitution, for Jefferson the main road lay in the calculation that only through union might the commerce of America be freed from the British yoke.

The threat to the interior also weighed on Jefferson's mind during this period. As a delegate to the Continental Congress, Jefferson had played a major role in drafting the Northwest Ordinance of 1784, which opened to settlement the territory beyond the Appalachians and established the principle that, once a certain level of population had been reached, new states would be formed and incorporated into the Union on an equal basis.[5] He credited all the rumors of British machinations among Indian tribes in the Ohio country, and he was appalled to learn of the terms of Jay's negotiation with Gardoqui: "The act which abandons the Mississippi," he warned, "is an act of separation between the Eastern and Western country. It is a relinquishment of five parts out of eight of the territory of the United States, an abandonment of the fairest subject for the paiment of our public debts, and the chaining those debts on our own necks in perpetuum." Worse, it would inevitably bring on disunion and, ultimately, a war. The settlers beyond the Appalachians would inevitably think that their own interests had been sacrificed. Were they to declare themselves a separate people, it would be absurd to think that the citizens from the eastern states would by force seek to retain them in the Confederacy and thus "cut the throats of their own brothers and sons." No. The more likely result was that the westerners themselves would open by force the navigation of the Mississippi and seize New Orleans, bringing on a war between them and Spain, and raising the "question with us whether it will not be worth our while to become parties with them in the war, in order to reunite them with us, and thus correct our error?" Jefferson thought it likely that "the inhabitants of the U.S. would force their rulers to take the affirmative of that question."[6] These were dark forebodings; they raised the specter of war, with which Jefferson was never comfortable, even when he acknowledged its necessity.

Jefferson's expectation that the adventurers populating the western country could themselves seize New Orleans from Spain indicates the low opinion he entertained of Spanish power, then as later. Indeed, it was not Spanish strength so much as Spanish weakness that worried him. "Our confederacy," he wrote in 1786, "must be viewed as the nest from which all America, North & South is to be peopled." The only real barrier that might stand in the way of the vision lay in the prospect that Spain, the sick man of the Americas, "might be too feeble to hold them till our population can be sufficiently advanced to gain it from them piece by piece."[7] The real danger, in other words, was that Britain or France might take territory in Spanish America before the United States was in a position to do so. This idea would never leave Jefferson. It lay behind the alarm with which he viewed the retrocession of Louisiana to France in the early years of his presidency, and it formed the basis of the "large policy of 1808," in which the Jefferson administration, in its last year, made known to influential persons in Cuba and Mexico its caution over the recognition of

Spanish American independence as well as its "strongest repugnance to see you under subordination to either France or England, either politically or commercially."[8] Ultimately it would form the basis of the Monroe Doctrine. It is interesting at this stage of Jefferson's life because it indicates the extensive character of his ambitions for the United States even at a time when the integrity of American independence itself had come increasingly into question.

The threats to American commerce and territory were related in Jefferson's mind, and not only because the most serious of these threats stemmed from the same source—Great Britain. Securing the great interior and breaking down the restrictions that fettered American commerce were equally necessary if Jefferson's vision of domestic society was to be realized. America could remain uncorrupted only if it remained a largely agricultural society, and it could remain agricultural only if "an immensity of land courting the industry of the husbandman" were available for settlement and if the farmer could find a ready outlet for his surplus produce. Free trade and territorial expansion were the means by which the United States might escape the curse of modernity itself, "expanding through space" rather than "developing through time," and in the process creating a civilization immune to the corruptions of Europe.[9]

His vision of America was thus a kind of a Vergilian paradise magnified a thousand times,[10] shorn of manufacturing establishments and the mobs of great cities that inevitably accompanied them. These added "just so much to the support of pure government, as sores do to the strength of the human body." The cultivators of the earth, by contrast, were "the most virtuous and independent citizens," the "chosen people of God, if ever he had a chosen people," as Jefferson wrote in his Notes on the State of Virginia, "whose breasts he has made his peculiar deposit for substantial and genuine virtue."[11]

Jefferson's thought on the relationship between agriculture, commerce, and manufactures was not without ambivalence. In the 1780s, he flirted with the ideal of total isolation, both political and commercial. Were he to indulge his "own theory," he told a European correspondent eager to know his views on commerce, he would wish America "to practice neither commerce nor navigation, but to stand with respect to Europe precisely on the footing of China. We should thus avoid wars, and all our citizens would be husbandmen." Jefferson quickly added that it was "theory only, and a theory which the servants of America are not at liberty to follow." [12] But the difficulty of strictly pursuing the ideal of isolation rested not merely, as Jefferson claimed, on the "decided taste" of the people for navigation and commerce. In truth, he did not yet know his own mind on the question. On one side lurked the danger of the foreign wars that an extensive commerce might bring on the United States; on the other lay the possibility that the American husbandman would be unable to find an outlet for his produce and would turn to manufactures. Both alternatives were potential sources of corruption, yet one or the other appeared inescapable.

This dilemma would remain with Jefferson throughout his life; it came to perplex and even haunt him in the course of his presidency, as his diplomacy came to be ever more deeply committed to the protection of American neutral rights. By the late 1780s, his thinking had reached at least a temporary resolution, and he threw his support behind the promotion of American commerce. In the event of another European war, he now thought it advisable to "become the carriers for all parties as far as we can raise vessels," thus fattening on the follies of a Europe at war. [13] He insisted to Washington that he understood the dangers of his chosen course. He was "decidedly of the opinion we should take no part in European quarrels, but cultivate peace and commerce with all." Yet who, he asked, "can avoid seeing the source of war, in the tyranny of those nations who deprive us of the natural right of trading with our neighbors?" The language of natural rights recalled the heady struggles of the Revolution, and it was a pointed reminder that the aims of the Revolution had yet to be achieved. To secure them might yet require force of arms. The danger that the products of the United States would soon exceed the European demand was no longer a distant one; it bore immediately upon the foreign policy of the new republic. For what, Jefferson asked, would "be done with the surplus, when there shall be one? It will be employed, without question, to open by force a market for itself with those placed on the same continent with us, and who wish nothing better." It was possible, to be sure, that the United States might avail itself "of the wars of others to open the other parts of America to our commerce, as the price of our neutrality." Still, the danger remained that war might follow inexorably from the commitment to commerce, and the new constitution was valuable for that very reason, for "war requires every resource of taxation and credit." [14]

On the eve of his assumption of the post of Secretary of State, Jefferson had given as much thought to foreign policy as any American. He had a clear idea of what he wanted, but he had yet to resolve in his own mind what he might have to pay in order to get it. Two great objects dominated his outlook. One was security and prosperity of the interior, which required from Spain the recognition of the 31st parallel and the opening of the Mississippi to American citizens and from Britain the evacuation of the Northwest posts in accord with the treaty of peace. The other was "throwing open all the doors of commerce & knocking off its shackles." [15] Access to the West Indian colonies of the European powers was the first desideratum of this policy; breaking free of the English connection by encouraging the direct trade to Europe in American ships was the second. He had embraced these objects, he often said, because the interests of the Confederacy required that the claims of each section be regarded as legitimate and not be sacrificed to the partial advantage of another; but they also conformed to his vision of America as a "great agricultural republic" that would spread out over the continent. Commerce, he knew, was the handmaiden of agriculture: the latter could be encouraged, and manufactures discouraged, only if a vent was given to its surplus productions.

Ideas proliferated in his mind on how these objectives might be achieved. The idea of playing the rival states of Europe against one another and making them bid for American neutrality had already occurred to him; so, too, had the notion of meeting European—and especially British—commercial restrictions with countervailing prohibitions on the part of the United States. But he had yet to discard the idea that war was probably inevitable for America, and he had not yet come to a full realization of what war might mean for the republic. This would come only in the context of his opposition to Hamilton and Federalism, only after he had come to appreciate how war might confirm the Hamiltonian system and bring to America the enlargement of executive and national power that he quickly learned to dread.

5

The Rival Systems
of Hamilton and Jefferson

I T IS ONE OF THE MORE REMARKABLE features of the party rivalry of the 1790s that it seems not to have been anticipated by its great leaders on either side. Hamilton and Jefferson had never met; their backgrounds were vastly different. But both had rendered important service to their country in securing independence from Great Britain, and there is no evidence that either man anticipated any particular problems in working with the other as the two leading members of Washington's cabinet. But disagreement quickly surfaced. Before Jefferson had even assumed his post as Secretary of State (which he did in February 1790), Hamilton had worked to defeat James Madison's proposals in the House of Representatives to introduce discriminatory tariff and tonnage provisions, aimed principally at Great Britain, against nations not in commercial treaty with the United States. A year later, when a clash between Great Britain and Spain over Nootka Sound on the northwest coast of North America threatened to produce a European war, Jefferson and Hamilton differed in the cabinet over the response that the United States ought to give to the expected British request to cross American territory with forces based in Canada in an attack on Spanish New Orleans. The crisis passed, and the request was never made. But it signaled the profound differences that divided the two men in their approach to foreign policy and that would badly split the nation when global war did come in early 1793. By 1791 the quarrel was public, and it grew steadily in ferocity thereafter.

In early 1792, Washington had become so alarmed at the division within his cabinet and within the nation that he asked for an accounting from the two men. But the answers tendered by both showed merely that their differences were irreconcilable, with Jefferson claiming that the history of the Secretary of the Treasury, "from the moment at which history can stoop to notice him, is a tissue of machinations against the liberty of the country which has not only received and given him bread, but heaped it's honors on his head." [16]

Hamilton's system rested on the funding and assumption of the revolutionary war debt, a program of external and internal taxes, the creation of a national bank, and the encouragement of domestic manufactures. In his celebrated reports to Congress on his financial system at the outset of the 1790s, Hamilton argued that each of these elements was indispensable to national power in the modern world. By establishing the credit of the United States on a firm foundation, the United States would restore the confidence of foreign and domestic lenders and enable the United States to borrow immense sums in the event of war. The pool of capital thus created would provide a circulating medium, answering the want of specie in America, and encourage investment in large-scale economic enterprise. The taxes laid to support the program would call forth the powers of the new government while binding creditors to its support. [17]

Hamilton's system had the advantage that it did not require the United States, a weak and fledgling power, to seek to overturn the system that regulated "the general policy of Nations." The prevailing system of exclusion among the mercantilist nations of Europe made it dangerous to depend too heavily on their markets; the alternative was to encourage the development of "an extensive domestic market" based on manufactures and capable of absorbing agricultural surpluses. He acknowledged that there existed a "want of reciprocity" in the treatment of American commerce by foreign nations. So long as the economy rested on agriculture, the United States would not be able to "exchange with Europe on equal terms"; but this he regarded as a fact which American prohibitions and discriminations were powerless to alter. That it might be contrary to the interest of the European powers he readily conceded, but he thought it belonged to them to judge "whether, by aiming at too much they do not lose more than they gain. 'Tis for the United States," he contended, "to consider by what means they can render themselves least dependent, on the combinations, right or wrong, of foreign policy."[18]

The Republican system rested on wholly different assumptions and principles. Rather than adjusting to "the general policy of Nations," Jefferson and Madison sought to overturn it. They nursed from the days of the Revolution the belief that the Anglo-American commercial tie was inherently unequal and exploitative, and they believed that in Great Britain's dependence on American commerce they had a weapon well suited to the task of reformation. In the commerce between the two countries, as Madison put it, Great Britain supplied the United States "chiefly with superfluities"; in return, the trade with America "employs the industry of one part of her people, sends to another part the very bread which keeps them from starving, and remits, moreover, an annual balance in specie of ten or twelve millions of dollars." The mass of English workers that would be thrown out of employment by American nonimportation, alongside the impending starvation of planters and slaves in the British West Indies, reflected the two pressure points where Britain was most vulnerable. In peace, Britain would have no choice but to open its ports and colonies to American goods and shipping; at war, her dependence on the United States would compel a scrupulous respect for neutral rights.[19]

Underlying this view was an intricate theory of what political economists now call "asymmetrical interdependence," in which American strengths were perfectly matched against British liabilities, placing the country in a position of enormous strength.[20] Federalists cautioned that this interdependence worked both ways, and that the United States would feel the dreadful effects of a commercial war far sooner than Great Britain. Whereas Britain's trade with the United States was only about one-sixth of its total imports and exports, "full one-half the value of our whole trade is with Great Britain and her dependencies." Tax revenues depended to a large degree on customs duties, three-fourths of which was derived from commerce with the British.[21] Yet Republicans were confident that in "a contest of self denial" the United States would be sure to prevail, and they were equally confident that such a contest would not lead to war. "Every consideration of interest," Madison held, must prevent Britain from declaring war against the United States. "Would war employ her starving manufacturers? Would war furnish provision to her West India Islands, which in that case must also starve? Would war give employment to the vessels that had formerly imported luxuries to America?" Every counterregulation Britain might offer "would be a stroke against herself."[22]

The threat of "peaceable coercion" was the centerpiece of the Republican system in foreign policy. It promised to avoid war without the sacrifice of American rights; it, far more than any other aspect of Jeffersonian foreign policy, was the means by which Jefferson proposed to overcome the dilemma that had disturbed him in the 1780s, by breaking the intimate connection that had always existed in his mind between commerce and war. Peaceable coercion reflected a curious combination of "republican" and "liberal" principles. America would prevail, in part, because it was more virtuous than corrupt old England, because it could well dispense with the "geegaws" and "luxuries" on whose export to America the prosperity of Britain depended. Though America would be governed by virtue, however, Britain would be governed by interest; its choice in a conflict that pitted its avarice against its pride or ambition was a foregone conclusion.[23]

This is not the only instance where Jefferson and Madison appeared, simultaneously, to embrace "republican" and "liberal" precepts, and it suggests that the tendency among historians to set these ideas in opposition—at least in the realm of foreign policy— is more misleading than enlightening. In their confidence in the efficacy of economic sanctions, in their distrust of standing military institutions, in their common rejection of debt and taxes, in their faith in public opinion, and in their insistence that morality and reason would prevail over the dictates of power, the Jeffersonians articulated an outlook whose roots were republican but whose broad themes became a central part of the liberal tradition in American foreign policy.[24] Opposing that tradition has always loomed the figure of Hamilton, "a collossus to the anti-republican party . . . a host within himself,"[25] whose own outlook provided the main foundation for American conservatism.

In each of these rival systems, foreign and domestic policy were so closely linked that it is difficult, even in retrospect, to tell where one ended and the other began. Just as the Jeffersonian system rested on the expansion of markets for American produce, Hamilton's depended on the creation and mobilization of capital and

the confidence of international creditors. Breaking free of the connection with Great Britain required, for the Republicans, the cultivation of close ties with France. The workability of Hamilton's
system, on the other hand, rested on avoiding a great rupture with
Great Britain. In his efforts to strengthen the American state,
Hamilton accommodated himself to the practice of European nations, for the purpose of rendering American policy less dependent
on events abroad. Jefferson, by contrast, had a vision of domestic
society that led him to reject the European example, but which
depended for its viability on his ability to effect peaceful change
in the external world.

Underpinning each system were profoundly different views of
the nature of republicanism and modernity. Hamilton's comprehensive program of economic development bore remarkable re-
semblances to that which had emerged in England after the Glo-
rious Revolution of 1688, a point well understood by advocates
and critics alike. All of the institutions against which England's
country opposition had railed in the eighteenth century—the "paper system, stockjobbing, speculations, public debt, moneyed interest, etc."—thus made their appearance on the American scene,
where they were denounced in remarkably similar terms.[26] The
fact that members of both houses of Congress purchased government securities was seen in the most ominous terms by the Republicans, who viewed it as the means by which Hamilton maintained
a "corrupt squadron" and a pernicious "influence" over the legislative branch. For Madison and Jefferson, the same financial system that had corrupted the balanced constitution in England was
now being introduced in America, and they fought it with every
weapon at their disposal.[27]

Just as the Republicans felt betrayed by Hamilton's opposition
to retaliatory legislation against Great Britain and saw in his financial system a conspiratorial design to overthrow the Constitution, Hamilton found barely comprehensible the opposition that
Madison and Jefferson made to his measures. The financial institutions he erected represented for him a way of coming to terms
with the requirements of a modern commercial society, and he
utterly rejected the idea that the modern commercial republic might
find its model "in the simple ages of Greece and Rome." "We
might as soon reconcile ourselves," he had once said, "to the Spar-
tan community of goods and wives, to their iron coin, their long

beards, or their black broth."[28] That a national bank was indispensable to the effective management of the government's taxing and spending operations seemed a proposition so obvious that he could not help but see the opposition to it as "subversive of the principles of good government and dangerous to the union, peace and happiness of the Country."[29]

These contrasting understandings of the meaning of republicanism were also in part responsible for the profoundly disparate estimates of national power. Far from seeing England's national debt as a source of strength, the country opposition had always seen in it a symptom of national weakness and decay; so, too, did Jefferson and Madison. Thinking that the whole system of credit rested on a precarious foundation, Jefferson and Madison believed in the imminence of an English financial crisis throughout the 1790s, and when the Bank of England suspended specie payments on the debt in 1797, they believed that the crisis they foretold was at hand.[30] Hamilton, by contrast, was far more impressed with the stability of the English financial system. "If, on the one hand, [Great Britain] owes an immense debt, on the other, she possesses an immense credit, which there is no symptom of being impaired. . . . How long it may last, how far it may go, is incalculable. But it is evident, that it still affords prodigious resources, and that it is likely for some time to come, to continue to afford them."[31] Hamilton saw in the rejection of "English" institutions a perverse misreading of the means by which the nation might become stronger and a failure to grasp the most elementary principles of political economy.[32]

Equally irreconcilable were the views taken of the power granted by the Constitution to the national government. Hamilton argued, on behalf of the bank, "that every power vested in a Government is in its nature *sovereign,* and includes by *force* of the *term,* a right to employ all the *means* requisite, and fairly *applicable* to the attainment of the *ends* of such power; and which are not precluded by restrictions & exceptions specified in the constitution." This principle, he contended, was *"inherent* in the very *definition* of *Government* and *essential* to every step of the progress to be made by that of the United States."[33] In the Republican view, such a broad construction of the constitution would ultimately prove fatal to liberty. In his opinion challenging the constitutionality of a national bank, Jefferson held that "to take a single step beyond the boundaries" of powers specifically delegated to Congress was

"to take possession of a boundless field of power, no longer susceptible of any definition."[34] Just as Congress lacked the power
to charter a bank, so the executive was without authority to issue
a proclamation of neutrality. It made no difference that the osten-
sible purpose of the proclamation was the preservation of peace.
Such attempts to "remove the landmarks of power" would ultimately make more likely the advent of war. It was "morally certain," Madison held, "that in proportion as the doctrines make
their way into the creed of the government, and the acquiescence
of the public, every power that can be deduced from them, will be
deduced, and exercised sooner or later by those who may have an
interest in doing so."[35]

Both systems were profoundly pacific, though their reasons for
avoiding armed conflict differed significantly. Each was wholly
persuaded that the adoption of the adversary's system would lead
to war. Hamilton, believing that the commercial connection between England and America was in no way "unnatural," early on
appreciated the possibility that the Republicans' policy of commercial discrimination might have disastrous consequences for his
financial system. Believing, as he did, that the accumulation of
injuries produced by an Anglo-American commercial war would
in all likelihood end in a real war, he came to fear "that in the
fermentation of certain wild opinions, those wise just and temper-
ate maxims which will for ever constitute the true security and
felicity of a state would be overruled and that a war upon credit,
eventually upon property and upon the general principles of public order might aggravate and embitter the ordinary calamities of
foreign war." No less than the Republicans (albeit for different
reasons), Hamilton feared the domestic consequences of the resort
to force. These constituted, for him, one of the most persuasive
arguments for a pacific policy on the part of the United States.[36]

For the Republicans, war held out the danger of confirming
the enlargement of national and executive power for which Hamilton had boldly contended. It would fasten upon the United States
an immense debt, heavy taxes in perpetuum, and a standing army
whose control would rest in the hands of those with views fatal
to the preservation of the Constitution. "War," as Madison argued in this *Letters of Helvidius,*

> is in fact the true nurse of executive aggrandizement. In war,
> a physical force is to be created; and it is the executive will,
> which is to direct it. In war, the public treasuries are to be

unlocked; and it is the executive hand which is to dispense them. In war, the honours and emoluments of office are to be multiplied; and it is the executive patronage under which they are to be enjoyed. It is in war, finally, that laurels are to be gathered; and it is the executive brow they are to encircle. The strongest passions and most dangerous weaknesses of the human breast; ambition, avarice, vanity, the honourable or venial love of fame, are all in conspiracy against the desire and duty of peace.[37]

Once the Hamiltonian system had revealed itself to Jefferson and Madison, they swiftly jettisoned the limited concessions to military power each had made in the 1780s. In the previous decade, both had supported the establishment of an American navy. The sea, Jefferson argued in the *Notes on Virginia*, "is the field on which we should meet an European enemy." The naval force he then contemplated was not a large one: "To aim at such a navy as the greater nations of Europe possess, would be a foolish and wicked waste of the energies of our countrymen. It would be to pull on our own heads that load of military expence, which makes the European labourer go supperless to bed, and moistens his bread with the sweat of his brows." Circumstances, as it happened, were such that no large navy was required, even as against the maritime powers of Europe: "Providence has placed their richest and most defenceless possessions at our door; has obliged their most precious commerce to pass as it were in review before us." Since the maritime powers of Europe would not risk the whole of their fleets against America, they would attack "by detachment only; and it will suffice to make ourselves equal to what they may detach." Still, what Jefferson was then calling a small navy would have represented a substantial force at sea, and he wished that it would be able to contend not only with Barbary corsairs but with the great navies of Europe. He then thought that around eighteen ships of the line and twelve frigates would be both necessary and sufficient. That was approximately the size of the force for which the Federalists contended in the 1790s, though they never came close to obtaining it.[38]

These ideas were abandoned in the 1790s. In 1790, on the floor of Congress, Madison opposed the creation of a navy. He wanted merely to encourage the development of resources that might be converted into a naval force in a war, and he fell back on the idea that privateers which could disrupt European shipping would be entirely adequate to the task. Jefferson spoke in the same terms in his report on commerce that was submitted to Congress in his final act as Secretary of State: the protection of American commerce required the possession of "a respectable body of citizen seamen, and of artists and establishments in readiness for shipbuilding." But the reference to citizen seamen indicates that Jefferson had moved away from the idea of building a navy composed of ships of the line and frigates and was thinking now simply in terms of privateers. In the course of the decade, as the Federalists threw their support more and more openly behind the construction of ships of the line prepared to do combat with the navies of the European powers, the Republicans hammered away at the enormous expense such a project would entail, the vast amount of patronage it would open up to the executive, and the danger that it would provoke a war on the high seas.[39]

A similar transformation occurred with respect to the establishment of a standing army. Jefferson's experience as Governor of Virginia during the darkest period of the revolutionary war had once convinced him of the inadequacy of militia forces. Militia forces were "expensive to the public" and "distressing and disgusting to individuals." Men called to the militia served for such brief periods that their training and experience were wasted, and they often made off with the firearms with which they had been provided. At this desperate juncture, Jefferson hoped that the Virginia assembly would consider raising and maintaining a sufficient number of regulars to carry on the war.[40] In the aftermath of the war, he called for "magazines and manufactures of arms," and he wrote often of the virtues of preparedness.[41] Yet by the 1790s a standing army had reemerged as an object of detestation and danger. Though he made no formal opposition to plans to raise a small regular force on the frontier, Jefferson was convinced of the superior utility of militia for Indian warfare, and he consistently opposed Federalist attempts to enlarge the army in reaction to the war scares with European powers that arose during the decade— with England in 1794 and with France in 1798.

Many Republicans had predicted that the government would soon be forced to institute a standing army to enforce their onerous system of taxation, fears that were heightened by the administration's move to crush the Whiskey Rebellion. The Federalists argued, as Jefferson had earlier, that the complexities of the military art required skill in commanders as well as discipline and training among recruits. Militia forces provided neither. By this time, however, the Republicans had become so thoroughly convinced that the standing army desired by the Federalists was to be the instrument of domestic repression that they opposed not only the efforts to create a body of regular soldiers but also all attempts to bring the state militias under national control. By the end of the decade, Jefferson's position was fixed. For internal defense, he proposed relying "on our militia solely, till actual invasion, and for such a naval force only as may protect our coasts and harbors from such depredations as we have experienced." He opposed "a standing army in time of peace, which may overawe the public sentiment"; and he rejected the "expenses and the eternal wars" that a navy would bring.[42]

Jefferson's passion for peace did not prevent him from waving the threat of war before his adversaries; nor did his aversion to becoming embroiled in the wars of Europe disable him from holding out dangerous diplomatic combinations against the powers of Europe. As Secretary of State, he conjured up for Spain a flurry of inevitable wars and overwhelming diplomatic pressures; but the Spaniards hung on to their position in the Southwest, apparently oblivious to the mortal danger that might soon befall them. Jefferson, in truth, had no intention of making war or contracting alliances for his objectives on the Mississippi. In the 1790 crisis over Nootka Sound, the British were to be told that they might have American neutrality "if they will execute the treaty fairly and attempt no conquests adjoining us." The Spaniards, on the other hand, were to be threatened with an Anglo-American alliance unless the navigation of the Mississippi was secured and an entrepôt was granted at the mouth of the river.[43] As these threats canceled each other, it may safely be concluded that Jefferson believed in the utility of threatening wars and alliances but was exceedingly reluctant to resort to either possibility in fact. To do so, he became more and more convinced, would only give a handle to the "monocrats" and "Anglomen," the better to raise armies and increase debts. Only against the prospect that the British would take both Louisiana and the Floridas and thus surround the United States on sea and land did Jefferson's voice take on a note of urgency and peril; under those circumstances, he advised the President, "we ought to make ourselves parties in the *general war* expected to take place, should this be the only means of preventing the calamity" of a British encirclement. But here, too, he wanted to "defer this step as long as possible";[44] as Secretary of State, no less than as President, the advantages of delay and the arts of procrastination were central elements in his diplomatic armory.

For Hamilton, war was an inescapable fact of political life. It was something for which every nation must prepare. War was to be avoided not through the absence of preparedness but through the moderation of diplomatic ambition. His outlook in this respect was thoroughly classical. He thought it "impossible to read the history of the petty republics of Greece and Italy without feeling sensations of horror and disgust at the distractions with which they were continually agitated," and he looked for inspiration to the wise example of Rome in the age of the Antonines—who, as Gibbon said, "preserved peace by a constant preparation for war" and who "were as little disposed to endure as to offer an injury." Unlike Jefferson, Hamilton was never a man for making threats he was unwilling to carry out, and his diplomatic style was always distinguished by a frank statement of the American position. He was temperamentally opposed to bidding for the favors of other nations, as was Jefferson's wont. His system was "ever to combine energy with moderation," and he sought to limit the pretensions of the United States toward foreign powers while strengthening the military and financial power of the country. The disjunction between national ambition and the rejection of the power state so characteristic of the Republicans' outlook was tempered on both ends in Hamilton's system.[45]

In the early numbers of *The Federalist,* Hamilton had rejected the idea that republics were in practice less addicted to war than monarchies. "Are not the former administered by *men* as well as the latter? Are there not aversions, predilections, rivalships, and desires of unjust acquisitions that affect nations as well as kings?"[46] By the 1790s, by contrast, the Republicans had come to endorse fully the view that republics were naturally peaceful, monarchies naturally aggressive. Jefferson applauded Paine's *Rights of Man,* a pamphlet notable not only because of its assault on Burke's *Reflections on the Revolution in France* but also because of the philosophy of international relations it contained. For Paine, as for Jefferson, war stemmed from the very nature of despotic government. It was "the common harvest of all those who participate in the division and expenditure of public money, in all countries. It is the art of *conquering at home;* the object of it is an increase in revenue; and as revenue cannot be increased without taxes, a pretense must be made for expenditures." Paine's belief that "taxes were not raised to carry on wars, but that wars were raised to carry on taxes" held a powerful appeal for the Republicans. Its implication was that the causes of war lay not in external conflicts of interest among states but in the domestic necessities of the power state itself. The monstrous aristocracies and bloated monarchies of Europe could maintain themselves only through war. To abolish the one would be to abolish the other.[47]

A similar analysis was made by Madison in his essay on "Universal Peace." Rousseau's plan for a "confederation of sovereigns, under a council of deputies, for the double purpose of arbitrating external controversies among nations, and of guaranteeing their respective governments against internal revolutions," was for Madison both undesirable and impossible—"preposterous" as well as "impotent." So long as war depended

> on those whose ambition, whose revenge, whose avidity, or whose caprice may contradict the sentiment of the community, and yet be uncontrouled by it; whilst war is to be declared by those who are to spend the public money, not by those who are to pay it; by those who are to direct the public forces, not by those who are to support them; by those whose power is to be raised, not by those whose chains may be riveted, the disease must continue to be *hereditary* like the government of which it is the offspring.

The regeneration of existing governments was the first, and indispensable, step toward a cure; the absurdity of Rousseau's plan for perpetual peace lay in the fact that it not only failed to allow for such regeneration but created a mechanism that might actively prevent it. In the confederacy of kings that had raised its head in opposition to the French Revolution, Madison could see the confirmation of his argument (as Hamilton saw in the conduct of French diplomacy the confirmation of his).[48]

Jefferson's initial attitude toward the Revolution in France had been a skeptical one. Believing that France was not yet suited for republican government, he hoped that a constitutional monarchy would emerge from the crisis of 1789. He urged moderation on all sides. On his return to America, however, he advanced side by side with the torrent of the Revolution, which moved with dizzying rapidity toward war, regicide, and terror. When his protégé William Short wrote him of the awful scenes through which the Revolution was progressing, Jefferson shot back a reproving letter: "The liberty of the whole earth was depending on the issue of the contest," he told Short, and he doubted that such a prize was ever "won with so little innocent blood." His own affections were "deeply wounded by some of the martyrs to this cause, but rather than it should have failed, I would have seen half the earth desolated. Were there but an Adam & an Eve left in every country, & left free, it would be better than as it now is."[49] His fear that the failure of republicanism in France would gravely wound it in America was deepened by the outbreak of war, for he saw in the European coalition against France a desire to snuff out liberty in the Old World as a prelude to extinguishing it in the new, and to this prospect no American could be indifferent. Therefore, he considered the establishment and success of republican government in France "as necessary to stay up our own, and to prevent it from falling back to that kind of half way house, the English constitution."[50]

Jefferson's sympathy toward France was the obverse of his hatred and fear of England. If he celebrated every success of French arms in the 1790s, as he did, it was because France stood in the way of the hostile designs that he believed England continued to entertain against America. In the late 1780s, he had insisted that, despite the treaty of peace, the English remained "our natural enemies, and as the only nation on earth who wished us ill from the bottom of their souls."[51] Once free from European entanglement, with an appetite whetted by success, England's resentment and hostility would again find release in America. "One thing," as Madison would later say, is "certain and conclusive": "Whilst the war against France remains unsuccessful the United States are in no danger from any of the Powers engaged in it."[52]

Hamilton's understanding of the European war was far different, though at the outset of the conflict it was not simply the mirror image of the Republicans'. From its earliest beginnings, to be sure, revolutionary France was for him an object of horror, and he had no sympathy for the violent and irregular proceedings of the successive governments that rose to power in that country. He admired, moreover, the principles of the British constitution, even those elements of it condemned as corrupt by the classical republicans, believing that England's balanced frame of government embodied a successful reconciliation of the competing claims of liberty and order.[53] At the outset of the Anglo-French wars, however, he wished simply that the United States could immunize itself from the contagion. Unlike Jefferson, he saw no compelling stake for the United States in the outcome of the contest. As yet, France did not pose the great threat to the balance of power that it would in coming years. It had made enemies of virtually the whole of Europe, and Hamilton dwelt frequently on the possibility that the First Coalition might overthrow the usurpers. Only if the United States threw itself recklessly into the contest would it court danger from abroad; thus his main concern was to ensure that the widespread sympathy for France among the American people did not lead the country to embark on a crusade for liberty on behalf of France.

Even before the outbreak of the European war, the rival systems of the Federalists and the Republicans were already fully formed and stood in implacable opposition to each other.[54] Much had changed since Hamilton and Madison had collaborated as Publius in justifying the new federal Constitution. Hamilton had retreated from the idea of employing the commercial weapon against Great Britain, an idea that had received his apparent sanction in *The Federalist* No. 11. Jefferson and Madison, on the other hand, drew back in horror from Hamilton's program of national consolidation, part of which they, too, had previously embraced. Perhaps of greatest significance was the attitude the Republicans now took toward the prospect of foreign war. In the previous decade, military preparations to meet that eventuality had seemed reasonable and prudent; by 1793, however, they appeared as the entering wedge of a program of domestic tyranny and monarchical consolidation. The traditional program of commercial discrimination now emerged not simply as a device to break open the closed trading system of Great Britain but as a whole new way of conducting statecraft, one capable of dispensing with the fleets and armies that could only imperil the domestic experiment in republican government. When Jefferson learned of the outbreak of war in 1793 and thought that Britain might strike at American trade, he observed that it would be a justifiable cause of war if there ever was one. But war, he hoped, would not be the response of the legislature. On the contrary, the occasion furnished the United States, as he told Madison, "a happy opportunity of setting another precious example to the world, by showing that nations may be brought to do justice by appeals to their interests as well as by appeals to arms." He hoped "that Congress, instead of a denunciation of war, would instantly exclude from our ports all the manufactures, produce, vessels and subjects of the nations committing this aggression, during the continuance of the aggression, and till full satisfaction made for it. This would work well in many ways, safely in all, and introduce between nations another umpire than arms."[55]

6

Neutrality and
the Law of Nations

THROUGHOUT THE EARLY PERIOD of American diplomacy, the law of nations exerted a powerful hold on all the leading statesmen of the country, Republican and Federalist alike. They differed often on the proper interpretation of this law, and each side charged the other with perversions of its meaning. But they were intimately acquainted with the works of the publicists—Grotius, Pufendorf, Wolf, Burlamaqui, Bynkershoek, and Vattel—and they shared the assumption that the law of nations was authoritative. "Ever since we have been an Independent nation," as Hamilton once observed, "we have appealed to and acted upon the modern law of Nations as understood in Europe. Various resolutions of Congress during our revolution—the correspondencies of Executive officers—the decisions of our Courts of Admiralty, all recognized this standard." So, too, did the "executive and legislative Acts and the proceedings of our Courts under the present government."[56] Jefferson and Madison were part of this larger consensus regarding the obligatory character of the law of nations; insofar as theirs represented a new diplomacy, it did not consist of a revolutionary repudiation of the "jurisprudists" scorned by many in France.[57] The Republicans, however, did wish to further the tentative efforts to reform that law, particularly in securing a wide berth for the rights of neutrals.

At no time during the early period of American diplomacy did the law of nations assume a more important place than in 1793, when the overthrow of the French monarchy and the outbreak of war between England and France presented a vast range of difficult questions to the American government. The precedents then established were of signal importance not only with respect to the next two decades of nearly continual warfare, but also in regard to the subsequent diplomatic stance of the nation and the respective constitutional prerogatives of the executive and legislative branches of the government. The central question facing Washington's cabinet lay in determining the respective rights and duties enjoined by a status of "fair neutrality." Less important at the time, though of considerable importance to posterity, were the rules laid down governing the law of recognition and intervention.

In response to the anxious queries of Gouverneur Morris, then American minister at Paris, Jefferson set forth the basic rule governing the recognition of new governments. "It accords with our principles," he wrote in his instructions of November 7, 1792, "to acknowledge any Government to be rightful which is formed by the will of the nation, substantially declared." Later, Jefferson wrote to Morris again, holding that the United States "surely cannot deny to any nation that right whereon our own government is founded—that every one may govern itself according to whatever form it pleases, & change these forms at its own will; & that it may transact its business with foreign nations through whatever organ it thinks proper, whether king, convention, assembly, committee, president, or anything else it may chuse. The will of the nation is the only thing essential to be regarded."[58] As subsequently understood in American diplomatic practice, Jefferson's dicta here amounted to a rule that conferred recognition on any government in effective possession and control of the governmental power and territory of a nation. Although opaque, Jefferson was surely not here enunciating a policy in which recognition would depend on a standard of democratic legitimacy.[59]

The abolition of the French monarchy on August 10, 1792, was the event that prompted Jefferson's instructions, and in setting forth this rule of recognition he has often been thought to have enunciated a wholly new principle in international law,[60] and one, moreover, which Hamilton did not share.[61] Neither assertion will bear scrutiny. Vattel, the most authoritative of eighteenth-century publicists, had recognized that the "nation alone" had a right to decide "disputes over the fundamental laws, over the public administration, or over the rights of the various powers which have a share in it." The British government had long subscribed to the same doctrine, sharply distinguishing between the internal character and the external behavior of foreign states. In the modern law of nations as it had developed in the eighteenth century, such a principle of recognition followed directly from the insistence on the perfect sovereignty of the state, and in laying it down as a rule of American policy Jefferson did not depart from the accepted understanding.[62]

Nor did Hamilton dissent from Jefferson's principle of recognition. In urging the conditional reception of Edmond Genet, on his way to Philadelphia in April 1793 as the representative of Girondist France, Hamilton acknowledged that "a Nation has a right to manage its own concerns, as it thinks fit, and to make such changes in its political institutions as itself judges best calculated to promote its interests." Hamilton denied, however, such a maxim meant also that "it has *therefore* a right to involve other nations, with whom it may have had connections, *absolutely* and *unconditionally,* in the consequences of the changes, which it may think proper to make." In conformity with this view, Hamilton proposed receiving Genet "in the character which his credentials import," while leaving to future determination the question whether the 1778 treaties with France "ought not to be deemed temporarily and provisionally suspended." Hamilton sought to distinguish between "the acknowledgement of a Government by the reception of its ambassador, and the acknowledgement of it, *as an ally.*" He would do the one but not the other. Thus he argued for the strict payment of the American debt to the governing power of France— to those who had acquired *"possession* of its political power" and had become "Master of its *goods.*" But in the expanded war that now pitted France against nearly the whole of Europe, he advised a position of strict neutrality for the United States.[63]

Both Jefferson and Hamilton propounded the basic norm in the law of nations prohibiting interference in the internal affairs of other states. This norm was of critical importance for Hamilton, and his animus toward France owed much to the contempt with which French diplomacy flouted the very principle—"an interference by one nation in the internal Government of another"— that it had most complained of. France "gave a general and just cause of alarm to Nations" by its decrees of November 19 and

December 15, 1792. In the former, the Convention, in the name of the French nation, declared "that it will accord fraternity and assistance to all peoples who shall wish to recover their liberty." In the latter, the Convention announced "that it will treat as enemies the people who, refusing liberty and equality, or renouncing them, may wish to preserve, recall, or treat with the prince and the privileged castes." Taken together, Hamilton argued, these decrees were "little short of a declaration of War against all Nations, having princes and privileged classes," equally repugnant "to the general rights of Nations, to the true principles of liberty, [and] to the freedom of opinion of mankind." France's "open patronage of a Revolution" in England, its incorporation of the territories over which French arms had temporarily prevailed, and its forced opening of the navigation of the Scheldt, contrary to treaty—all this breathed a spirit of defiance toward the public law of Europe.[64] Not incidentally, such conduct also showed that France's war was an offensive one, thus relieving the United States of any obligations it might have under the 1778 defensive alliance, even supposing it were still in effect.

Jefferson's stance toward intervention was more ambiguous. As Secretary of State, he frequently upheld the principle of nonintervention, which was closely linked to the maintenance of American neutrality. At the outbreak of the war, he disclaimed any intention "to meddle with the internal affairs of any country, nor with the general affairs of Europe. Peace with all nations, and the right which that gives us with all nations are our objects."[65] As nonintervention should be the rule for American policy, so it should also have formed the line of conduct for France. To his son-in-law, he confessed in June 1793 that the French "have been guilty of great errors in their conduct towards other nations, not only in insulting uselessly all crowned heads, but endeavouring to force liberty on their neighbours in their own form."[66] This bout of disaffection with France, however, was atypical: Jefferson's dominant mental picture of the European war was one that continued to see a "confederacy of princes" warring "against human liberty."[67] Responsibility for the war, he believed, lay with the despots of Europe, not the republicans of France, as did the first violation of the principle of nonintervention. As an American statesman, committed to neutrality, he would not indulge these sentiments in public, and his state documents contain no record of them. Ex officio, however, he still thought France and liberty were as one, and when he left office at the end of 1793, his revolutionary fervor was rekindled and passions repressed by office liberated. He then declared his conviction that France would triumph completely over its enemies, expressing the "hope that that triumph, & the consequent disgrace of the invading tyrants, is destined, in the order of events, to kindle the wrath of the people of Europe against those who have dared to embroil them in such wickedness, and to bring at length, kings, nobles & priests to the scaffolds they have been so long deluging with human blood."[68] Still later, in 1795, he rhapsodized about the possibility of a French conquest of England and spoke of "dining with Pichegru in London," so as to better "hail the dawn of liberty and republicanism in that island."[69]

Jefferson initially opposed a proclamation of neutrality. He doubted the executive had the constitutional authority to issue one, and he deemed it inexpedient to make known American intentions without first receiving explicit recognition from Great Britain that it would concede, as the price of it, "the *broadest privileges*" to neutral nations.[70] Washington overruled him on this point, and a proclamation was issued that warned American citizens "carefully to avoid all acts and proceedings whatsoever which may in any manner tend to contravene . . . a conduct friendly and impartial toward the belligerent powers." Those citizens who committed hostilities against the powers at war would be subject to punishment; shippers that carried articles "deemed contraband by the modern usage of nations" would not receive the protection of the United States.[71] Jefferson was able to strike mention of the word "neutrality" from the proclamation (a hollow victory, because it soon became known as such), and he inserted the word "modern" in the section dealing with contraband. Otherwise, the proclamation conformed to Hamilton's ideas on the subject. Madison, when he heard of it, thought it an abomination: "It wounds the national honor, by seeming to disregard the stipulated duties to France. It wounds the popular feelings by a seeming indifference to the cause of liberty. And it seems to violate the forms & spirit of the Constitution, by making the executive Magistrate the organ of the disposition the duty & the interest of the Nation in relation to War & peace, subjects appropriated to other departments of the Government."[72]

Left uncertain by the proclamation was the status of the French alliance, contracted at a critical moment in the American war for independence. The 1778 treaty obliged the United States to guarantee "from the present time and forever . . . the present Possessions of the Crown of France in America as well as those which it may acquire by the future Treaty of peace."[73] Since Britain would in all likelihood strike at the French West Indies, the alliance threatened to drag the United States into the conflict. Jefferson, however, argued in the cabinet for the retention of the alliance, insisting that the treaties were not "between the U.S. and Louis Capet, but between the two nations of America & France." He dismissed the prospect that they would draw the United States into the war. Posing a string of rhetorical questions to the President, he cast doubt on the proposition that France would ever invoke the guarantee of its West Indian islands (just as the United States had not requested French aid in securing the removal of British forces from the Northwest posts). Apart from the manifest inability of the United States to supply France a naval force, Jefferson believed that "we shall be more useful [to the French] as neutrals than as parties by the protection which our flag will give to supplies of provision." And though he declared that the danger of involvement in war "is not yet certain enough to authorize us in sound morality to declare, at this moment, the treaties null," he left the impression that such a danger, if extreme, imminent, or probable, would indeed be grounds for renouncing the alliance.[74]

The belief that American neutrality would be more useful to the French than active participation in the war was reinforced by the reports of famine in France; only if that famine worsened, Jefferson thought, would the revolution be in danger.[75] On these assumptions, a position of "fair neutrality" was not difficult to square with his twin convictions that France was the natural ally of the United States and that the survival of republicanism in America depended on its survival abroad. Nor, in fact, had he renounced in his own mind that the United States ought to make good on the territorial guarantee to France, even though Genet had not invoked it.[76] One of his purposes in introducing the word "modern" into the proclamation, he later recalled, was "to lay the foundation of the execution of our guarantee, by way of negotiation with England." His reasoning was as follows: "By the ancient law of nations, e.g., in the time of the Romans, the furnishing a limited aid of troops, though stipulated [by treaty], was deemed a cause of war. In latter times, it is admitted not to be a cause of war. This is one of the improvements in the law of nations. I thought we might conclude, by a parity of reasoning, that the guaranteeing a limited portion of territory, in a stipulated case, might not, by the modern law of nations, be a cause of war."[77] Consistent with this understanding, he told Madison, after his retirement, that "whatever doubts may be entertained of the moment at which we ought to interpose, yet I have no doubt but that we ought to interpose at the proper time, and declare both to England and France that these islands are to rest with France, and that we will make a common cause with the latter for that object."[78] Jefferson did not reveal his view of the guarantee to his colleagues because he "apprehended it would occasion the loss of the word." Nor did he mean to use military force: he believed throughout that the guarantee might be honored and the rights of neutrals secured "in a peaceable way."[79]

Genet's instructions were not consistent with this strategy, nor did the passionate French diplomat understand what Jefferson was up to. Though not immediately invoking the guarantee, he did act so as to badly compromise American neutrality. He distributed commissions to fit out French privateers in American ports, established French prize courts on American soil, and raised a little army for an attack, originating from the United States, on New Orleans and the Floridas, then in the possession of Spain. None of these actions was compatible with American neutrality. Though Jefferson winked at Genet's plans to liberate New Orleans,[80] he nevertheless found himself in a war of words with the French minister over a range of maritime issues. As a consequence of Genet's repeated defiance of the American government, Jefferson ultimately had no choice but to join in the demand for the recall of Genet, whose appointment he had come to regard as "calamitous" for Franco-American relations and whose outrageous behavior seriously compromised the Republican position at home.

In the course of the conflict with Genet, and with new questions, many of a highly technical nature, forcing their attention on the American government week after week, Washington's cabinet hammered out a policy of neutrality, though one that was fully satisfactory to neither Jefferson nor Hamilton.[81] France retained the right, granted by Article XVII of the Treaty of Amity and

Commerce, to admit its prizes and privateers into American ports; the same privilege, however, was denied to Britain (under terms of Article XXII of the same treaty). These two stipulations apart, however, the neutrality policy was perfectly equal to the two warring parties. Treaty obligations required the United States to admit French vessels of war into American ports, but as the treaty did not positively forbid the entrance of British warships they were granted the same privilege.[82] The "usurpation of admiralty jurisdiction by the consuls of France" was disallowed by the American government, as was the right of France to fit out privateers in American ports or to issue commissions to American citizens.[83] Those that were illegally commissioned were banned from American ports. Restitution was to be made to Great Britain of all prizes taken and brought into American ports, subsequent to June 5, by such privateers.[84] On balance, Britain's minister to the United States, George Hammond, had "every reason to be satisfied with the conduct of the *federal government*."[85]

The collision with France prompted by Genet's conduct was quickly parried. Of greater importance in the long run was the three-cornered dispute among the United States, France, and Great Britain over the rights (as opposed to the duties) of neutrals. The conflict, at home and abroad, over the exact latitude of these rights ebbed and flowed over the next two decades, but never disappeared. Disputes over their precise character brought Britain and the United States to the verge of war in 1794, produced a quasi war with France during the Adams administration, led to the embargo of 1807, and ultimately issued in Mr. Madison's War of 1812. By leaning toward the French view, the United States risked war with Britain; by leaning toward the British view, the United States risked war with France. The dilemma was complete, and it was exacerbated by the fact that the two great parties at home were by no means of one mind on the question.

The traditional policy of the United States was first set forth in 1776, in the Model Treaty adopted by the Continental Congress. That instrument, drafted mainly by John Adams, adopted the "liberal" and "enlightened" policy that "free ships made free goods." This doctrine, which would be incorporated in the 1778 Treaty of Amity and Commerce with France, was supported by a host of states in Europe, which formed the Armed Neutrality of 1780 in support of it. Under this principle, sea power could not be employed to harass or interdict the shipping of neutral nations, whose flags on the ocean were an extension of territorial sovereignty at home. The right contended for was that of a general freedom of trade in wartime, save in contraband articles and to places blockaded or besieged. Both in the Model Treaty and in the 1778 Treaty of Commerce with France, the definition of contraband was a narrow one, expressly excluding provisions and naval stores (both products of peculiar importance to America).[86] But though seeking to gain the widest possible acceptance of these rights, the United States was careful to limit its obligations to support them by force, and the Continental Congress, accordingly, had declared, in June 1783, that though it was willing to recognize neutral rights in the final peace treaty, it wished to "avoid accompanying them by any engagements which shall oblige the contracting parties to support those stipulations by arms."[87]

By the outbreak of the War of the French Revolution, these principles had received the assent of virtually all states comparatively weak in naval power. But they were evidently in opposition to the interests of the dominant naval power, and Great Britain had never accepted them as part of the necessary or customary law of nations (i.e., as that part binding irrespective of treaty obligations with particular states).[88] Were the principle of free ships, free goods "once really established and honestly observed," John Adams would later recall, "it would put an end forever to all maritime war, and render all military navies useless." For this reason, Adams thought, it would never take place, "however desirable this may be to humanity, how much soever philosophy may approve it and Christianity desire it."[89] Britain could not assent to the principle without surrendering the key advantage afforded by its naval power. In practical terms, it meant that the United States might enter fully into France's carrying trade with the French West Indian colonies. Nor had Britain accepted as part of the modern law of nations the exclusion of provisions and naval stores from contraband lists. On the outbreak of war in 1793, Britain sought (by an order of June 6, 1793) to deny corn, meal, and flour from landing in France and justified its action by declaring "that, by the law of nations, as laid down by the most modern writers, it is expressly stated, that all provisions are to be considered as contraband, and as such, liable to confiscation, in the case where the depriving an enemy of these supplies, is one of the means intended to be employed for reducing him to reasonable terms of peace."[90]

Neither Jefferson nor Hamilton accepted the British claim that provisions might be treated as contraband. Before news arrived of the June 8 order, Jefferson had told Pinckney that the seizure of American provisions on their way to France "would be so un-equivocal an infringement of the neutral rights, that we cannot conceive it will be attempted." When Britain did attempt it, Hamilton strongly protested to the British minister, telling Hammond that the order was "harsh and unprecedented," militating against "the principal branch of the present American exports."[91] Jefferson also vigorously protested the order, calling it "manifestly contrary to the law of nations." Were the United States to acquiesce in the British pretension while continuing to allow exports to Great Britain, France would have every right to complain of partial and unneutral treatment, an injury that might lead to war.[92]

On the question of free ships, free goods, however, the opinion of Washington's advisors was divided. Hamilton did not regard the assertion of this principle as worth a commercial or military conflict with England, a position consistently maintained by the Federalists in the course of the decade.[93] Jefferson's posture was more ambitious. On learning of the outbreak of war, he had instructed Thomas Pinckney, American minister to Great Britain, to press for British recognition that the principles embodied in the 1778 Franco-American treaty "should form the line of conduct for us all."[94] By inserting the word "modern" into the neutrality proclamation Jefferson had meant to hint that the United States should require Britain's acquiescence in the principle of free ships, free goods "as the condition of our remaining in peace."[95] In the course of the conflict with Genet, however, those broad neutral principles were soon abandoned as untenable by Jefferson himself. It cannot be doubted, he told Genet, "but that, by the general law of nations, the goods of a friend found in the vessel of an enemy are free, and the goods of an enemy found in the vessel of a friend are lawful prize." The neutral's vessel, in other words, did not cover the goods of an enemy; free ships did not make free goods. In the aftermath of the slave revolts on Santo Domingo, the British acted on this principle in seizing, from American ships, the property of French émigrés fleeing from the slaughter, and Jefferson told Genet that he "should be at a loss on what principle to reclaim it."[96] The same doctrine, at greater length, was presented to Gouverneur Morris in the letter requesting Genet's recall.[97]

Jefferson had not in fact renounced the attempt to secure British acquiescence in the principle of free ships, free goods, though many observers, both at the time and subsequently, thought that he had done so.[98] He had been drawn into the statement by Genet, who threatened that if the United States "did not resent the conduct of the British in taking French property in American bottoms, and protect their goods by *effectual measures* (meaning by arms), he would give direction that the principle of our treaty of goods following the bottom, should be disregarded." At the same time, the French were disregarding in fact the very principle Genet contended for, and it became necessary, Jefferson recalled, to oppose Genet's construction. He could do so only "by placing it on its true ground, to wit: that the law of nations established as a general rule that *goods should follow the owner,* and that the making them *follow the vessel* was an exception depending on special conventions only in those cases where the convention had been made: that the exceptions had been established by us in our treaties with France, Holland, and Prussia, and that we should endeavor to extend it to England, Spain, and other powers; but that till it was done, we had no right to make war for the enforcement of it." The denouement was ironic and, in Jefferson's view, most regrettable. Genet "had obliged us to abandon in the first moment the ground we were endeavoring to gain, that is to say, his ground against England and Spain, and to take the very ground of England and Spain against him."[99]

Hamilton's and Jefferson's understanding of the rights and duties of neutrality, as well as the obligations entailed by the French alliance of 1778, though not opposed on all questions, were evidently very different. Hamilton proposed the abandonment of the French alliance, which he saw as incompatible with a position of strict neutrality. Jefferson wished to retain the alliance, including the territorial guarantee, without, however, entering the war as an active belligerent. Hamilton would not risk a conflict with Britain over free ships, free goods, whereas Jefferson's strategy lay in forcing Britain to choose between acceptance of the principle and the loss of American markets and supplies.

Despite these differences, both men were advocates of the modern law of nations, which aimed at mitigating the rigors of war and reflected a marked departure from ancient practice. Jefferson could not recollect "a single species but man which is eternally and systematically engaged in the destruction of its own species. What is called civilization seems to have no other effect than to teach him to pursue the principle of *bellum omnium in omnia* on a larger scale, and in place of the little contests of tribe against tribe, to engage all quarters of the earth in the same work of destruction." He joined with Madison in the wish to see man "softened."[100] Hamilton, too, thought that the rigors of war might be mitigated "by the softening and humanizing influence of commerce" and believed that, in Roman times, "the world was yet too young—moral science too much in its cradle—to render the Roman jurisprudence a proper model for implicit imitation."[101] Here, as well, however, there was a subtle difference of emphasis. Hamilton invoked the "softening and humanizing influence of commerce" to argue against the sequestration and confiscation of private British debts, a weapon of commercial warfare that the Republicans were unwilling to relinquish.[102] And whereas Jefferson invoked "the exchange of wants" facilitated by commerce by way of arguing for the greatest possible latitude for neutral rights, Hamilton desired an understanding of these rights commensurate with the limitations of American power and compatible with the avoidance of war with Great Britain. The tendency of Hamilton's thought was to impose duties on ourselves; that of Jefferson, to impose duties on others.

Hamilton's outlook also differed from Jefferson's in the greater weight it gave to considerations of relative power and usage, as against the tendency to reform that law in accordance with an ideal standard founded in "nature and reason." Hamilton was not a positivist; he, too, believed in "the natural or necessary law of nations" and "the eternal principles of morality & good faith." But the inclination in his thought was to rely more on considerations of usage and accepted practice; these were, for him, of the "highest authority."[103] Jefferson, by contrast, though he always invoked usage as precedent when convenient for his argument, was led by his desire for a change in international norms to greater reliance on "the natural or necessary law of nations," as discovered by the exercise of dispassionate reason. Hence he could write, as Hamilton never would have done, that "our geographical peculiarities may call for a different code of natural law to govern our relations with other nations from that which the conditions of Europe have given rise to there."[104]

In searching for that new code of natural law, Jefferson frequently had recourse to what he called "the dictates of national morality." His emphasis on moral principle coincided with the vigorous defense of American commercial rights, gratitude toward France, support for republicanism abroad, the navigation of the Mississippi, and the right of deposit in New Orleans. Indeed, there was scarcely a single question affecting the foreign policy of the United States that Jefferson did not convert into a moral question. This was fully consistent with, and indeed followed from, his rejection of the distinction between private and public ethics so characteristic of the *ancien régime* and which in Europe was closely identified with the doctrine of reason of state. There was, he thought, "but one code of morality for men, whether acting singly or collectively."[105]

National morality required, first of all, gratitude toward France for its aid in securing American independence. To deny this sentiment as a proper motive of national conduct was "to revive a principle which has been buried for centuries with its kindred principles of the lawfulness of assassination, poison, perjury, etc. All of these were legitimate principles in the dark ages which intervened between ancient and modern civilization, but exploded and held in just horror in the eighteenth century."[106] Hamilton, as usual, took a different approach. In defending the Proclamation of Neutrality, he did not "advocate a policy absolutely selfish or interested in nations." He did, however, insist that "a policy regulated by their own interest, as far as justice and good faith permit, is, and ought to be, their prevailing policy: and that either to ascribe to them a different principle of action, or to deduce from the supposition of it arguments for a self-denying and self-sacrificing gratitude on the part of a Nation, which may have received from another good offices, is to misconceive or mistake what usually are and ought to be the springs of National Conduct."[107]

Hamilton's position was singularly austere. Jefferson's, by contrast, was more delicate. In placing such stress on the ties of gratitude to France, he laid himself open to the charge that he was betraying the interest of the United States. In fact, however, he wanted to do no such thing. Gratitude was merely one among many considerations that made France the centerpiece of his foreign policy, and it was surely not the most important one. His doctrine is nevertheless significant and revealing, for it displays his propensity to see moral significance in virtually all questions of

American foreign policy. As Secretary of State, Jefferson located American rights sometimes in nature and reason, at other times in treaty law or customary usage, and frequently in all. In seeking to secure the navigation of the Mississippi from Spain, he appealed to the law of nature and nations, "as we feel it written in the heart of man," and asked "what sentiment is written in deeper characters than that the ocean is free to all men, and their rivers to all their inhabitants? Is there a man, savage or civilized, unbiased by habit, who does not feel and attest this truth?" If the right of American citizens to navigate the Mississippi was unquestionable, so, too, was their insistence on securing the right of deposit in New Orleans: "A right to a thing gives a right to the means without which it could not be used."[108] America's interest in opening the European West Indies to American ships and produce was something that could not be validated by appeal to treaty obligation or customary usage; Jefferson found the right instead in moral principle: "An exchange of surplusses and wants between neighbor nations," he held, "is both a right and a duty under the moral law."[109] The rights of neutrals were more generally established and had the support of the weak naval powers of Europe; here he made his appeal to both "reason and usage." Together, they established "that when two nations go to war, those who choose to live in peace retain their natural right to pursue their agriculture, manufactures, and other ordinary vocations, to carry the produce of their industry for exchange to all nations, belligerent or neutral, as usual, to go and come freely without injury or molestation, and in short, that the war among others shall be, for them, as if it did not exist."[110] The states of Europe, he knew, often held to different principles; "with respect to America," however, "Europeans in general, have been too long in the habit of confounding force with right."[111]

The principles for which Jefferson stood and which he identified with the natural rights of all men posed no serious obstacle to the pursuit of America's interests. Indeed, they posed no obstacle at all.[112] The moral order he embraced was simply the order of his expectations, an order which in its scope and magnitude was breathtaking. Yet it would be wholly misleading to conclude from this fact that Jefferson was merely a clever or cynical advocate—though surely he was sometimes both. The emphasis on moral principle and natural rights was always a factor of profound significance in shaping his outlook toward the external world; so, too, was the radical distinction he drew between the unjust pretensions of the European powers and the natural rights of the United States.

In so closely identifying the pursuit of the national interest with the vindication of natural right, Jefferson invested American pretensions with a kind of sacred character. To give them up, even temporarily, could easily be seen as a betrayal of republicanism and a surrender of independence. That made compromise difficult, particularly with the British, against whom claims of the same character had been made before. Jefferson's moralism also had the effect of magnifying his deep distrust of the motives of other powers. Great Britain's refusal to concede its maritime pretensions became easily linked with its malevolent hatred of America and republicanism. From this premise, to which Jefferson often recurred, it was easy to conclude that concessions would do nothing to appease Britain's appetite but only whet it further. Psychologically, it converted the reformation of the international system demanded by his statecraft into a defensive action; and it transformed Britain's efforts to preserve its maritime dominion and Spain's efforts to preserve its empire in the Americas into grave and menacing assaults on the moral order.[113]

By the time Jefferson retired as Secretary of State, no consensus existed within the nation over the meaning of the European struggle, and there still continued a fundamental difference between the emerging parties over the reciprocal rights and duties of the belligerent and neutral powers—conflicts that would continue for the duration of the Anglo-French wars. Each side appealed to the already traditional American policy of nonentanglement, neutrality, and nonintervention. In 1793, however, it was Hamilton and the Federalists who had the better claim to it. Jefferson's sentiments toward the French alliance and his fervent hope for the success of republicanism abroad made his commitment to an impartial neutrality doubtful, at the least. That commitment, however, would not be tested in the early 1790s. Jefferson's resignation tilted the balance within the cabinet ever more toward Hamilton, and the Virginian sped away toward Monticello, damning the miseries of public life as he went.

Soon after Jefferson left office, Britain and the United States were brought to the verge of war. Under the terms of an Order in

Council of November 6, 1793, the British swooped down on un-suspecting American vessels in the West Indies, taking over 300 American merchants vessels as prize. America's Mediterranean trade was also dealt a serious blow by a British treaty with Portugal that left American ships unprotected from Barbary corsairs. In the Old Northwest, new evidence of British machinations among the Indians came to hand, all of which added up to a menacing pic-ture of British hostility. In accordance with their respective sys-tems, Federalists called for military preparations; Republicans, for commercial retaliation. In a last desperate attempt to avoid war through negotiation, Washington sent John Jay to London in early 1794. The treaty he produced had a profound impact on the de-velopment of political parties at home and dominated American politics until the end of the decade. It also produced a wholly new situation in American relations with France and Spain.

7

The Diplomacy of
Federalism

THE GREAT ACCOMPLISHMENT of Federalist statecraft was the vindication of America's effective title to the eastern Mississippi valley. In the Jay Treaty of 1794 and the Pinckney Treaty of 1795, Washington's administration made peace with, respectively, Britain and Spain. From Britain Jay secured the evacuation of the Northwest posts and the removal of a potent source of discord; from Spain Pinckney obtained recognition of the 31st parallel as the boundary between the United States and Spanish Florida, the navigation of the Mississippi, and the right of deposit in New Orleans. The two settlements ensured that the American Union would extend to the Mississippi, which had been highly doubtful only recently; and such was the dominant position of the United States if it remained at peace that there could be little question that the Floridas would one day fall in its grasp. Hamilton's judgment of the Jay Treaty's significance was surely correct: "The most important *desiderata* in our concerns with foreign powers," he argued in defense of the treaty, "are the possession of the Western posts and a participation in the navigation of the River Mississippi. More or fewer of Commercial privileges are of vastly inferior moment. The force of circumstances will do all we can reasonably wish in this respect, and in a short time without any steps that may convulse our trade or endanger our tranquillity will carry us to our goal."[114]

Hamilton's judgment was borne out in both respects. American commerce prospered in the years that the Jay Treaty was in effect. The Federalists' acquiescence in certain aspects of British maritime practice dealt nothing approaching the blow to it that the Republicans prophesied. Nor can there be doubt that the interior was the place where the Union ran the most serious risk of foreign complication and possibly disunion. Before the two settlements, there had existed a potent danger of an Anglo-Spanish combination covering the whole of the interior; were the views of these powers to be united, the security of the interior would have been badly compromised. The British had not abandoned their attempt to detach Vermont or their project of an Indian barrier in the Old Northwest; the Spaniards retained numerous friendships with adventurers in the western country who had no real attachment to the eastern states. The French Revolution had thrown Britain and Spain suddenly into alliance, a reversal of a European diplomatic constellation that dated from 1585, and which was as unexpected as it was portentous. The essential precondition of American security in the interior lay in the division of these two powers; the "dissolution or prevention" of the community of views between Britain and Spain, as Hamilton rightly argued, was "a point of the greatest moment in our system of national policy."[115]

The Jay Treaty secured this dissolution. It did more. Hamilton's confidence that an understanding with Great Britain would render the Spanish position untenable in the Southwest was soon borne out by events. Pinckney's arrival in Spain coincided with a rising fear in Madrid that its war on France had been a dreadful mistake. Undertaken as a gesture of solidarity with Louis XVI, the deposed cousin of Charles IV, Spain soon came to regret its war against France. England's blows against French maritime power— symbolized by the burning of the French fleet at Toulon (instead of its transfer to the Spanish Navy)—were quickly seen as a threat to Spain's imperial position in the Americas. That Spain intended to sign a separate peace with France and give Britain a pretext of war made it all the more fearful, and when the news of the Jay Treaty reached Don Manuel de Godoy, *El Príncipe de la Paz,* it became apparent that the territorial demands of the United States would have to be recognized and the right to navigate the Mississippi surrendered. The nature of the agreement between America and Britain was not precisely understood at Madrid, but Spanish fears were eminently justified. Were the quarrels of the interior to fester, there was little reason to doubt that the Federalists might seek an understanding regarding military cooperation with Great Britain. That the two powers combined might deprive Spain not only of the Floridas and Louisiana but of a good part of its empire in the Americas made it imperative to avoid all cause of friction. Pinckney, who knew well the terms of the Jay Treaty, was careful to avoid any mention of its terms in his negotiations. When he left San Lorenzo, he had secured virtually everything for which Jefferson had contended in his negotiations with Spain.

The Jay Treaty thus remains the central episode in the diplomacy of Federalism. On the merits of the treaty largely turn the political judgment of Hamilton's diplomacy (for Hamilton was indisputably the central figure in the diplomatic design that produced it, despite the fact that the burning suspicions entertained of him by the opposition made him incapable of playing a public role). In the agreement, Britain surrendered the Northwest posts that had been held since the conclusion of the War for Independence, while the United States agreed to submit the claims of British creditors, long frustrated by the inaction of state courts, to a mixed commission of arbitration. Also submitted to a joint commission were American claims for spoliations committed by British cruisers and privateers (most of which centered around the notorious Order in Council of November 1793). In thus securing compensation for spoliations and in removing the potential menace of a British-American war in the interior, Jay secured the main objects of his mission. With regard to maritime rights, however, he was far less successful. American desires for a liberal regime of neutral rights were brushed aside by Grenville, the British Foreign Secretary, and one article of the treaty which admitted American ships of less than seventy tons burden into the British West Indies was rejected by the Senate. But the agreement did secure the peace. Under the circumstances facing the country, peace was an object, as Hamilton said, "of such great and primary magnitude," that it was not to be given up "unless the relinquishment be clearly necessary to Preserve Our Honor in some Unequivocal Point, or to avoid the sacrifice of some Right or Interest of Material and Permanent Importance."[116]

Absorbed in the cultivation of his peas and clover at Monticello, Jefferson made no extensive comment on the terms of the treaty, contenting himself with denouncing it in the harshest possible terms. It was a "monument of folly and venality," an "infamous act, which is nothing more than a treaty of alliance between England and the Anglomen of this country against the legislature and people of the United States."[117] He could not abide the treaty's acquiescence in the English system of maritime domination. Hamilton pressed the argument that the United States could not give the law to Great Britain, that Britain's "very existence as an independent power" rested upon the superiority it enjoyed at sea, that the United States, if it adhered to the Republican position, was asking the British to give up what they had resolved never to relinquish.[118] Jefferson nevertheless found this acquiescence galling. Three betrayals—against republicanism, neutral rights, and France—oppressed him; that the treaty might even be considered confirmed the malignant weight of England's influence in America. It was a craven surrender to Britain's maritime pretensions, while it took out of America's hands the commercial weapons that might secure justice in the future. Finally, it promised to badly compromise American relations with France and would in fact, he thought, lead to war with that power. He was convinced that that objective now lay at the heart of Federalist diplomacy.

Madison and Gallatin drove these points home on the floor of Congress. By the treaty, Gallatin declared, "we had promised full compensation to England for every possible claim they might have against us, . . . we had abandoned every claim of a doubtful nature, . . . and we had consented to receive the posts, our claim to which was not disputed, under new conditions and restrictions never before contemplated." Having made an unequal treaty on the issues that had grown out of the 1783 Treaty of Peace, Jay had gone on to agree to commercial articles that secured to the country "no commercial advantages which we did not enjoy before." No security against future aggressions or in favor of freedom of navigation had been obtained by Jay, yet he had "parted with every pledge we had in our hands, with every power of restriction, with every weapon of self-defence, which was calculated to give us any security." This last was, perhaps, the worst blow of all. To surrender the threat of commercial restrictions against Britain (which the treaty did by its most favored nation clause), to give up permanently the threat of sequestering and confiscating British private debts, meant that the United States could take no measures short of war to enforce its rights against Great Britain.

Still, Gallatin was for the postponement, not the complete rejection of the treaty. Conscious that a complete rejection might lead to a further retention of the Northwest posts, no reparations for spoliations on American trade, and continued uncertainty in Anglo-American relations, he would not give a final negative to the resolutions for carrying the treaty into effect. But Britain would have to disavow the new hostilities it had undertaken since the conclusion of the compact—particularly its continued impressment of American seamen and its provision order of April 25, 1795—before the ratifications might be exchanged.[119]

Gallatin's support for postponement and not rejection reflected the qualities of temperament that would make him such a formidable opponent of Federalist measures in 1798. Madison's position, by contrast, was less compromising, his rejection of the treaty final and irrevocable. Everything was wrong with it. The articles settling the execution of the 1783 Treaty of Peace represented "the grossest violation" of the principle of reciprocity, as did those that regulated Anglo-American commerce and that dealt with the "great points in the Law of Nations." Of all nations, he thought, the United States "ought to be the last to unite in a retrograde effort" against the principle that free ships make free goods. The situation of the United States "particularly fits them to be carriers for the great nations of Europe during their wars. And both their situation & the genius of their Government & people promise them a greater share of peace and neutrality than can be expected by any other nation." The treaty, Madison thought, would strike a mortal blow at American commerce and navigation. By allowing that provisions might in some circumstances be deemed contraband of war, Jay had introduced a principle in violation of "the fundamental rights of nations & duties of humanity," one that had as its purpose "the horrible scheme of starving a whole people out of their liberties." Under these circumstances, an American rejection of the treaty was imperative. "Unequal in its conditions, . . . derogatory to our national rights, . . . insidious in some of its objects, . . . and alarming in its operation to the dearest interests of the U.S. in their commerce and navigation," there was no plausible argument for its acceptance, least of all that a failure to ratify might lead to an Anglo-American war. The idea of a war with Britain "was too visionary and incredible to be admitted into the question," whereas the prospect of war with France as a consequence of such an unequal treaty was a very real one. Appeasement, in any case, was more likely to lead to war than a stance that did "justice to all nations" and sought it from them "by peaceable means in preference to war." For there was no evidence, "either in History or Human nature," for the proposition that nations could be "bribed out of a spirit of encroachment & aggressions by humiliations which nourish their pride, or by concessions which extend their resources & power."[120]

The debate over the treaty ultimately turned on two questions: whether the compromises with Britain required too great a sacrifice of American interests, and whether the failure to compromise would have led to war with Great Britain. Madison's position reflected not only the belief that compromise was tantamount to the rankest appeasement but also a thoroughgoing denial that war alone could bridge the immense gulf his uncompromising stance would leave between the parties. In both respects, it seems clear, his position was fundamentally misguided. Though he spoke often of reciprocity in negotiation, his conception of what reciprocity required was in practice indistinguishable from a British capitulation. There remains little question, moreover, that his own favored strategy would rapidly have led to war between the two countries. Though it is possible that Jay might have wrung more concessions out of the British, there is no reason to believe that Pitt's ministry would have consented to anything approaching what was demanded by Madison; ultimately the Republicans would have had to make good their threat to resort to economic sanctions, whose domestic consequences would likely have been just as severe as the ill-fated embargo of 1807. The Republicans knew that Great Britain was loath to begin another war with America, especially given its precarious position in Europe. What they failed to appreciate was that Britain was equally loath to surrender its maritime pretensions. In all likelihood, the "unsettled state of things" that would have ensued from the failure of Jay's mission, as Hamilton argued, "would have led to fresh injuries and aggravations, and circumstances too powerful to be resisted would have dragged us into War."[121] The failure of the commercial propositions Madison advocated to secure the capitulation of Great Britain would have left only the alternatives of war and disgrace. That was precisely the dilemma of Madison's presidency from 1809 to 1812, and it was a predicament resolved in the same fashion that Hamilton, writing seventeen years before the event, said it would be.[122]

Such a rupture might also have seriously prejudiced the negotiation with Spain. It is, in any case, highly unlikely that Republican diplomacy in the mid-1790s would have secured a more favorable result in the Southwest. Spain, it is true, was moving toward a rupture with Great Britain in any case. Spain had awakened suddenly to the realization that the security of its empire depended on the French Navy, as it long had. With northern Spain occupied by French troops, Spain saw no alternative but to appease France. The Anglo-American reconciliation was nevertheless an important factor in producing Spanish concessions—first in July 1794, when the news of Jay's mission to England arrived in Spain and Godoy agreed to accept American demands for the boundary and the navigation on condition of a treaty of alliance and reciprocal guarantees of territory in America; then again in the late summer of 1795, when the Spanish demands for an alliance and territorial guarantees were dropped. Had the United States failed to reach an accommodation with England while Spain moved toward a rapprochement with France, the distribution of power in North America would clearly have been less favorable to the United States than it was. Under those circumstances, the United States would have been reduced to prying concessions from Madrid through the good offices of France. The Republicans staked much on this possibility; the truth was that at this very moment France was seeking to gain from Spain the retrocession of Louisiana, and was promising to close the Mississippi in order to stop the inexorable expansion of the United States.[123]

The Republicans were right in thinking that the treaty would badly compromise American relations with France. Federalists had denied the danger. In avoiding Scylla, Hamilton held, there was little danger of running upon Charybdis: "No cause of umbrage is given to France by the Treaty, and it is as contrary to her interest as to her inclination wantonly to seek a quarrel with us."[124] That prediction was wide of the mark. Severely displeased with Anglo-American reconciliation, France charged the United States with willful violations of the 1778 Franco-American treaties and proceeded to unleash its cruisers and privateers on American shipping, going in fact well beyond the declared policy of treating "the flag of neutrals in the same manner as they shall suffer it to be treated by the English."[125] That reaction prepared the ground for the "crisis of '98" and deepened still further the division between the parties at home. Federalists became vulnerable to the charge of coveting a war with France; Republicans, of inviting the very depredations on American commerce they had foretold.

Ultimately, the issue turned on those controverted points in the law of nations that had been agitated since the beginning of the war; here, at least, the Federalists stood on reasonable ground. The United States had no obligation, as the French charged, to support by force and as a general code of maritime conduct the principle of free ships, free goods, explicitly surrendered for the duration of the war in Article XVII of the Jay Treaty. As Jefferson had himself acknowledged, this was no part of the customary law of nations. In making it the rule of the 1778 treaty of commerce, the United States had obliged itself to respect the rule when it was a belligerent and France a neutral. It had not undertaken a positive stipulation to force England to observe it against the United States. And though the balance of obligations created by the 1778 and 1794 treaties was indeed unequal against France in the present war—in that France was forbidden to take English goods from American vessels, whereas the English were allowed to take French goods from American vessels—the American government offered to change the stipulations in the 1778 treaty if France so desired, eliminating all the inequalities of which France complained.

Seconded by the Republicans, France charged as well that, by including naval stores and timber for shipbuilding, the Jay Treaty had "swelled" the list of contraband in violation of the "least contested" rights of neutrals; not content with this "unbounded condescension" to England, the United States had also tacitly consented to treat provisions as contraband. Here, too, however, the American position did not constitute a violation of its obligations as a neutral. Though the interests of the United States dictated the exclusion of naval stores from contraband lists, it had no positive obligation to support this exclusion either by the existing law of nations or by treaty. Nor did a violation of neutrality arise from the part of Jay's Treaty relating to the trade in foodstuffs to France and its colonies. Article XVIII acknowledged "the difficulty of agreeing on the precise cases in which alone provisions and other articles not generally contraband may be regarded as such," and provided that when such goods were seized they should not be confiscated, England agreeing instead to pay "the full value of all such articles," together with a reasonable mercantile profit, freight, and demurrage. The clause, as written, was not unfavorable to

France, in that it encouraged American shippers to send goods to French ports, knowing that, if seized by Britain, they might still make a profit on the trade. And though Great Britain renewed, by its order in council of April 25, 1795, its policy of seizing American provisions bound to France, the United States protested against this construction of the treaty and the order was subsequently repealed.[126]

The Federalist interpretation of the rights and duties of neutrality, though it assuredly yielded to the British interpretation on some key points, thus did not wholly acquiesce in British doctrine, either in Jay's Treaty or subsequently. Indeed, with the exception of naval stores (which many publicists, including Vattel, had quite reasonably considered as contraband), that policy preserved the right of the United States to carry on the trade to which it was accustomed in peace, most particularly in the right to trade with France in the produce of the American soil. At the same time, it did not press these claims so far as to take advantage, under cover of neutral rights, of that large "share of peace and neutrality" promised by Republican policy. That Virginians, led by Madison, would wish to push the carrying trade to the limit, while the mercantile classes were generally content with a less ambitious stance, was an irony much commented on at the time, and it was one that would arise again when the Jeffersonians took power.[127]

Nor is this all. It is often said of Jefferson that the key to his own diplomatic outlook lay in his confidence that Europe's distress would ultimately redound to America's advantage, above all in securing the American claim to the great interior. The theme was central to the writings of Samuel Flagg Bemis, the outstanding historian of early American diplomacy, who thought it encapsulated the essence of Jefferson's method in foreign policy:

> Out of future chapters of circumstances, the precise nature of which he could not foresee, but to the general character of which he confidently trusted, [Jefferson] awaited the opportune moment when the United States, a nation of no appreciable organized military or naval power, could by a threat of potential strength and by astute diplomacy force from Spain, harassed by troubles in Europe, the free navigation of the Mississippi River and the recognition of the southern United States boundary of 31 degrees north latitude . . . ; could compel Great Britain to get her troops out of our

northern territory and admit our commerce into her remaining American possessions.[128]

Yet the most striking feature of the Jeffersonian system is how ill-suited it was to the attainment of each of these objects. The whole calculation rested on the view that England, so dependent on the commerce of the United States, would have no choice but to submit to American demands. There is little question, however, that the calculation was mistaken, and that the Republican system would have led to a war with Great Britain. Jefferson was always alive to the possibility that America might profit from Europe's quarrels, and the theme has powerful support in his correspondence. Yet his conception of how this might be done was defective. As the diplomacy of Federalism showed, the true path of securing America's advantage from Europe's distress lay in an accommodation with England—an accommodation that, because of its retroactive effect on Spain, secured the American claim to the great interior. This was a fact which Jefferson did not as yet understand and which his own system did not allow for.

8

Toward the Republican
Triumph of 1800

FEW EPISODES IN AMERICAN HISTORY have the intrinsic drama
of the great crisis that afflicted the country as it approached
the dawn of a new century. The European war had unleashed tre-
mendous passions and dreadful energies: nations disappeared, gigantic projects were afoot. The victories of French arms on the
continent and the possibility that Britain might go down under the
onslaught of naval mutinies, national bankruptcy, domestic rebellion, and French invasion changed the face of world politics and
had extraordinary repercussions at home. In reaction to the Jay
Treaty, France revealed a face of unmitigated hostility to the United
States, instituting decrees that threatened American neutral shipping wherever on the ocean it might be found. The idea that France
aimed at universal empire and that, having finished with Great
Britain, America was next on the list of victims, took hold of the
Federalist imagination and led to domestic measures that would
have been deemed impossible only a few years before. A wave of
indignation broke over the country when news arrived that agents
of Talleyrand (Messrs. X, Y, and Z) had approached the American peace commissioners in Paris and demanded, as a condition
of good relations, a substantial *douceur* as well as a subsidized
loan. In response to these humiliating conditions, public opinion
swung toward the Federalists, who pushed through large increases
in land and naval forces in addition to repressive legislation (the
Alien and Sedition Acts) aimed at foreign and domestic subver-

sives. Republicans were convinced that Federalist policy had as its secret intention the provocation of a war with France, with its awful train of domestic consequences; Federalists, that a seditious faction at home, in league with the most powerful nation on earth, aimed at overthrowing the government by force and introducing republicanism *à la français*. Depending on the witness, the nation tottered on the brink of anarchy, civil war, foreign invasion, or despotism.

Civil war, however, was avoided, as was war with France. The Directory was content with despoiling American trade, and most Federalists (including Adams and Hamilton) were unwilling to pass beyond a state of mitigated hostility with France. For a variety of reasons, the crisis ultimately broke in favor of the Republicans, ensuring Jefferson's election in 1800. Unbeknownst to Jefferson, divisions had been multiplying within the ranks of the Federalist party—over the kind of military preparations the nation should undertake, over the proper handling of negotiations with France, and over the command of the substantially enlarged army. When Adams announced a new peace mission to France in February 1799, division broke into the open, with Hamiltonians regarding the mission as a gratuitous and avoidable concession inconsistent with the dignity of the government, and Adams prizing the measure as a bold stroke against a secret cabal in the government intent on a war with France. This internal division within Federalism coincided with a reversal of French fortunes in Europe—beginning with Napoleon's defeat in the Battle of the Nile in the summer of 1798. Those reverses, together with the conciliatory gestures of the French government, dimmed the prospect of a French invasion and made the Federalists' defense preparations appear unnecessary at best, malicious at worst (even to the President himself). The "war fever" thus steadily cooled, the doctor curing it, as Jefferson said, in the guise of the tax gatherer.

None of the great men who dominated the crisis—Adams, Hamilton, and Jefferson—played a particularly creditable role. Adams acted a very inconsistent part, blowing hot and cold over relations with France. His later recollection, which portrayed him as consistently aiming at peace against the war party in his cabinet, is not supported by the record, which was one of dizzying inconsistency.[129] Hamilton, though surely right in urging preparedness on the nation and in concentrating his attention on a possible French occupation of Louisiana, nevertheless erred badly in seriously entertaining a project for the liberation of Spanish America, to be carried out, in the event of a decisive rupture with France, by American land forces in conjunction with the British fleet. Adams's later comment on this plan—that he had seen too many revolutions to "be desirous of engaging myself and my country in most hazardous and bloody experiments to excite similar horrors in South America"—was surely just.[130]

Nor did either man act to stem Federalist sentiment for the Alien and Sedition laws in the Congress, though both had reservations over the propriety of the measures Congress passed. Under authority of the Sedition Act, prosecutions were undertaken against printers for opinions that, though assuredly false and even scandalous, were not too dissimilar from those entertained by a large number of Republican leaders. It was bad policy and contrary to the genius of American institutions to combat such opinions through the courts. "The American Government," as Gallatin held, "had heretofore subsisted, it had acquired strength, it had grown on the affection of the people, it had been fully supported without the assistance" of such laws. "To repel opposition by the single weapon of argument" was the only policy appropriate to the crisis, and in departing from it the Federalists not only failed to secure their object but indeed laid the groundwork for the ruin of their own party.[131]

Jefferson's reaction to the Alien and Sedition laws was scarcely more creditable. In June 1798, he counseled patience to those contemplating a dissolution of the Union in opposition to Federalist measures. The "reign of witches," he told John Taylor, would soon pass over. The people would rapidly recover "their true sight, restoring their government to its true principles." The "evils of a scission" would be endless. This famous letter to Taylor was one of Jefferson's greatest despatches; it was full of wise sayings on the inevitability of party conflict and on the utmost danger that might ensue were the Union to be broken: "Seeing that we must have somebody to quarrel with, I had rather keep our New England associates for that purpose, than to see our bickerings transferred to others."[132] In the event, however, his own line of conduct wavered from this sound advice. His draft of the Kentucky resolutions, completed in November 1798, contained much in the way of reckless utterance. It was one thing to declare the repressive legislation of the Federalists as unconstitutional infringements on powers explicitly delegated by the Constitution, quite another to urge the propriety of the direct nullification of federal authority by individual states. No rampart, he thought, now remained "against the passions and the powers of a majority in Congress." He deprecated the constitutional remedy whereby the people themselves, by free election, might turn out of the national government those who had pushed through such measures. That remedy was appropriate only "in cases of an abuse of the delegated powers." In the present case, however, "where powers are assumed which have not been delegated, a nullification of the act is the rightful remedy." The general government was a creation of the several states, which had made a compact among themselves: "In cases not within the compact, (casus non fœderis)," every state had a natural right "to nullify of their own authority all assumptions of power by others within their limits." Thus he hoped that "the co-States, recurring to their natural right in cases not made federal, will concur in declaring these acts void, and of no force, and will each take measures of its own for providing that neither these acts, nor any others of the General Government not plainly and intentionally authorized by the Constitution, shall be exercised within their respective territories." [133]

That these were dangerous and inflammatory utterances is conceded even by Jefferson's friendly biographers. The doctrine which Jefferson presented "without express qualifications or safeguards," as Dumas Malone has well said, "could have paralyzed the general government if carried to its logical conclusion"—as it was, of course, a generation later.[134] Happily, the Kentucky legislature, under the direction of John Breckenridge, excised the most extreme passages, and Madison took even more careful ground in directing the Virginia resolutions through the state legislature— moderating tendencies to which Jefferson made no objection. A comparable sequence occurred the following year, when Jefferson's proposals for radical language were again rejected by Madison, again with Jefferson's acquiescence. Even so, the appearance of the 1798 resolutions—alongside reports that Virginia was arming to forcibly obstruct the measures—would prompt Hamilton's suggestion that a "clever force" be collected and "drawn towards Virginia for which there is an obvious pretext." Then, he thought, measures might "be taken to act upon the laws & put Virginia to the Test of resistance."[135] A wrong step at this critical juncture by either of the parties might well have brought matters to a forcible decision.

It is a curious fact that the same specter—unlimited and arbitrary power, insatiable in its aims, destructive to liberty—haunted both sides in the great crisis of '98. For Jefferson, that danger arose at home; for most Federalists, it came from abroad. France now emerged, in the thinking of the high Federalists (or Hamiltonians), as "the most flagitious, despotic, and vindictive government that ever disgraced the annals of mankind . . . a government marching with hasty and collosal strides to universal empire, and, in the execution of this hideous project, wielding with absolute authority the whole physical force of the most enthralled, but most powerful nation on earth." So Hamilton wrote in *The Stand,* a series of newspaper pieces published in the spring of 1798. "Swelled to a gigantic size and aping Rome, except in her virtues," France plainly aimed at "the control of mankind."[136] The possibility that Britain might be overthrown and the likelihood that France would succeed to the Spanish empire in the Americas alike pointed to serious danger. For Hamilton, there were two main implications.

One was that Britain, by virtue of its naval superiority, was a valuable shield against French ambition; the other was the indispensable necessity of defense preparations at home. The British Navy, on this view, was as much a barrier against France as a menace to American security, partly friend and partly foe. Britain, Hamilton argued, "has repeatedly upheld the balance of power [in Europe], in opposition to the grasping ambition of France. She has no doubt occasionally employed the pretense of danger as the instrument of her own ambition; but it is not the less true, that she has been more than once an effectual shield against real danger." This was the case in the Wars of the Spanish Succession, when Britain had ranged itself against the ambitions of Louis XIV; it was assuredly the case now.[137] The recognition that American security would be gravely imperiled by the overthrow of the European system not only dictated cooperation with the English; it also underlined the necessity of permanent preparations for defense at home. "It can never be wise," he wrote at the end of the year, "to vary our measures of security with the continually varying aspect of European affairs." In consideration of "the rapid vicissitudes, at all times, of political and military events" and "the extraordinary fluctuations which have been peculiarly characteristic of the still subsisting contest in Europe," the obvious policy was "to place our safety out of the reach of the casualties which may befal the contending parties and the powers more immediately within their vortex." Only by unremitting vigilance and exertion would the United States be able to secure its own destiny. "Standing, as it were, in the midst of falling empires, it should be our aim to assume a station and attitude which will preserve us from being overwhelmed in their ruins." [138]

Republicans denied both cardinal points of Federalist doctrine. Gallatin, now the leading spokesman for the Republicans, called the idea of a French invasion "a mere *bugbear*," designed to foist a standing army upon the people but otherwise to serve no legitimate purpose of national defense. The French, he conceded, were ambitious. Intoxicated with their victories, and feeling their own great power, they were a menace to surrounding nations. Nevertheless, "it was neither their interest nor in their power to effect an invasion of this country." Their lack of power was shown by the small force—"two or three frigates and a few hundred men with arms"—that they had been able to put into the West Indies during the present war. Too many other interests, of much greater magnitude than a punitive expedition against the United States, would occupy France's limited marine resources. Hence, so long as the European war continued, "they are either unable or unwilling to make any great exertions at a distance from Europe." [139]

Gallatin also questioned whether the balance of power was of any real concern to the United States. In a powerful speech on a bill providing for the salaries of foreign ministers, he argued that "we have no interest whatever in that balance, and by us it should be altogether forgotten and neglected." Recalling the old British debate over "continental connections," he held that the balance had served as a cause or pretense for many useless wars. Recollect, he asked the House, "at what late period the British Ministry wanted to involve the British nation in a war with Russia for the purpose of preserving that balance, which might, in their opinion, be affected by the transfer of Oczackow, situated as it is in a remote corner of the Black Sea, from the hordes of Tartars which rule Turkey, to the Tartar hordes which inhabit Russia." Nor was the cause of freedom itself one that should alter the foreign policy of the United States. "We may lament the fate of Poland and Venice"—victims, respectively, of each of the rival coalitions—"and I never can myself see, without regret, independent nations blotted from the map of the world. But their destiny does not affect us in the least."[140]

These debates were indicative of a general movement in the position of the parties since the outset of the European war. Hamiltonians moved from a posture of nonalignment and neutrality toward a position of much closer cooperation with Great Britain, approximating in all but name a condition of alliance with the British. At the least, there was in 1798 an explicit recognition of American dependence on British sea power, together with the sense that French ambitions would inescapably draw the United States into the vortex of European politics. The Republican leadership, on the other hand, moved in the opposite direction. The sympathies for the republican cause in France weakened in the face of France's ambitions in Europe and the accumulation of French injuries against the United States. Though Jefferson continued to believe that Federalist policy since the Jay Treaty was responsible for the state of mitigated war that came to exist between France and the United States, his basic instinct was to withdraw from the fray. Britain's insults against American commerce had not been deemed worth a war by the Federalists in 1794; it was absurd, he thought, to go to war against France for comparable violations in 1798. He hoped for peace in Europe and thought it would be the only thing that could save the country from plunging into the abyss.[141]

The quasi war with France also prompted a reassessment by Jefferson and other Republican thinkers of the role of the carrying trade. Only a few years before, Madison had warned that the Jay Treaty would deliver a crushing blow against American commerce and navigation. By the end of the decade, however, Republican thinkers arrived at the conclusion that the carrying trade was dangerously overextended. Jefferson's fears of the 1780s, which equated the carrying trade with continual warfare, now revived. The country, he wrote Joseph Priestley, "was running navigation mad, and commerce mad, and navy mad, which is worst of all."[142] These sentiments echoed the conclusions of other Republican pamphleteers (such as Thomas Cooper and Priestley himself), and they represented a great departure from Jefferson's earlier hopes that the country might become the carrier for all parties in the European war. That now seemed the road to ruin.[143]

Jefferson's thoughts on the balance of power and the value of the French alliance also underwent a significant change. In 1797, he was still describing his system as "justice to England, gratitude to France, and subservience to none."[144] He cheered Bonaparte's victories in Europe and the blows that England was suffering— only that change, he thought, prevented the "war party" at home from attempting bolder measures. Even then, however, he began to have second thoughts. The subjugation of England, he acknowledged in early 1798, "would indeed be a general calamity." Though Jefferson considered that eventuality to be impossible, he was slowly arriving at a more even-handed view of the European war. Of equal importance was his growing sense of the destructive impact of the European war on politics at home: "Our country-men," he told Elbridge Gerry in 1797, "have divided themselves by such strong affections, to the French & the English, that nothing will secure us internally but a divorce from both nations."[145] The war fever of the following two years confirmed him in these sentiments. The reviving sense of isolation derived partly from his conception of the direction the Federalists were taking the nation. "I am not," he told Gerry in 1799, "for linking ourselves by new treaties with the quarrels of Europe; entering that field of slaughter to preserve their balance, or joining in the confederacy of kings to war against the principles of liberty." That warning, directed against the possibility of an alliance with Great Britain, still left room for the treaty with France, and indeed Jefferson had objected to Congress's unilateral renunciation of the French alliance in 1798. Nevertheless, by April 1799, he could speak of abjuring "all political connection with every foreign power."[146]

The developing Republican rejection of all things European was reflected in the debate over diplomatic establishments that occurred in the late 1790s. John Adams, in supporting a modest foreign establishment, had declared that

> although it is very true that we ought not to involve ourselves in the political system of Europe, but to keep ourselves always distinct and separate from it, if we can; yet, to affect this separation, early, punctual and continual information of the current chain of events, and of the political projects in contemplation, is no less necessary than if we were directly concerned in them. It is necessary, in order to the discovery of the efforts made to draw us into the vortex, in season to make preparations against them.[147]

Jefferson had laid down a similar rule in the 1780s. Noting that the Turks' ignorance of Europe had exposed them to annihilation, Jefferson had warned that "while there are powers in Europe which fear our views, or have views on us, we should keep an eye on them, their connections and opposition, that in a moment of need, we may avail ourselves of their weakness with respect to others as well as ourselves, and calculate their designs and movements, on all the circumstances under which they exist."[148] These prudent views, however, came under attack in the late 1790s, not least by Jefferson himself. In his "profession of political faith" to Gerry, Jefferson now declared himself in favor of "little or no diplomatic establishment." He would speak, in the first year of his presidency, of calling in "our diplomatic missions, barely keeping up those to the more important nations."[149]

This growing sense of isolation and separateness among the Republicans was reflected, finally, in the now common rejection of the idea that the cause of liberty at home was intimately bound up with the survival of republican government in France. Gallatin had pointed the way toward this rejection, and Jefferson now accepted the proposition himself. Even before Napoleon's coup d'état of 18 Brumaire, Jefferson could write that "tho I cordially wish well to the progress of liberty in all nations, and would forever give it the weight of our countenance, yet they are not to be touched without contamination from their other bad principles."[150] The news that a "Dictatorial consulate" was established in France completed the disaffection. Following an old script, which Jefferson knew by heart, Napoleon had "transferred the destinies of the republic from the civil to the military arm." Some, he told Samuel Adams, "will use this as a lesson against the practicability of republican government. I read it as a lesson against the danger of standing armies." It was of immediate relevance to America, for the same ambitions lurked in the breast of "our Buonaparte"—Alexander Hamilton—who, if war came, would "step in to give us political salvation in his way."[151] If the Republicans could only muddle through the current crisis, however, none of this would matter. "Our vessel is moored at such a distance, that should theirs blow up, ours is still safe, if we will but think so."[152]

The outlook that Jefferson brought to the presidency in 1801 was significantly different from the views he once had entertained. The opposition to Hamilton and Federalism had converted him into a radical opponent of the power state, so much so that he had embraced the idea that the Union ought to be dissolved if the price of its preservation lay in submission to unlawful power. Armies, navies, and diplomatic establishments, the necessity of which he had acknowledged in the 1780s, were now thought to be potent instruments for the wreck and ruin of the republican experiment at home. Most of this was already clearly apparent by the early 1790s; the crisis of '98 forged them ever deeper in his consciousness.

For France, his ardor had risen to a high pitch with the outbreak of the European war but had finally cooled by the end of the decade. Now neither the cause of liberty abroad nor the alliance with France formed a part of his system. That the Adams administration had succeeded in terminating the alliance with France (which it did with the convention of Mortfontaine in 1800) was fully consistent with his new thinking, which called for "peace, commerce, and honest friendship with all nations—entangling alliances with none."[153]

Napoleon's coup d'état and the termination of the Franco-American alliance thus caused no loss of optimism. On the contrary, it served to confirm and strengthen Jefferson's sense of American exceptionalism. As he entered the office of the presidency, he enjoyed the governance of "a rising nation, spread over a wide and fruitful land, traversing all the seas with the rich productions of their industry, engaged in commerce with nations who feel power and forget right, advancing rapidly to destinies beyond the reach of mortal eye." He expected a "perfect consolidation," but it was one of sentiment, not of power. It was to be found in the people returning to their senses after being "hood-winked" by Federalism: "Now they see for themselves." The republican experiment, he thought, was just beginning, and as he reflected on its implications he could barely contain his enthusiasm. "A just and solid republican government maintained here," he told John Dickinson, "will be a standing monument & example for the aim & imitation of the people of other countries; and I join with you in the hope and belief that they will see, from our example, that a free government is of all others the most energetic; that the inquiry which has been excited among the mass of mankind by our revolution & its consequences, will ameliorate the condition of man over a great portion of the globe."[154]

III

THE DIPLOMACY OF
EXPANSION (1801–5)

9

The Nature of
Jefferson's Success

THE TRIUMPH OF THE PARTY of Jefferson in the election of 1800 was one of the great turning points in the political history of the United States. For the first time, a political opposition gained control of the national government. In taking power from the Federalists, the Republicans displaced those who were largely identified with the founding of the state. The fall of the Federalists was more than the defeat of a party that had governed from the moment of creation. It was also seen as the defeat of a vision of society that had inspired the ruling group and of the principles by which this vision was to have been realized. Among participants in the political struggle of the time, few held this view more strongly than did Thomas Jefferson. To the leader of the Republicans, the election of 1800 constituted a revolution, and one, as he wrote in later years, that "was as real a revolution in the principles of our government as that of 1776 was in its form."[1]

The characterization had merit. In their respective conceptions of the role the national government should play in domestic affairs, between Federalists and Republicans there appeared to be little common ground. Jefferson and his allies were pledged to the strict limitation of this role. The sum of good government, Jefferson declared in his First Inaugural Address, was one "which will restrain men from injuring one another, shall leave them otherwise free to regulate their own pursuits of industry and improvement, and shall not take from the mouth of labor the bread it has earned."[2] What few positive tasks government was considered to have were to be undertaken in furtherance of a society devoted primarily to agriculture, one in which commerce played no more than an auxiliary role and industry played almost no role at all. Given a sufficient supply of open land to absorb a rapidly growing population and an international environment affording the opportunity for disposing of agricultural surpluses, there was little need of the intrusive hand of government. In these circumstances, the national government was seen to hold out much more of a threat than a promise.

There was no greater manifestation of such threat in Republican eyes than debt and taxes. Jefferson came to office committed to do everything in his power to remove these seeds of corruption. In the course of his first term as President, the Republican administration largely redeemed this pledge, doing away with almost all internal taxes. At the same time, it devised and implemented a scheme for paying off the public debt in a period of sixteen years. Taken together, these measures appeared at the time a startling realization of Republican orthodoxy. They were made possible only by a further reduction in military expenditures and by the continuation of a high level of imports, since customs duties accounted for the bulk of government revenues. Thus dependent on peace and a favorable environment for American commerce, the administration's fiscal reforms could not be separated from developments in the external world. In the Jeffersonian calculus, American democracy depended on a particular sort of external environment.

The record of Republican accomplishment in domestic affairs was such as to secure and, indeed, to enhance the party's hold on power. Yet it was not in domestic affairs but in foreign policy that Jefferson and his administration enjoyed their greatest triumph. In comparison with the acquisition of Louisiana all else seemed to pale in significance. Almost half a continent had been gained and at a price that was negligible; without a resort to force, the France of Napoleon was kept out of the Mississippi valley. Only exceptionally was the attempt made, and then mostly by political adversaries, to distinguish between the fortunate development and the diplomacy that attended it. The Louisiana Purchase conferred on those participating in it one of the great political windfalls in American history.

"Among the various acts of political wisdom, for which the present administration of the United States has been distinguished, the late cession of Louisiana to this country, by the exercise of a magnanimous policy on the part of our government, may be classed as the most pre-eminent."[3] So began one of the better known pamphlets celebrating the purchase. What Allan Magruder considered the most preeminent act of political wisdom on the part of the Jefferson administration, David Ramsay found an act without historical parallel. "History affords no example," that remarkable historian of the American Revolution declared, "of the acquisition of such important benefits, at so moderate a price, and under such favorable circumstances."[4] With the exception of the establishment of independence and the making of the Constitution, Ramsay insisted, "the acquisition of Louisiana is the greatest political blessing ever conferred on these states." Louisiana was a great political blessing—even more, a triumph of statecraft—not only because of what we got but for the way we got it. "We have secured our rights by pacific means," wrote the editors of the administration's most faithful newspaper, "truth and reason have been more powerful than the sword."[5]

This initial reaction to Louisiana has stood the test of time remarkably well. Now as then, the acquisition of Louisiana is generally seen not only as a great political blessing but as a triumph of statecraft. If there is a significant difference, it is in finding Jefferson's diplomacy rather more sophisticated than contemporaries had been inclined to view it. In his masterful account of the *Mississippi Question, 1795–1803*,[6] Arthur Whitaker argued that since the recourse to force as a means for resolving the crisis brought on by the retrocession "was out of the question," Jefferson had to rely on diplomacy. But diplomacy depended on circumstance and this meant, in effect, that Jefferson had to wait until France and England again went to war. "Procrastination was, therefore, the keynote of his policy." This did not mean inaction; nor was Jefferson passive during the crisis. On the contrary, Whitaker wrote, he did "all that it was possible for him to do." When Spain suspended the right of deposit at New Orleans, Jefferson kept the reaction of the West from getting out of hand while he negotiated with Spain and France. At the same time, he made use of the domestic outcry to persuade the representatives of these two states that war was imminent and that he was the only force standing between war and peace. The envoys, persuaded by this performance, responded as Jefferson hoped they would.

The high marks Whitaker thus gave to Jeffersonian diplomacy in the Louisiana crisis are also given by the author of the leading biography of the nation's third president. Like Whitaker, Dumas Malone finds that the key to Jefferson's policy was in playing for time. Jefferson "followed a policy of patience," Malone writes, and "events were to vindicate his judgment." Malone acknowledges that in Jefferson's diplomatic triumph "there was a considerable element of luck." But the mark of an effective, and even a great, statecraft is not that it is unattended by fortunate circumstances; it is instead to be seen in the ability to adapt to, and make use of, such circumstances when they do arise. "The wisdom of its flexible policy was demonstrated by events, and if its measures had been described as feeble by some partisan critics they were shown to have been vigorous enough." In Malone's view, Jefferson was "ready to use force eventually if his peaceful efforts to meet the imperative needs of his country should fail." Luck notwithstanding, "the administration was alert and skillful, and, except in minor details, one wonders how its procedure could have been improved upon."[7]

The outstanding biographer of Jefferson, Merrill Peterson, is at one with Malone in this assessment. To Peterson as well, Jefferson's cautious diplomacy of watching and waiting for renewed war in Europe paid off. Indeed, outbidding all in his judgment of Jefferson's Louisiana diplomacy, Peterson writes that "Jefferson played the game to perfection":

> In the two-year-long campaign, he never overextended himself, never cut off his lines of retreat, never risked the consequences of an armed encounter, yet kept the objective steadily in view. Inaction was as much a part of his strategy as action, and he knew when to wave the sword and when to sheath it. . . . The entire proceeding was an impressive vindication of the ways of peace in the conduct of American affairs. In the final analysis, of course, he was saved by European war. But this was not simply a piece of dumb luck. The prospect of war, like the prospect of French defeat in Santo Domingo, entered into his calculations. He correctly weighed the imponderables of the European power balance, shrewdly threatened to throw his weight into the British scale,

gauged the effect of renewed war on Napoleon's imperial design, and prepared to take advantage of the *démarche* when it came.[8]

There have always been those who have dissented from the prevailing judgment of Jefferson's diplomacy in the affair of Louisiana, though they have never comprised more than a small minority. To Jefferson's admirers their criticism has nearly always seemed partisan in character. Certainly this is true of the President's first—and greatest—detractor. Almost from the outset of the Louisiana crisis, Alexander Hamilton was critical of what he regarded as the passivity with which the Jefferson administration responded to the rumored retrocession. In early 1803, Hamilton urged the administration to pursue a more "energetic" policy to meet the growing prospect of French control of Louisiana.[9] In the circumstances, he insisted that only two courses were possible: "First, to negotiate and endeavor to purchase, and if this fails to go to war. Secondly, to seize at once on the Floridas and New-Orleans, and then negociate." In Hamilton's view, the proper course to follow was the second. Consistent with his conviction that in such matters "energy is wisdom," he favored the path of war and to this end urged immediate increases in the army and navy.

The startling news of the Louisiana Purchase prompted an immediate response from Hamilton.[10] Acknowledging that the purchase would undoubtedly "give eclat" to Jefferson's administration, he went on to insist "that the acquisition has been solely owing to a fortuitous concurrence of unforseen and unexpected circumstances, and not to any wise or vigorous measures on the part of the American government." Having remained passive before the threat raised by Napoleon, Hamilton wrote, "we were about to experience the fruits of our folly" when Great Britain once again moved toward war with France. Even while at peace, however, France had wasted its opportunity by attempting to subjugate Santo Domingo, "and the means which were originally destined to the colonization of Louisiana, had been gradually exhausted by the unexpected difficulties of this ill-starred enterprise." "To the deadly climate of St. Domingo, and to the courage and obstinate resistance made by its black inhabitants are we indebted for the obstacles which delayed the colonization of Louisiana, till the auspicious moment, when a rupture between England and France gave a new turn to the projects of the latter, and destroyed at once all her schemes as to this favorite object of her ambition."[11]

Hamilton's view has since been dismissed by most historians as an all too obvious case of sour grapes. Still, it is a view that has been occasionally embraced. One historian to have done so, though without ever acknowledging that Hamilton first gave expression to it, was Henry Adams. Adams, too, attributed the outcome to fortune, to what Hamilton termed "a fortuitous concurrence of unforeseen and unexpected circumstances," rather than to diplomatic skill. Moreover, the nature of those circumstances was largely the same for Adams as it had been for Hamilton. For both, Santo Domingo was the great obstacle that drained Napoleon's resources and delayed the French occupation of Louisiana. It was the growing difficulties Napoleon met with in his attempt to reimpose French authority on the island that eventually turned him away from his grand scheme of empire in the West and back to a renewal of the conflict with England. Only the prejudice of race, Adams wrote, "blinded the American people to the debt they owed to the desperate courage of five hundred thousand Haytian negros who would not be enslaved."[12]

That the events on Santo Domingo affected the outcome of the Louisiana crisis has of course always been acknowledged, by both Jefferson and subsequent historians. But there is a considerable difference between the emphasis Hamilton and Adams placed on the role of Santo Domingo in the affair of Louisiana and the customary recognition that events on the island did affect the outcome of this affair. In the prevailing view, and it is the view that Jefferson himself initially intimated, Santo Domingo is only one factor among several that determined the outcome. Even more, it is only one element among several that Jefferson made use of in fashioning a diplomatic strategy for successfully countering French ambitions in Louisiana. In this view, the decisive consideration was the strategy and not any of the several elements of which this strategy made use. In the view of Hamilton and Adams, however, it was the element, that is, Santo Domingo, that largely determined the outcome. The strategy is seen as incidental at best and counterproductive at worst. Hamilton thought it was counterproductive, for the most part, and, here again, Adams followed his lead. "The essence and genius of Jefferson's statesmanship," Adams insisted, "lay in peace."[13] To this, all else was subordinated. Yet a diplomacy so guided was placed at an inherent, and perhaps fatal, disadvantage, particularly when dealing with a Napoleon. In the Louisiana crisis, Adams suggested, circumstances conspired to ward off what might well have been a catastrophe.

Hamilton and Adams to the contrary notwithstanding, the view persists that Jefferson was a "superb diplomatist" in the Louisiana affair. Whether he did indeed play the game to perfection and whether the acquisition owed more to diplomatic prowess or the arbitrament of chance—or, more simply put, to skill or dumb luck— are questions that will be examined at length in the pages that follow. Their significance resides in their capacity to illuminate the nature of Jefferson's reason of state. That Jefferson saw the necessity which the retrocession of Louisiana to France imposed on him is clear; that he was prepared to accept the consequences of this necessity—in the shape of either war with France or alliance with England—is much less so. In truth, he withdrew so much from the prospect of war with France that he was prepared to accept the French military occupation of Louisiana, though he knew that such an occupation would pose a severe danger to the Union. He never once, moreover, had in his mind a conception of alliance with Great Britain that the latter was prepared to accept, and for much the same reason that he waved the sword "yet never risked the consequences of an armed encounter." Alliance with England, like war with France, would do violence to his vision of domestic society.

His diplomacy in the Louisiana affair was thus not that of a man firmly committed to a clear strategy, but of one who feared that he might be impaled on the horns of a dilemma. Jefferson never resolved this dilemma in his own mind; indeed, he hoped beyond hope that he would not have to face it. This, above all, accounts for the special place that time enjoyed in his diplomatic scheme. His faith that time would bring him deliverance was rooted more in temperament than in solid expectation, and it was badly shaken on more than one occasion. That his faith was seemingly vindicated was, however, among the more important consequences of the Louisiana affair. He had escaped, he thought, the perils of reason of state. Necessity had reared its ugly head, yet he had resisted the temptation to give in to it. His success, he concluded, lay as much in the means he had employed—or rather not employed—as in the great end he had accomplished.

Yet the denouement was ironic. He had labored to avoid war and its corruptions not out of any sentimental attraction to pacifism but largely because he feared that it would wreck his constitutional edifice, in which the powers of the national government were strictly circumscribed by explicit delegations of authority. The manner in which he brought Louisiana into the Union, however, was fatal to this constitutional vision; it required, Jefferson believed, an act beyond the Constitution. The incorporation and governance of Louisiana could be justified only on the basis of constitutional doctrine in direct contradiction to prohibitions that Jefferson had long held inviolable—prohibitions, as Henry Adams said, that were "the breath of his political life."

Nor is this all. A diplomacy that had begun by trumpeting the virtues of peace and persuasion soon found itself incapable of resisting the attractions held out by force and fraud. The claim to West Florida that the administration set forth was of a highly dubious nature, though the administration quickly conferred upon it the character of certainty and clothed it in the language of high moral principle. Spain, moreover, soon found itself the object of reflections by Jefferson and Madison of the most traditional realpolitik. Threats deemed imprudent in the case of France were viewed differently in the case of Spain. The episode showed that the attraction of the administration for the new as against the old diplomacy was far more a function of relative power than of anything else.

The consequences of the Louisiana affair were thus extraordinary, and not least of these was the effect it had on the President himself. It enlarged his appetite while confirming his own system. It fortified his conviction that the key to policy with both Britain and Spain lay through France; it confirmed his faith that time was on his side; and on more than one occasion it made him confuse might and right, the deadly sin that he attributed to the *ancien régime*. By the end of Jefferson's first administration, he was already committed to the diplomatic design that would end in the ruin of the second.

10

The Significance
of the Mississippi Valley

THE IMMEDIATE SIGNIFICANCE of the Louisiana purchase was
that it finally resolved the issue of the control of the Mississippi valley and, in particular, of the great river that formed the
West's major avenue of communication with the outside world.
From the time independence and union were achieved, the security
and prosperity of the United States had been seen as inseparable
from the right to navigate the Mississippi, for without this the
loyalty of the "men of the western waters" would, it was thought,
have been greatly weakened. The vital connection was affirmed
early in the administration of President Washington when war between Great Britain and Spain threatened and fear arose over the
future status of the Mississippi. Jefferson, then Secretary of State,
took the position that freedom of navigation of the river was a
necessity for the United States, one that could be denied only at
the price of independence and union. To refuse to vindicate our
interest in the river, if necessary by going to war, he insisted, not
only would result in the loss of the western lands and peoples but
would lead ultimately to the sacrifice of the original states of the
Union as well ("Were we to give up half our territory rather than
engage on a just war to preserve it, we should not keep the other
half long").[14] Jefferson's emerging adversary, Alexander Hamilton, did not differ in assessing the significance of the Mississippi.
Responding to a query of the President on the passage of British
troops through American territory, the Secretary of the Treasury

declared that securing the navigation of the Mississippi was "essential to the unity of the Empire." The failure to ensure the right of navigation there would lead, Hamilton insisted, to the dismemberment of the western country with all its attendant consequences.[15]

At the beginning of the 1790s the equation drawn by both Jefferson and Hamilton—between independence and union, on the one hand, and the unimpeded use of the Mississippi, on the other hand—would not have been accepted in much of the North. By 1800, however, northern opposition to a "Mississippi interest" had sharply declined as the growing commercial importance in the trade of the river became apparent. By the time Jefferson became President the equation elicited something that approached consensus. Political adversaries might and did differ over the means by which the American interest in the Mississippi could be best secured. On the need to preserve this interest in the Mississippi there was now a near unanimity of view.

The logic of this need was simple yet compelling. It went roughly like this: the unimpeded use of the Mississippi was indispensable if the lands between the Appalachian Mountains and the river were to be secure against the danger of dismemberment or secession. To obtain an outlet for their produce, the people of the Mississippi valley would have to come to terms with the power that dominated the great waterway. A hostile power in control of the river would represent, at the very least, a barrier to further expansion on our part. More likely, if it displayed sufficient energy, such a power would' act as a magnet on those ready to exchange allegiance for protection. Spain in control of the Mississippi was tolerable, if barely so, only because the level of energy it was capable of displaying represented no serious barrier to future expansion. Given the near contempt that Spain's presence generally evoked, the prospect that Spanish power could serve as an effective magnet was negligible, so much so that Jefferson, when Secretary of State, had secretly applauded a Spanish plan to open lands to American immigrants.[16] But the same attitude could not be entertained toward France or Great Britain. Either in control of the Mississippi and its outlet, New Orleans, would directly threaten, in Hamilton's words, the unity of the American empire.

In 1801, when the threat of French control of New Orleans and the Mississippi suddenly appeared imminent, this view had come to represent almost a commonplace of political discourse. Of critical importance was the equation drawn between security and expansion. Security required expansion, the reasoning went, since only through expansion could the prospect of dangerous neighbors be removed. In turn, the removal of dangerous neighbors would obviate any need to enter into relationships of alliance that might compromise independence. For with the removal of dangerous neighbors the need to enlist the power of others to balance that of a neighbor would no longer obtain. If the desire for greater security pointed to expansion, so too did the desire for greater independence.[17]

The dimensions of the consensus over the Mississippi interest were borne out in the congressional debate of early 1803, which arose after news arrived that the Spanish Intendant at New Orleans had suspended the right of deposit, in apparent violation of the 1795 treaty with Spain. The West raged against the prospect that Americans might be unable to ship produce by way of the outlet at New Orleans, and the nation as a whole followed the lead of the West. The resolutions sponsored by Senator James Ross, a Federalist from western Pennsylvania, authorized the President to occupy New Orleans in order to ensure the right of deposit, and provoked a heated debate in the Senate.[18] Yet on the necessity of controlling the great waterway there was no disagreement among the supporters and opponents of the resolutions. "We can never have permanent peace on our Western waters," White of Delaware declared in support of the resolutions, "till we possess ourselves of New Orleans, and such other positions as may be necessary to give us the complete and absolute command of the Mississippi." Jefferson's defenders in the Senate did not deny the vital character of the interest. Breckenridge of Kentucky considered the right of navigation "so all-important, so indispensable to the very existence of the Western States, that it was a waste of words and time to attempt to portray the evils which a privation of it would produce." The only difference in the Senate, as well as in the country as a whole, concerned "the proper means to obtain this great end." The Republicans preferred the course of negotiations; the Federalists, of immediate war.

Though it was the suspension of the deposit that deeply stirred public opinion in the United States, the reaction owed much to the knowledge that Louisiana had been ceded to France. For it was generally assumed that the suspension had to have been the work of France, not Spain. The contempt with which Spain was held in American eyes made it seem incredible that Madrid would hazard such a step. But France was another matter and the suspension was seen as an indication of what was in store for the country once the French established their hold over New Orleans.

In fact, the suspension of the deposit was not the act of France but of Spain. A desire to embarrass the French and a sense of frustration over American smuggling had prompted the government in Madrid to act as it did.[19] This consideration apart, the story of Louisiana is, above all, a story of French aspirations and policy.

It was the French who provoked the crisis with the signing of the secret Treaty of San Ildefonso on October 1, 1800, by which Spain conditionally agreed to retrocede Louisiana to France.[20] And it was the French who brought the crisis to an end when Napoleon, disgusted by his ruinous reverses on Haiti, startled the American negotiators in Paris by his offer to sell the whole of Louisiana to the United States. By contrast, the continued presence of Spain in the Mississippi valley posed no real danger to the interests of the United States. Spanish rule in Louisiana was steadily declining in effectiveness. The widespread expectation was that the days of the Spanish empire in Louisiana, and in the Floridas as well, were numbered, and that the United States would be the inevitable beneficiary. But the prospect of French intervention put an entirely different face on matters. France in possession of New Orleans held out a serious threat to the integrity of the union. As Jefferson summarized the danger:

There is on the globe one single spot, the possessor of which is our natural and habitual enemy. It is New Orleans, through which the produce of three-eights of our territory must pass to market, and from its fertility it will ere long yield more than half of our whole produce and contain more than half our inhabitants. France placing herself in that door assumes to us the attitude of defiance. Spain might have retained it quietly for years. Her pacific disposition, her feeble state, would induce her to increase our facilities there, so that her possession of the place would be hardly felt by us, and it would not perhaps be very long before some circumstance might arise which might make the cession of it to us the

price of something of more worth to her. Not so can it ever
be in the hands of France. The impetuosity of her temper,
the energy and restlessness of her character, placed in a point
of eternal friction with us, and our character, which though
quiet, and loving peace and the pursuit of wealth, is high-
minded, despising wealth in competition with insult or injury, enterprising and energetic as any nation on earth, these
circumstances render it impossible that France and the United
States can continue long friends when they meet in so irri-
table a position.[21]

These words could not have come easily to Jefferson, given his
longstanding friendship for France. Indeed, they are prefaced by
the reminder that "we have ever looked to her [France] as our
natural friend, as one with which we never could have an occasion
of difference. Her growth we viewed as our own, her misfortunes
ours." The marked departure from a relationship of friendship in
the late 1790s had been seen by Jefferson and most Republicans
as the work of Federalists and not as something dictated by the
interests of the nation. In coming to office, Jefferson's hope and
expectation had been that his administration would once again
enjoy good relations with France. The quasi state of war between
the United States and France had come to an end in the closing
period of the Adams administration. The convention of Mortfontaine, concluded on September 30, 1800, had restored relations
between the two countries. Though the agreement scarcely resolved all of the issues that had led to hostilities, it was not unreasonable to see in it the basis for a new and improved relationship.

The retrocession of Louisiana necessarily put an end to this
expectation. Republicans and Federalists agreed with Jefferson that
"perhaps nothing since the revolutionary war has produced more
uneasy sensations through the body of the nation," though they
did so for different reasons. The Federalists saw in the retrocession
of Louisiana to France the first act in the dismemberment of the
Union. Yet the impending retrocession also was seen as providing
an opportunity, perhaps the most promising they would have, to
recapture power from the Republicans. The Republicans, certainly
Jefferson, found in the retrocession the prospect of losing power
to the Federalists, a prospect viewed almost as seriously as the
threat to their dreams of expansion.

In the circumstances, neither set of expectations was unreasonable. The retrocession of Louisiana to France, if consummated, would have given the United States a dangerous, because a powerful, neighbor. Such a neighbor would have barred the way to further expansion. In control of the Mississippi, it might have acted as a powerful magnet on the western peoples. To counter its power might require abandonment of a policy of independence for that of alliance. Jefferson was right to insist that the cession "completely reverses all the political relations of the U.S. and will form a new epoch in our political course." [22]

11

Napoleon's
Colonial Design

I N THE HISTORY OF the United States, the acquisition of Louisiana is an event that ranks in importance below only the achievements of independence and union. So Henry Adams concluded and his view has been generally shared by historians.[23] Although the story of Louisiana has been told many times, to this day it remains unclear in several significant respects. This uncertainty, moreover, must be expected to persist, for it is largely rooted in the refractory character of the two men who dominated the event from beginning to end, Napoleon Bonaparte and Thomas Jefferson.

The importance of understanding Napoleon as a key to understanding French policy has always been appreciated. However mystifying his actions, the attempt to account for French policy of necessity leads to speculation about motivation, given Napoleon's domination over policy. The lament of Robert Livingston, the American minister in Paris, may be recalled: "There is no people, no Legislature, no counsellors. One man is everything. He seldom asks advice, and never hears it unasked. His ministers are mere clerks; and his Legislature and counsellors parade officers."[24] Bonaparte was the great autocrat of the age, just as Jefferson was the great democrat. Bonaparte took it as self-evident that he, the sovereign, was the final judge of his powers, whereas Jefferson was committed to the belief that no government—including the government of a republic—should be the final judge of its own powers. Even so, in the conduct of foreign policy, differences that were otherwise considerable became much less so. Here Jefferson too could behave, and not infrequently did behave, with a degree of secrecy and discretion that resembled the most traditional representative of the *ancien régime*.

The move by Napoleon to return to the Mississippi valley could not have come as a great surprise to Jefferson.[25] It was no secret that the French had never given up their colonial aspirations in North American. Once the sovereign of Canada and the Mississippi valley, they had lost both at the end of the Seven Years' War in 1763. From that time until the years of Spain's retrocession of Louisiana the prospect of France's return to the Mississippi valley had been one of recurring interest to French statesmen and publicists. In the late 1780s, the French representative in the United States, a certain Moustier, had gone so far as to prepare for his superiors a lengthy memoir on the Mississippi valley and the advantages, commercial and otherwise, it would hold out in French hands.

The French revolutionaries continued and, indeed, intensified this interest in Louisiana. Edmond Charles Genet, the Girondist minister to the United States in 1793, had planned to enlist American support for a joint Franco-American intervention against Spain in Louisiana. Although Genet was disavowed and censured by Paris, his abortive scheme for recapturing Louisiana for France nevertheless provoked a sympathetic reaction at home. Two years after Genet, in 1795, the French tried to secure the retrocession of Louisiana from Spain during peace negotiations with that country. While this attempt too proved unsuccessful, France receiving the Spanish part of the island of Santo Domingo instead of Louisiana, it signaled the beginning of a sustained attempt by the government of the Directory to obtain Louisiana.

The continuity and persistence of French aspirations in Louisiana, then, could almost be taken as a political given by an American government. Napoleon simply acted on the basis of these aspirations, though he did so with his characteristic energy and determination. The retrocession of Louisiana was sought as part of a larger plan for the reconstitution of French power in North America. By 1800, after a decade of revolution and war, France had come close to losing almost all of its colonial possessions in the New World. Only French Guiana, on the South American mainland, remained firmly under French authority. The rest were either under the control of the English or in various stages of achieving independence from the mother country. Of the latter, Santo Domingo was the most important. The richest of the French colonies, Santo Domingo had achieved something close to independence as a result of a slave revolt that had begun in 1791. Led by the extraordinary black leader Toussaint L'Ouverture, the revolution had destroyed the power structure on the island but had stopped short of openly breaking away from French authority. Seeing on Santo Domingo the center of the reconstituted colonial system, Napoleon made the reassertion of French rule over the island his highest priority. It was here, he determined, that French power must first be employed.[26]

Although the decisive first step, the restoration of French power over Santo Domingo was insufficient, even when extended to encompass all of the French colonial possessions in the Caribbean. These possessions could never be secure, the French believed, so long as they remained dependent on outside sources for their supply of essential provisions. Throughout the preceding century, France had sought in vain to exclude foreigners from trading with its Caribbean possessions. Particularly in the case of Santo Domingo, foreign trade not only persisted but steadily increased with Britain's American colonies. It did so because this trade was indispensable to the life of the island, France being unable to supply it with essential foodstuffs and lumber.[27]

The importance of trade with the American mainland was especially apparent in time of war when France's merchant marine was driven from the seas by England and the island was utterly dependent for the necessities of life on the English colonists, who, despite the opposition of London, continued to trade with the enemy. Even in peace, this dependence was very considerable and French statesmen believed it threatened the hold of the metropole over the island. In the manner of his predecessors, Napoleon dreamed of ridding France of this dependence. In Louisiana, the architects of the new empire were persuaded, everything needed for the sustenance of these colonies could be found. With Louisiana in French possession, provisions that had otherwise been supplied mainly by the United States—and that in the case of Santo Domingo had recently sustained those rebelling against French authority—would now be secured from sources under French control.[28]

The significance of Louisiana in this scheme of empire was not confined simply to the role of supply depot. In the manner of his predecessors, Napoleon also thought Louisiana might serve alternately as a magnet and a barrier. If France were to control the outlet of the great waterway, the prospects of secession in the West could not be discounted. In the words of Laussat, the new colonial prefect of Louisiana, "A just distinction, if possible, must above all be introduced between the interests of the Anglo-American states identifiable here with us, and the interests of the merchant marine and all British commerce of the Anglo-Americans of the coasts."[29] Were France successful in thus dividing the allegiance of the westerners from their brethren across the Appalachian Mountains, Louisiana might then serve as a barrier to the further expansion of the United States. The French had long cast themselves in the role of protector of the Spanish empire, an empire centered in Mexico. At the time of the negotiations over the retrocession of Louisiana, much was made of this role that France would presumably play on Spain's behalf. A French presence in the Mississippi valley would constitute, in Talleyrand's phrase, a "wall of brass" in the way of the Americans.[30] If Spanish governments were not always reassured by this offer of protection, it was because they suspected, and with reason, that the French increasingly saw themselves as the successors to the lands making up the Spanish empire in North America.

Regarding one portion of this empire, France had made its aspirations quite clear. In the negotiation of the Louisiana retrocession, the French representatives had pressed Spain to include all or part of the Floridas, a concession Charles IV was unwilling to make. The reasons for Napoleon's interest are obvious. Although New Orleans was of great significance as a port for the Mississippi River traffic, Louisiana did not possess a good naval harbor. In possession of the long Florida coastline, and with it the harbors of Mobile and Pensacola, Napoleon would have held a commanding position in the Gulf of Mexico. Until Napoleon decided to sell the whole of Louisiana, the French continued to hope that Spain might be persuaded to part with the Floridas. In fact, there were no discernible limits to France's colonial ambitions in North America. There had seldom been any clear limits to these ambitions and Napoleon was not one to introduce limits where few had before existed.

The French plan above all required peace with England for its successful execution. In war, the far superior naval power of England would frustrate France's colonial aspirations in this hemisphere. The weakness of the French Navy as an obstacle to implementing Napoleon's scheme had been demonstrated by the abortive attempt in 1800 to send an armed expedition to Santo Domingo.[31] The failure of that effort had persuaded Napoleon to suspend further efforts to reestablish French authority over the island until peace was made with Great Britain. More than a year later, on October 1, 1801, the preliminaries of peace were finally concluded and the major obstacle in Bonaparte's path of empire in the West was removed. Even before signing the treaty, the French had satisfied themselves that England would view with tolerance their efforts in the Caribbean. The British government was informed of the great armada for Santo Domingo being prepared at French ports and, in response, indicated its approval of the venture.[32]

The French plan also counted on America's tolerance, if not its active support, for the effort in Santo Domingo. The Americans could not, of course, be expected to view this effort with favor if they were once to appreciate its full implications. It was imperative, then, that the significance of Santo Domingo for French aspirations on the mainland of North America be kept hidden from them as long as possible. The day before the French signed the secret treaty with Spain at San Ildefonso, they had signed an agreement with the United States ending the quasi state of war. The Convention of Mortfontaine (September 30, 1800) had been entered into in part to facilitate the grand scheme of empire. At peace once again with France, the United States would no longer have an incentive to aid those who placed themselves in defiance of French authority, as it had during the years between 1798 and 1801 when the Adams administration had facilitated the provisioning and arming of Toussaint's forces.[33] The French calculation was that by the time the United States might appreciate the close relationship between Santo Domingo and Louisiana its support would no longer be needed.

This was the simple reasoning on which the French plan was founded. In proceeding on the basis of it, Napoleon and his ministers were aware that time was a critical factor. French authority would have to be reasserted soon in Santo Domingo else what little was left of the French position on the island might be lost entirely. Then, too, it would presumably have to be reasserted before the Americans understood the logic of the French design. For once it was realized that the French campaign on Santo Domingo was a step preparatory to the occupation of Louisiana, the Americans might throw up obstacles to the campaign on Santo Domingo, or, more directly, act on their natural advantages and move to occupy New Orleans before the French could firmly establish themselves there. And looming above and beyond these considerations was the prospect of the peace between France and England breaking down before the various parts of the design had been executed, and the British Navy putting a sudden end to the best laid of plans.

In moving to gain the approval of both England and the United States for his actions in Santo Domingo, Napoleon enjoyed the great advantage of dealing with governments that were prepared to go to considerable lengths to avoid war. In England, the Addington ministry had already indicated in a number of ways in 1801 its desire for peace. Exhausted after a long period of war and burdened with a large national debt, it had remained passive toward affairs on the continent. In the peace negotiations with France, it had given up virtually all the conquests made in the war. The Jefferson administration not only matched but exceeded the Addington ministry in its desire for peace. The Republicans had come to office pledged to reduce the national debt and to rid the nation of the burden of internal taxation. This pledge could not be carried out unless the level of military preparedness was allowed to suffer accordingly. War threatened the heart of the Republicans' program.

There was yet another advantage that Napoleon enjoyed. In reasserting French authority over Santo Domingo, Bonaparte portrayed the effort as one that served the interests of all of the "civilized" states and not merely France. The portrayal was largely successful, for the black rebellion on Santo Domingo did strike at the interests of others. The statement made in the secret instructions to General Leclerc, the commander of the French expedition to the island—"The Spaniards, the English, and the Americans look upon the Black Republic with equal anxiety"[34]—was simple fact. The English government saw in the rebellion a threat to a number of that country's colonies. That the blacks were supposedly Jacobins as well only added to the interest of the Addington ministry in having them suppressed. The American government had an even greater interest in their suppression, given that government's responsiveness to the interests of the southern states. The President had himself confirmed this interest in the course of a conversation with the French Chargé in July, 1801.[35]

Napoleon had every reason, then, to credit the expressions of understanding and even of support that he received from the English and American governments for his move to restore French authority in Santo Domingo. That he deceived these governments about the larger purposes informing his Santo Domingo expedition is quite true. Still, this consideration does not alter the conclusion that Bonaparte took his first steps to subjugate the island only after he had received what amounted to a green light from England and the United States. The simple sequence of events supports this conclusion, as do the tone and contents of the secret instructions from the First Consul to the commander of the expedition.[36]

Thus it was that in mid-December 1801, one of the largest armed forces ever to set sail for the New World left the French port of Brest, arriving off the coast of Santo Domingo at the end of January 1802. Up to this point everything seemed to be in place diplomatically for Napoleon to undertake the first and indispensable step in the execution of his plan, which lay in the subjugation of Haiti. This, Napoleon confidently assumed, would require at most a few months. Once the Spanish reluctance to transfer title to Louisiana was overcome, the way would be clear for sending a force of occupation to New Orleans.

12

War and Alliance
in Republican Diplomacy

O N SEPTEMBER 28, 1801, the Secretary of State forwarded a
letter of instructions to the newly appointed Minister to
France, then awaiting passage to Bordeaux.[37] The minister's attention was directed to the information that a "transaction" had likely
been concluded between France and Spain involving "the mouth
of the Mississippi, with certain portions of adjacent territory." The
impending change of neighbors, Madison wrote Livingston, "is of
too momentous concern not to have engaged the most serious attention of the Executive." Livingston was instructed to take up
the subject with the French government and to express the anxiety
of the United States over the prospects and consequences of the
rumored transfer of territory. The "tendency of a French neighborhood," the government in Paris was to be delicately reminded,
must "inspire jealousies and apprehensions which may turn the
thoughts of our citizens towards a closer connexion with her rival,
and possibly produce a crisis in which a very favorable part of her
dominions would be exposed to the joint operations of a naval
and territorial power." Should, however, Livingston find the cession a fait accompli, or virtually so, he was to make every effort
to preserve those rights of trade and navigation then obtaining
between the United States and Spain. Additionally, he was to see
whether France could not be induced to cede the Floridas—or at
least West Florida—to this country, assuming these territories to
be included in the cession. Such cession by France would at once
prove that nation's good will and serve to reconcile the United

States to an arrangement so much "disrelished" by them. But should the Floridas neither have been ceded to France nor their cession contemplated by France, efforts were to be made to dispose Paris in favor of "experiments on the part of the United States, for obtaining from Spain the cession in view."

Thus France was to be warned that a return to the Mississippi valley could result in an alliance between this country and England, an alliance that might well result at some future date in France being driven altogether from North America. But the threat was to remain less than explicit, lest it unnecessarily arouse the French and thereby jeopardize American rights in the use of the river. Alternatively, France might be induced to give part of the territories ceded to it in order to gain the good will of this country. The cession of West Florida to the United States would reconcile the latter to the French presence elsewhere, while relieving France of spoliation claims by American citizens. On the other hand, if Spain still possessed the Floridas, France might nevertheless be disposed to assist us in obtaining these lands in return for our favorable disposition toward its presence here. How France might undertake this role was not elaborated.

These instructions, drawn up at the outset of the Louisiana crisis, form a succinct summary of the principal features of Jeffersonian diplomacy, not only in the Louisiana crisis but also in the subsequent efforts to obtain the Floridas. Essentially the same considerations, but with greater urgency, were pressed by Jefferson in his famous letter to Livingston in April 1802. Although once our "natural friend," Jefferson wrote, in possessing New Orleans France must become "our natural and habitual enemy." By doing so, he warned, France must seal its own fate, for in taking New Orleans it would not only force America into alliance with Great Britain but require the United States "to turn all our attention to a maritime force, for which our resources place us on very high grounds." When war broke out again in Europe, the United States would seize the opportunity to tear up any settlement France had made, and would then hold "the two continents of America in sequestration for the common purposes of the united British and American nations." In such a contest of arms, France would immediately lose New Orleans. "For however greater her force is than ours compared in the abstract, it is nothing in comparison of ours when to be exerted on our own soil."[38]

The April letter stands as the fullest and most considered expression Jefferson gave to the crisis brought on by Louisiana. Against the background of a steadily worsening atmosphere, the President decided to convey to Bonaparte the concern of the American government over French actions and to make clear the likely consequences of these actions if they were continued. A Frenchman, Du Pont de Nemours, was entrusted with this and a later letter for Livingston, letters that he might read before delivering them in Paris to the American minister. Well known in both countries, De Pont was to serve as the unofficial channel of communication between the two governments.

There is no evidence that the contents of Jefferson's April 18 letter to Livingston were ever conveyed to the French government.[39] Its carrier, Du Pont, was of the view that the letter could only offend Napoleon, who would see in its threats a challenge to his honor. Du Pont suggested that an offer of purchase would instead be the most promising method of obtaining the ends Jefferson sought.[40] While the President was quick to reply that Du Pont had misunderstood him, and that he had not intended to threaten France but only to point out the unavoidable consequences of that nation's apparent course of action,[41] the letter has nevertheless been seen as laying out the essential diplomatic strategy Jefferson pursued in the crisis. The heart of the strategy was the putative alliance with Great Britain.

In the spring of 1801, prior to the Peace of Amiens, Great Britain had in fact shown a disposition to oppose the French colonial scheme. Responding to the rumors of the retrocession of Louisiana (thought to include the Floridas) to France, the British Foreign Secretary, Lord Hawkesbury, had informed the American minister of London's strong opposition to this extension of French power. Because the acquisition would reverse the results of the Seven Years' War, threatening Canada and the British West Indies, "England must be unwilling that the territory should pass under the dominion of France." Rufus King had replied that his government was content to see the Floridas "remain in the hands of Spain, but should not be willing to see them transferred, except to ourselves."[42]

Madison, in commenting to King on Hawkesbury's statements, expressed himself in stronger terms. "Considering the facility with which her extensive Navy can present itself on our part, that she already flanks us on the North, and that if possessed of

Spanish countries contiguous to us, she might soon have a range of settlements in our rear, as well as flank us on the South also . . . she is the last of Neighbors that would be agreeable to the United States."[43] Not unwisely, Madison was concerned that a unilateral British occupation of New Orleans or the Floridas, even if designed as a preemptive move against the French, might issue in the permanent settlement of an equally dangerous neighbor on the American frontier.

Almost two years later, in the spring of 1803, the British again displayed a tendency to oppose forcefully the French design on Louisiana. In April 1803, the Peace of Amiens was clearly breaking down. Once again a British cabinet minister, Addington, came to the American minister to London, now with a tentative plan to occupy New Orleans in the event of war. In reply, Rufus King came very close to repeating the views he had earlier expressed. Although the United States could not see with indifference Louisiana in the hands of France, King declared, "it would be contrary to our views, and with much concern that we should see it in the possession of England." To this concern, Addington assured King "that England would not accept the Country . . . that were she to occupy it, it would not be to keep it, but to prevent another power from obtaining it, and . . . that this end would be best effected by its belonging to the United States." Although King expressed agreement with Addington's last remark, he voiced opposition to any British occupation, even if followed by a transfer of the territory to the United States, for fear that the appearance of a British–American concert would embroil the United States in conflict with "another Power with which we desired to live in Peace."[44] As it had two years before, Washington echoed King's coolness toward any such action on the part of Britain. In a letter to Monroe and Livingston, written after the sale of Louisiana to the United States but before word of the sale reached Washington, Madison disparaged the prospect that England might make a gift of New Orleans to the United States.[45]

This skepticism with which English schemes of opposing the French in Louisiana were received is all the more striking in that it brackets a period of a year and a half during which American diplomacy toward France was very much dependent on the promise—or threat—of an alliance with England should France persist in its plans with respect to Louisiana. At the time King and Madison expressed their opposition to the English proposal, they did not know that Napoleon had abandoned his plan to send a military force to occupy New Orleans and that he had decided to sell Louisiana. For all they knew, then, a French occupation force would soon be sent to New Orleans to replace the small Spanish force there. Save for the English Navy, there was no impediment to the French force. Although the renewal of war between France and England was already a distinct prospect, no one could say when war would come. Yet rather than agree to an English occupation of New Orleans, the American government was apparently prepared to risk a French occupation.

These considerations are instructive when dealing with the Jefferson administration's alliance diplomacy during the Louisiana crisis. The marked diffidence shown by this country toward an English occupation of New Orleans, however attended by assurances that it would be only temporary, was reciprocated by the diffidence shown by London toward hints of alliance during the period of peace between Great Britain and France. The preliminaries of peace were concluded on October 1, 1801. The following month, the American minister found the English government cool and detached about France's plans in North America. When Rufus King asked Lord Hawkesbury whether England would cooperate with this country by using its influence at the peace negotiations to effect a restoration of Louisiana to Spain, the Foreign Secretary refused to give him an answer on this point, responding instead that the English government had taken no notice of the cession and did not think it would soon become important.[46] Six months before, the same Hawkesbury had declared his government's willingness to seize Louisiana.

The change that had occurred in the English position was subsequently underscored. The British government, even though it did not necessarily look upon the cession of Louisiana with favor, was unwilling to do anything that would prejudice the continuing peace discussions at Amiens. Rufus King concluded that "England abstains from mixing herself in it, precisely from those considerations which have led her to acquiesce in others of great importance to the Balance of Europe, as well as her own repose."[47] Nor was it clear that the English government, quite apart from those European interests that dictated abstention from the French adventure in the New World, did in fact look with disfavor on the cession of Louisiana to France. Although the English might view a French presence in Louisiana with some apprehension as a possible threat to Canada, they might also be expected to see this presence as guaranteeing them American dependence and friendship. They might be expected to see on the occasion of the Louisiana crisis what they had earlier been so singularly blind to when they had removed the French from the Mississippi valley at the end of the Seven Years War, thereby removing a central source of colonial dependence on England and setting the stage for the American Revolution. This lesson was reflected in the comment Lord Hawkesbury made in the spring of 1802 during a parliamentary exchange: "That it was sound policy to place the French in such a manner with respect to America as would keep the latter in a perpetual state of jealousy with respect to the former, and of consequence unite them in closer bonds of amity with Great Britain."[48] The *Annual Register* of 1802, in quoting Hawkesbury's remark, reported that it filled Americans with "indignation and abhorrence." It did so, however, because it pointed to a condition that Americans had long been at pains not to admit even to themselves, much less to others. The President was as reluctant as anyone to admit this condition of dependence on Great Britain. Yet there was no other meaning that could be given his words: "The day that France takes possession of New Orleans . . . seals the union of two nations who in conjunction can maintain exclusive possession of the ocean. From that moment we must marry ourselves to the British fleet and nation."[49]

While the French never succeeded in taking possession of New Orleans, save in a formal sense and even then only for the briefest of periods, until the time Napoleon decided to sell the colony American diplomacy had to prepare against that day. How serious were the American efforts and what effect did they have on the French? The first question can be answered with greater assurance than the second. To characterize the American efforts to propitiate Great Britain as serious is to take no little liberty with that term. It may be argued, of course, that the propitiation of British power was virtually impossible during the period England was at peace with France, that there was no incentive the Jefferson administration could hold out that would tempt London to jeopardize the peace by allying England with the United States in opposing French plans of empire in North America.[50] Even so, the question concerning the nature of the American efforts to enlist British support would remain. At best, those efforts must be characterized as halfhearted and without real significance.[51] Given the outlook of Republicans, this was understandable. If an alliance was almost by definition an illicit relationship, an alliance with Great Britain represented for Jefferson and his political associates something approaching a state of mortal sin. Perhaps this explains why Jefferson and Madison appear to have thought of an alliance with England not as a product of the union of political wills but as something that resulted from a kind of political immaculate conception. Whereas the benefits of alliance were desired, the act that produced those benefits was not. For that act meant, if it meant anything, that benefits had to be apparent on both sides.

There is no evidence that Jefferson ever seriously considered the prospect of an alliance from this perspective. His strategy was not directed to the end of actually concluding an alliance with Great Britain. Instead, it was one of appearing to be moving in this direction, with the expectation—or hope—that the object of this strategic feint would be persuaded that the appearance foreshadowed the reality. Thus the story of Jefferson's alliance diplomacy during the Louisiana crisis is a story of attempts to pressure the French into yielding by threatening that the United States would ally itself with England against France. On occasion, the French were told that Great Britain was eager to conclude an alliance with this country.[52] None of these threats reflected a discernible reality in the sense that the American government was, in fact, prepared to enter into an alliance or that the English were disposed to offer one. Instead, it was the threat that formed the sole reality of this diplomacy.

There is one apparent exception to this pattern and even it is not quite clear. In April 1803, when the crisis had moved close to its denouement, the President, with the support of a majority of his cabinet, decided that if the French proved intractable in the negotiations then in progress in Paris, Monroe and Livingston should be instructed to enter into discussions with the British government "to fix principles of alliance."[53] The instructions Madison wrote to the two envoys were based, in the main, on the supposition that "Great Britain and France are at peace, and that neither of them intend at present to interrupt it." In this situation, the instructions read, if the French government should either "meditate hostilities" or force the United States to initiate a war by closing the Mississippi, Monroe and Livingston were to communicate with the British government and "invite its concurrence in the war." It was to be made clear to Great Britain that although war depended on the choice of the United States alone, our choice of war in turn depended on Britain undertaking to participate in it. At the same time, Madison emphasized, the certainty of our choice of war "should not be known to Great Britain who might take advantage of the posture of things to press on the United States disagreeable conditions of her entering the war." Instead, the United States was to determine those conditions and they were to include the stipulations that neither party should make a separate peace and that Great Britain was to enjoy certain commercial advantages in the Mississippi beyond those already secured by existing treaties. There was to be no mutual guarantee of existing possessions or of conquests that might be made by the parties, however, as this would lead to undesirable entanglements in the wars and disputes of others. And the envoys were to reject any and all British attempts to extend that nation's domain on the west side of the Mississippi. Any disappointment England might have as a consequence of this rejection was to be alleviated by the promise that France would not be allowed to retain or acquire any territory from which Great Britain was excluded. Monroe and Livingston were to conduct these negotiations with England if France, by denying free navigation of the Mississippi, made war inevitable. If, however, navigation was not disputed but the deposit alone was denied, the envoys were to make no explicit engagement, leaving to Congress the decision between war or further procrastination.[54]

The instructions were addressed to a contingency that had already been excluded by France: closure of the Mississippi to navigation by the United States. The French had earlier undertaken to abide by Spain's obligations to this country once they succeeded to Louisiana.[55] These obligations included the right of navigation and of deposit. In this context, what is significant about the instructions was not so much that they provided for situations that were very unlikely to arise but what they had to say about the nature of the administration's alliance diplomacy. For the alliance Monroe and Livingston were to conclude was an alliance Great Britain could not be expected to make. The instructions simply brushed aside the experience of the preceding year and a half. Although assuming that England was neither at war nor intending to enter upon war, the envoys were nevertheless to conclude an alliance the real value of which to Britain was apparent only if the nation were at war. But if that were to occur—if, the instructions read, war has "actually commenced, or its approach be certain"— the American ministers were to avail themselves of this change of circumstances "for avoiding the necessity of recurring to Great Britain, or, if the necessity cannot be avoided, for fashioning her disposition to arrangements which may be the least inconvenient to the United States." In brief, they were either to make no alliance or to make one that would carry virtually no obligations for the United States.

Even if it is concluded that the alliance diplomacy of the Jefferson administration had no solid basis and that it rested on little more than a kind of incantation, it does not follow that the party to whom it was primarily addressed was unaffected. Whether this diplomacy was "real" or not, the result, after all, was what mattered. Did it impress the French sufficiently to lead them, when taken together with other considerations, to draw back or substantially alter a course they were otherwise intent on taking? This question cannot be answered in the manner many have sought to answer it, by reading the diplomatic dispatches of the French representative to the United States. There is no doubt but that Pichon was duly impressed by the threat that Jefferson and Madison voiced to him on more than one occasion. There is doubt whether Pichon's reports of this threat to form an alliance with Great Britain made a similar impression in Paris. Pichon was a young and impressionable Chargé d'Affaires. He was clearly taken with the President. He was opposed to the course Napoleon was intent on pursuing in North America. While intelligent and quick to detect French mistakes, he was not prone to look deeply and critically into the diplomatic methods of the Jefferson administration (unlike his successor, Turreau). The merest threat of alliance or war made to Pichon by the President or Secretary of State sent him scurrying to write an urgent dispatch to his superiors in Paris. But the latter were not so readily impressed.[56]

If Jefferson's alliance diplomacy was to affect French behavior, it was desirable—even imperative—that it do so during that critical period when France was moving militarily to occupy Louisiana and to build up the defenses of the colony. In the French plan this was the period when France would be at peace with England. The French calculation was that England would not break the peace over Louisiana. The calculation was sound. The English in entering into peace negotiations with France had deliberately put aside the issue of Louisiana in deference to Paris. That issue could not attract Great Britain into responding to the alliance diplomacy of the United States, and the French appreciated this very well.[57]

It was only if the peace broke down for other reasons that this diplomacy could evoke a British response. Indeed, in the event that peace broke down, the English did not need Jefferson's encouragement to prompt them to strike at France's colonial efforts both at sea and on land. Then this diplomacy would be directed to restraining England's efforts rather than to eliciting them and for fear that otherwise these efforts might pave the way for subsequent claims. It is quite true that when peace visibly began to break down, Napoleon was at pains to see that the United States did not ally itself with Great Britain. The abandonment of Louisiana readily accomplished that. Still, Napoleon did not abandon Louisiana because he was affected by Jefferson's alliance diplomacy, but for quite different reasons. Having once decided to abandon the colony, however, he did so in a manner he considered would best serve his interests. Not the least of those interests was the creation of a power in the New World that might in time provide a serious challenge to the maritime power of England.[58]

The question of Jefferson's alliance diplomacy is closely related to that of the President's use of the threat of war. The prevailing view among historians is that Jefferson was quite willing to use military power in pursuit of his diplomatic aim and that he effectively threatened France with war should it claim New Orleans.[59] Whether or not this view is well taken depends in no small measure on determining what constitutes a meaningful threat of force. The persuasion that Jefferson was quite willing to use military power is, in part at least, dependent on the fact—or what is so claimed—that he seriously threatened war. Did he?

A serious threat of war has the quality of imminence, if not of immediacy, and is dependent on the occurrence of a specified event. The significance of the temporal element is apparent. The longer the period of time that is entertained, the greater the possibility that changing circumstances may alter the will of the party making the threat. The element of specificity in defining the conditions or circumstances productive of war is also important. The more vague or ambiguous the conditions, the more difficult it is for a threat of war to serve as a deterrent (and that, presumably, is its principal purpose). Jefferson's threat of war satisfied neither of these conditions. Indefinite in time and vague in circumstance, his threats to use force were all of a contingent and hypothetical nature, dependent on future developments that were likely to put the French at sufficient disadvantage as to make the prospect of this country resorting to force against them a credible one. But these were not the kinds of threats likely to command Napoleon's respectful attention. He might be deterred by a meaningful threat of force in the here and now. He could not be deterred by a hypothetical threat addressed to circumstances in which he already knew that his empire would be put at risk.

There is one apparent exception to this pattern of Jeffersonian threats and it is to be found in the famous letter of April 18, 1802. Although clearly containing a threat of war, the threat was contingent on an alliance with Great Britain. The letter contrasts markedly with the official instructions Madison wrote at the time to Livingston. In the latter, Livingston was directed to state that "a mere neighborhood could not be friendly to the harmony which both countries have so much an interest in cherishing; but if a possession of the mouth of the Mississippi is to be added to the other causes of discord, the worst events are to be apprehended."[60] These threats of war, contingent and hypothetical though they were, undoubtedly reached the French, for as early as the beginning of 1802, Jefferson had told the French Chargé that a French occupation of New Orleans would lead to a rupture between the United States and France, an alliance between the United States and Great Britain, and, eventually, a war between this country and the occupant of New Orleans.[61]

While Jefferson was voicing this to Pichon, however, his minister in Paris was assuring the French government that "as long as France conforms to the existing treaties between us and Spain, the government of the United States does not consider herself as having any interest in opposing the exchange."[62] Madison's letter to Livingston of May 1, 1802, represented an abrupt change from the assurances the American minister had been giving the French, and Jefferson's private letter to Livingston amounted to an even more radical shift in policy.

The shift was more apparent than real, however, for it was not otherwise attended by a hardened diplomatic position. On the contrary, having delivered themselves privately and officially of hard-line positions, Jefferson and Madison seemed content to let matters rest. The impression was thereby given of a government that was not to be taken at its word, when that word was bellicose. This impression would have been confirmed had adversaries only been able to read the letter Jefferson wrote in October 1802 to Livingston.[63] Having told Livingston in April that the closest possible relationship must be formed between this country and England, in October the President had again seen "all the disadvantageous consequences of taking a side" between France and Great Britain. While acknowledging that we may yet be forced into taking sides by a "more disagreeable alternative," in that event, Jefferson insisted, "we must countervail the disadvantages by measures which give us splendor and power, but not as much happiness as our present system." And after insisting in the spring of 1802 that war must be the inevitable result of a French occupation of New Orleans, Jefferson insisted in the fall of that year that "no matter at present existing between them and us is important enough to risk a break of peace,—peace being indeed the most important of all things for us, except the preserving an erect and independent attitude."

These considerations notwithstanding, the view has been taken that Jefferson made the landing of French forces in New Orleans the occasion for war between the United States and France.[64] There is no evidence, though, to support the contention that the President threatened war if the French were to attempt the military occupation of Louisiana. Indeed, the threat would have been an anomaly in view of the fact that the cession of Louisiana to France was never as such considered a *casus belli*. Yet the cession was without meaning unless the new sovereign could effect the military occupation of the colony. From the outset, France showed every intention to occupy Louisiana as soon as the territory was formally surrendered by Spain and a military expedition could be mounted. In the end, these two conditions proved critical, since Spain refused to surrender Louisiana until October 1802, while the demands of the terrible struggle taking place on Santo Domingo resulted in delay after delay of the French expeditionary force. The decisive effect of these developments, though, must not be confused with the supposed effects of a policy that kept France from occupying Louisiana by making such act the occasion for war. This policy never existed.

Instead, the record shows that from early 1802, Livingston was advising the government in Washington that the French were intent on effecting the military occupation of their colony as soon as they could dispatch an expedition. In February, the minister reported that General Bernadotte was understood to be in command of the expedition and that he had asked for ten thousand troops.[65] In April, the number of troops reportedly assigned to the Louisiana expedition was reduced to between five thousand and seven thousand. "They will shortly sail for New Orleans," Livingston warned, "unless the state of affairs in St. Domingo should change their destination. You may act upon this information with absolute certainty."[66] In late May, he wrote to Madison that the Louisiana expedition had been delayed until September because of some difficulty that the Foreign Minister, Talleyrand, would not explain, "but which, I have no doubt, has arisen from the different apprehensions of France and Spain relative to the meaning of the term Louisiana, which has been understood by France to include the Floridas, but probably by Spain to have been confined to the strict meaning of the term."[67]

Thus through the winter and spring of 1802, the administration was warned of the impending military occupation of Louisiana. In the same period, one will search in vain for a threat from Washington to the effect that the landing of French troops at New Orleans must provide the occasion for war. The same lack of response attended Livingston's subsequent warnings in the late summer and fall of that year.[68] What disrupted the French plans time and again, causing delays in sending an expedition and reductions in the force that was to be sent, were the voracious demands of the campaign on Santo Domingo. Had it not been for these demands, there is every reason to believe that New Orleans would have been occupied by a French force by the end of 1802 and that this force would have been one of substantial size.

It was not the appearance of French forces in New Orleans that formed the *casus belli* for the administration, but the denial of free navigation of the Mississippi. This position is apparent in the general instructions March 2, 1803, given Monroe and Livingston.[69] The American commissioners were to try to buy New Orleans and the Floridas. They might offer close to ten million dollars for this as well as guaranteeing France commercial privileges. The commissioners, if pressed, might also offer as an inducement a guarantee of the west bank of the Mississippi. But if France refused to sell any of its territory under any conditions, the two envoys were to secure the right of deposit, and hopefully to improve on the old right. Should even the right of deposit be denied, Monroe and Livingston were to be guided by instructions specially adapted to the case.

The conclusion of Henry Adams that the instructions "offered to admit the French without condition" is difficult to resist.[70] The same must be said of the instructions Madison subsequently wrote Monroe and Livingston on April 18.[71] Unaware that Napolean had decided to sell Louisiana to the United States, Madison drafted the April instructions primarily to provide for the eventuality that France "should be found to meditate hostilities or to have formed projects which will constrain the United States to resort to hostilities." But war was to be judged inevitable only in the event that France "should avow or evidence a determination to deny to the United States the free navigation of the Mississippi." If France should not dispute the right of navigation and instead only deny deposit, the envoys were advised that "it will be prudent to adapt your consultations to the possibility that Congress may distinguish between the two cases, and make a question how far the latter right may call for an instant resort to arms, or how far procrastination of that remedy may be suggested and justified by the prospect of a more favorable conjuncture."

In the April instructions as well there was no serious consideration given to war, or, for that matter, to an alliance with Great Britain should France neither dispute the right of navigation nor deny the deposit. Napoleon had no interest in doing either. Quite the contrary, the evident interest of the French, once they were to become the masters of New Orleans and of the great river, was to grant both navigation and deposit to the Americans, thereby establishing the conditions that might in time secure the support and even the loyalty of the western states. The instructions addressed a contingency that was never intended by France and that, indeed, had been expressly disavowed.[72]

In sum, one searches in vain for serious threats of war by the administration. Such threats simply did not form a part of the President's diplomacy. They did not do so for the simple reason that Jefferson's diplomacy was guided above all by a commitment to peace. War was not absolutely precluded. But it was accepted only in the extreme and altogether unlikely event that France denied Americans the use of the Mississippi. In that event, Jefferson had no course to follow other than war. Short of this extreme contingency, though, he was prepared "to palliate and endure" until such time as France would be placed in a position so disadvantageous that it would no longer be able effectively to resist American claims to New Orleans and the great waterway.

The passivity and determination to avoid war that marked the diplomacy of the Jefferson administration during the Louisiana crisis were paralleled by a military policy that sought to give the appearance, though not the reality, of activity and strength. Even the appearance of military preparation was not sought after until the very late stages of the crisis. This simple fact is critical; it reveals more clearly than any other consideration the outlook and disposition of those responsible for the nation's security.

At the close of 1802, the administration did finally begin to take some measures of a defensive character. It did so largely in response to domestic pressures that were the result of Spain's suspension of the deposit. The action of Juan de Dios Morales, the Spanish Intendant of New Orleans, had triggered widespread public fears that the use of the Mississippi might be seriously threatened. Whereas before the suspension Jefferson had dealt with the crisis without interference, in the period immediately following the suspension he was confronted by the sudden rise of forces that threatened the complete control over policy he had theretofore enjoyed. In the West, the demand arose to seize New Orleans before it was occupied by the French. In Congress, a warlike mood emerged that was not confined to Federalists. The cry for stronger measures to defend American rights in the Mississippi River found a strong echo in much of the press and resounded in resolutions adopted by several state legislatures.

In these circumstances, Jefferson reluctantly responded to the rising demand for action. One response—indeed, the principal response—was the appointment of James Monroe as Minister Extraordinary to France and Spain. Although a modest act, and one from which the President expected little of consequence, it had the effect of calming the West.[73] If Jefferson's comments to Monroe are to be credited, this goal weighed more in recommending the mission than did considerations of strategic necessity. Monroe was seen by westerners as a firm supporter of their interests, and his appointment undercut the prospects of Federalist inroads into the Republican West.[74] Shortly after Monroe's appointment, the government received word that Talleyrand had pledged the French government to the strict observance in Louisiana of the treaties existing between the United States and Spain.[75] The news had a further calming effect not only in the West but elsewhere as well.

At the same time, the administration took certain measures, including limited preparation, for the possibility of armed conflict. These measures were taken against the background of a military policy that had reduced the army from an authorized strength of 5,500 to 3,300 regulars.[76] In his second annual message, on December 15, 1802, Jefferson had declared that no change in the nation's military establishment was "deemed necessary" and had contented himself with recommending to the Congress a "review" of the militia with the purpose of giving it "those improvements of which you find it susceptible."[77] Two weeks before, the Secretary of War had ordered the reenlistment of every valuable soldier whose period of service was expiring, a step, however, that added nothing to the existing force.

Beginning early in 1803, several measures were taken along the Mississippi frontier to increase the defensive capability of fortified positions there.[78] Of these posts, the largest and most important was Fort Adams, situated on the Mississippi River just above the border with Spain. It was Fort Adams that constituted the principal defensive bulwark against an offensive thrust from New Orleans, just as it was this post that would have provided the base for an assault on New Orleans. In March 1803, Fort Adams had a force of seven companies. In the same month two additional companies were ordered there. Since the companies were in all likelihood under strength, the size of the force at this critical post probably numbered no more than six hundred men. Elsewhere, the distribution of forces both along the southern frontier, where they would face the French, as well as on the northern frontier, where they faced the British, inevitably meant that the army was stretched very thin.

How significant were the various measures taken to improve

the readiness of the forces in the western posts and what do they indicate about Jefferson's intentions? Judged by the view that Jefferson himself seemed to have taken of his administration's efforts, they do not appear to have been more than of modest significance. Certainly, they scarcely seem to have been of such character as to have "left nothing to chance" in a showdown with Napoleon's forces.

The military measures the Jefferson administration took in the winter and spring of 1803 were not measures intended for offensive action against a French force of occupation. The American forces were intended to serve a defensive role. In the south, they were to block a French move against American positions along the Mississippi north of New Orleans. During the early winter of 1802-3, the expectation persisted that a French force would eventually be sent to New Orleans. There were no plans to contest its arrival by military means. Nor did the American government intimate, let alone declare, that the attempt by France to occupy New Orleans would be regarded by this country as a *casus belli*.[79] It is true that by late winter, American authorities entertained strong doubt that the French would in fact attempt the military occupation of Louisiana. But this doubt was not the result of the preparatory military measures taken by the administration. It arose instead from the repeated failure of the French to mount an expedition because of the continuing demands of the Santo Domingo campaign as well as from the growing realization that war would soon be renewed between France and England.[80]

13

"Playing for Time"

JEFFERSON WAS KEENLY AWARE of the interests at stake in the Louisiana crisis. To prevent the France of Napoleon from becoming the successor of Spain in the Mississippi valley formed at once his clear and compelling reason of state. Even so, he was not prepared to accept all the consequences this necessity might dictate. Although desperate, he was willing to give up very little for deliverance from peril. He wished, in particular, to avoid both war with France and alliance with England. He was loath to accept either prospect not because he failed to appreciate the interests that were at stake but because he was persuaded that either course of action would mean the sacrifice of his vision of domestic society. The one would threaten the corruption of the Republican experiment, which depended so heavily on avoiding the debt and taxes that war would bring; the other would compromise America's status of independence. His dilemma sprang from the fact that the refusal to accept either war or alliance might also lead to the same result.

Jefferson found the escape from his dilemma in the conviction that time was against the French plan. Time was against it, in part, because France was unlikely to be successful in quickly subjugating Santo Domingo. While engaged on the island, Napoleon would be hard pressed to undertake the occupation of Louisiana. Nor did time favor a permanent French occupation of Louisiana. Before France could develop its position sufficiently in Louisiana, it was likely to be at war again with England. And even if this did not prove to be the case, he argued, time would surely work against France in the sense that the Mississippi valley was being rapidly settled by Americans. Sooner or later, the sheer weight of numbers would operate to drive the French from the position of occupant over an ever greater alien and hostile population. In the meantime, as Jefferson wrote to a friend after the crisis, his hope was "to palliate and endure."[81]

Although the President claimed in the same letter that he "very early saw that Louisiana was indeed a speck in our horizon which was to burst in a tornado," Jefferson in fact came late to an appreciation of the impending crisis. Particularly, he failed to grasp at an early date the importance of Santo Domingo as a barrier to Napoleonic aspirations in Louisiana. In July of 1801, he had confirmed to the French Chargé that the United States wished success to Napoleon's efforts to reconquer the island. Jefferson pointed out to Pichon that trade with the island was of great importance to the United States, and that any attempt by the government to suspend it would "gravely compromise public opinion." Nevertheless, he responded favorably to the French request for cutting off provisions to the island's rebels, telling Pichon that, once France had made peace with England, "nothing would be more simple than to furnish your army and your fleet with everything and to starve out Toussaint."[82] In all likelihood the President made his promise to Pichon because he wanted to preserve the relationship with France set out in the convention of Mortfontaine and because he felt obliged to respond to the mounting fears of the southern slaveholders over the spread of the black rebellion on Santo Domingo, fears that he shared in part.[83]

In the ensuing months, however, the President failed to carry out the promise he made in July to the French representative. In the early fall of 1801, the French made peace with England, as Jefferson had urged them to do, but the American trade with the rebels of Santo Domingo continued. When Pichon went to the President early in 1802 to see if he might obtain a reversal of the policy that sanctioned this trade, a reversal the Secretary of State had earlier refused, he found the President reserved and cold.[84] The message was quite clear. The French could attempt to close down this trade by their own efforts, but the American government would do nothing to help them.[85]

One reason for Jefferson's reversal lay in his unwillingness to propose an embargo on the flourishing trade to Santo Domingo to Congress, where it might quickly be turned into a sectional issue. At best, cooperation with Napoleon promised embarrassment; at worst, a bitter division along sectional lines. Sometime in the winter, Jefferson came to realize that Santo Domingo was part of a larger scheme of which he had hitherto been unaware. In November 1801, the American minister in London, Rufus King, sent to Madison a copy of a treaty between France and Spain, concluded eight months earlier, in which the cession of Louisiana by a previous treaty was clearly set out.[86] During the same month, Livingston arrived in Paris to hear of the imminent French expedition to Santo Domingo, later speculating that some of these forces might proceed to Louisiana.[87] The apprehensions of the administration over the fate of Louisiana were only aggravated by the persistence with which the French government continued to deny, in the face of evidence now almost conclusive, that an agreement on retrocession had been made.[88]

The climax in this sequence of events came on the heels of the arrival, in February 1802, of the French expedition to Santo Domingo under the charge of General Leclerc, brother-in-law to the First Consul. Leclerc's instructions had led him to believe that he could count on American cooperation and assistance. Instead, he found that American merchants continued to trade with the rebels. When Leclerc, who was in need of supplies for his own forces, had tried to buy the cargoes of the American ships, the shipmasters had asked prices he considered extortionate. The French commander had thereupon seized the cargoes and, having found that some of the ships contained arms as well as food, imprisoned several shipmasters. In an atmosphere charged by increasing clashes between American interests on the island and French policies, Leclerc dismissed the American Consul, Tobias Lear, in early April.[89] The conduct of Leclerc, considered along with the growing evidence that part of his expedition was ultimately destined for Louisiana, had a further sobering impact on the administration.[90]

It was in these circumstances that Jefferson wrote his famous letter to Livingston. While placing primary reliance on the threats of war and alliance, such as they were, Jefferson also took note of the likely difficulties the French would face in subjugating Haiti. In Washington, the President wrote, "the idea . . . is that the troops sent to St. Domingo were to proceed to Louisiana after finishing their work on the island. If this were the arrangement, it will give you time to return again and again to the charge, for the conquest of St. Domingo will not be short work. It will take considerable time to wear down a great number of soldiers."[91] Of all the prophecies that regularly sprang from Jefferson's fertile mind, this one seems the most remarkable in retrospect. It is in striking contrast with the optimism that then prevailed in Paris. As late as the spring and early summer of 1802, the prospects for Napoleon's scheme of empire seemed quite bright to its architects. In fact, the stage was already being set for the disaster that would overtake the French by the late summer of that year. Napoleon had badly underestimated the requirements of the campaign on Santo Domingo, as the American Consul on the island had noted before his expulsion.[92] The expedition sent out under Leclerc was not nearly of the size that was needed to ensure against defeat. But to men who held the blacks in contempt, as did Napoleon and Leclerc, the prospect of defeat at their hands was not to be credited.[93] Besides, Leclerc had waged a determined and ruthless campaign, one that by April had left French forces in control of all the principal towns and ports of the island. Although the French had paid a very heavy price in casualties for their successes, this did not appear to detract from what they saw as a victory. In May, Toussaint had been deceived into giving himself up to the invaders. With his surrender, the way seemed open for a quick and successful pacification of the island. In June, Napoleon informed the Minister of Marine, Decrès, that the possession of Louisiana was to be undertaken without delay, that an expedition was to be organized in the greatest secrecy, and that it appear to be directed to Santo Domingo.[94]

The optimism that attended the initiation of these measures did not last long. In July, the storm suddenly broke over Leclerc's head. The French commander had been prudent enough not to attempt the immediate implementation of Napoleon's order restoring slavery. But rumors that slavery had been restored on Guadeloupe soon reached Santo Domingo, where they aroused the blacks to a fury that persisted until independence, and freedom, had been achieved.[95] By July, Leclerc's forces were weakened by an outbreak of yellow fever that grew in violence as the summer progressed. By late summer, a badly depleted force was clearly on the defensive. Before Paris could respond to Leclerc's frantic calls for reinforcements, he too had succumbed to yellow fever in November 1802.[96] Only with a new force of ten thousand troops did his successor, Rochambeau, manage to launch a new offensive that kept the French in the field as an effective fighting force.

Even if Jefferson was endowed with an unusual degree of prescience, it is unreasonable to assume that he clearly foresaw the various difficulties that would beset French plans. Certainly the disastrous decision to restore slavery on the island could not have been counted on. Nor was there reason to expect that yellow fever would strike with the intensity that it did.[97] Jefferson might have hoped that the black rebels of Santo Domingo would continue indefinitely to consume French forces until Napoleon abandoned his colonial ambitions in the New World, but he could not prudently *count* on this outcome. Nor did he. As it turned out, the demands imposed by Santo Domingo, formidable though they were, could not preclude Napoleon from sending a force to occupy Louisiana. He might well have done so in the fall of 1802 and very nearly did so in the winter of 1803, only to be blocked from acting by unusual weather.[98] For Jefferson, the expected difficulties on Santo Domingo provided time; but true deliverance from peril, he thought, lay elsewhere.

The central hope of Jefferson's Louisiana diplomacy was that Britain and France would again go to war. Had the French succeeded in occupying New Orleans, a renewed European war, which would have immediately isolated what forces France had deployed in the Western Hemisphere, might have allowed the United States to seize New Orleans at little cost. Or the United States might have acted in concert with Britain. Whether Jefferson would have employed force even under such favorable circumstances remains uncertain; he clearly hoped that the prospect of such an event would induce Napoleon to sell New Orleans to the United States, or at least to return Louisiana to Spain.

As events turned out, Jefferson's best hopes were realized. Napoleon did not succeed in sending an occupation force to New Orleans. By the time the weather permitted the departure of the military expedition, the English had blockaded it. Although still formally at peace with France, it was already apparent to the government in London that the die had been cast for a renewal of the hegemonic struggle. With the imminent renewal of the war, Napoleon had little to lose by getting what he could for Louisiana. Once at war, Louisiana could easily be taken from him by England and he doubtless thought that this loss was more than likely. By selling Louisiana to the United States, he would be able to finance a part of the coming war. Abandoning Louisiana, it is true, meant abandoning his grand plan of empire in the West. But that plan could not survive renewal of the war with England, and Napoleon appreciated this only too well.

The outcome of events has since obscured the risks inherent in Jefferson's dependency on European war. In retrospect, the eventual renewal of war between France and England was as inevitable as anything in politics could be. But the same cannot be said for the moment war would again be taken up. England, independently of the situation in Europe, was not disposed to break the peace over Santo Domingo or Louisiana,[99] and Jefferson was unwilling to pay a price that might have tempted England to do so—that is, an alliance which would entail benefits readily apparent to Great Britain. For the President, salvation was located in events beyond his control. "We did not, by our intrigues, produce the war," Jefferson wrote after the crisis, "but we availed ourselves of it when it happened."[100] In this respect, the President's fortune owed more than anything else to the restless nature of Napoleonic ambition. For almost three years, the First Consul had subordinated all else to the exigencies of his plan for empire in America. As long as the plan seemed to be working, he was reasonably careful to avoid acts that might provoke England to resume hostilities. In part, at least, Napoleon appears to have gone to war when he did because he had not succeeded in reducing Santo Domingo. A vital feature of his plan for empire in America had miscarried. Despite the huge reversals, though, the island had not been irrevocably lost; in the winter of 1803 the French position on the island had even stabilized.[101] But to bring the island under his control would have required another great investment of time and resources. By this time, however, the First Consul's patience had been worn thin; fresh visions of empire beckoned, requiring ready cash, of which Bonaparte was normally in short supply. It may be, as many have speculated, that Bonaparte had never been as attracted by and committed to the colonial scheme as some of his ministers, that this scheme was ultimately incompatible with his ambitions in Europe (and, for that matter, in the East) and that he was by nature incapable of contenting himself with what he must have disdainfully regarded, to use Henry Adams's phrase, as a shopkeeper's vision of empire—one that required the patient development of French naval and marine power.

The failure of grandiose schemes nearly always suggests a logical fulfillment. The plan for making France "preponderant in America," in the words of Napoleon's Finance Minister, Barbé-Marbois, was indeed grandiose. When failure did come, Napoleon abandoned the scheme as readily as he had taken it up. "Damn sugar, damn coffee, damn colonies!" he had exploded on learning of the disaster on Santo Domingo. But the suddenness with which he abandoned his scheme does not necessarily testify to a lack of commitment. Up to that point he had made great efforts toward the fulfillment of his colonial design. The First Consul's commitment, however, had never been unqualified. Discouraged by the slow pace of empire-building, Napoleon at last turned away from the West and focused his energies on Europe, on the war that would leave him, in four years, the master of the continent.

In spite of his faith that events in Europe might operate to remove the threat to New Orleans, Jefferson expected that at some point the French would take possession of Louisiana. His intelligence from Livingston, which he had no reason to suspect, warned him of continued French preparations, though the veil of secrecy that Bonaparte placed around the expedition made reliable forecasts difficult. The optimism aroused by French reverses on Santo Domingo had quickly cooled, largely as a result of the public furor arising from the revocation of the deposit at New Orleans.

Just as the initial stages of the crisis had been marked by contrasting perceptions in Paris and Washington, so too was the final crisis before Napoleon's renunciation of his claim to Louisiana. As Napoleon's depression over the failure of his western policy deepened, Jefferson became alarmed that the expedition to New Orleans was at last under way. By the winter of 1803, Jefferson appeared to have been resigned to a French occupation.

Against this prospect the President accelerated his program of buying land from the Indians with the idea of establishing a substantial white population on the Mississippi as rapidly as possible. The influx of settlers was expected to form a new line of defense against the French. "The crisis is pressing," Jefferson wrote to the

Governor of the Indian territory in late February, "what ever can now be obtained must be obtained quickly. The occupation of New Orleans, hourly expected, by the French, is already felt like a light breeze by the indians." To facilitate the purchase of lands from the increasingly recalcitrant Indians, Jefferson suggested the usefulness of seeing them run into debt at the government trading houses, noting that "when these debts get beyond what the individuals can pay, they become willing to lop them off by a cession of lands."[102]

Beyond forming a defensive barrier, this American population might at some future date be sufficient to drive the French from New Orleans. "Although I am not sanguine in obtaining a cession of New Orleans for money," Jefferson wrote on April 30, 1803, he was "confident in the policy of putting off the day of contention for it, till we are stronger in ourselves, and stronger in allies, but especially till we shall have planted such a population on the Mississippi as will be able to do their own business, without the necessity of marching men from the shores of the Atlantic 1500 or 2000 miles thither, to perish by fatigue and change of climate."[103] In the meantime, he would do what he could by negotiation.[104]

This was, indeed, Jefferson's true policy: to conquer without war or, if this proved impossible, to conquer without a costly war. If war was the nemesis that threatened to destroy everything the Republicans had achieved and still hoped to achieve, Jefferson's unwillingness to face squarely the prospect of war with France and to make serious preparations for it was surely understandable. At the same time, his policy involved considerable risk. It did so in the first instance because it was prepared to accept for the time being a French military presence in New Orleans. Once this force was there, its removal might well have proven to be a difficult task. There was no way of knowing how large this presence might ultimately become. Nor was the size the principal consideration. Instead, it was the commitment itself. Once New Orleans was occupied by French forces, it was reasonable to expect that the nature of the crisis would be transformed. With the military occupation, the French commitment in Louisiana would have deepened.[105]

Second, Jefferson's policy involved risk because it gambled with the sentiments of the western people. The argument that time would work increasingly against the French by virtue of numbers alone necessarily assumed that these numbers could always be counted on to support the American government's cause. This assumption was at odds, though, with the view that the very presence of a foreign power in the Mississippi valley would raise the specter of secession. Earlier, Jefferson himself had given expression to this dread prospect; he had done so in circumstances that were far less ominous than those faced in Louisiana.[106] Moreover, even if in the long run time did work against the French, there might still be an enormous price to pay before the demographic fact worked its inevitable way. And once it had done so, it might still have left a legacy of disunion in its wake.

The purchase of Louisiana has often been regarded as one of the greatest triumphs of American diplomacy. Half of a continent was gained without war. The gauge of Jefferson's success has been measured almost as much in the means as in the end. Foremost among the means, presumably, was a brilliant diplomacy the central feature of which was to play for time. For time, Jefferson quickly sensed, was the great enemy of Napoleon's design. In time, the terrible cost of the ill-conceived campaign on Santo Domingo would become apparent to all and would operate to constrain even a Napoleon. In time, the struggle between England and France, implacable adversaries as they were, was bound to be renewed. And in time, the American position in the Mississippi valley could only become stronger. The essence of Jefferson's strategy was to wait for these developments, to play for time until they could work their expected result. In this, historians have generally concluded, Jefferson was right. He used conditioning circumstances to his great advantage, just as he used to his advantage the threat of forming an alliance or of going to war. And this capitalizing on circumstances, it is argued, testified above all to his insight and skill rather than to his good fortune.

The favorable assessment of Jefferson's statecraft has often been further underlined by comparing it with the course of action recommended at the time by Alexander Hamilton. Two options, Hamilton declared, were open to the United States: "First, to negociate and endeavor to purchase, and if this fails to go to war. Secondly, to seize at once on the Floridas and New-Orleans, and then negociate," a course of action that required an immediate increase in the army and militia as well as the full cooperation and support of Great Britain.[107] For this counsel, the leading Federalist opponent of Jefferson has since been criticized. Hamilton's advice, this criticism runs, would have involved us in a war with both France and Spain, a war for which we were utterly unprepared. In this war, we could not count on the support of Great Britain, then at peace with France. Hamilton was too intelligent not to realize that the course he advocated could result only in disaster. This being the case, critics have concluded, it represented little more than a political maneuver made to embarrass the administration of the day.[108]

Clearly, Hamilton's advice was intended to put the administration in an unfavorable light. Was this advice so manifestly misguided, however, that it does not warrant serious consideration? A positive response can scarcely be supported by pointing out that Hamilton's position rested on the assumption that France would never sell. Whatever his hopes might have been, Jefferson was often as skeptical of the chances for purchase as was Hamilton. But Jefferson stopped well short of accepting the proposition that in the event efforts to purchase failed, the alternative was war. The President did not accept Hamilton's first course. The war that for Jefferson would result from the failure of purchase negotiations was not a war in the here and now, as it undoutedly was for Hamilton. Instead, it was a war projected into the future and one that was dependent on the fulfillment of certain conditions. The immediate prospect and reality for Jefferson was not war but palliating and enduring until these conditions were met. But this course, to repeat, carried its own risks. It meant allowing a French occupation of New Orleans. And it meant relying on the cooperation and help of the British.

In the circumstances of the winter of 1803, prudence dictated Hamilton's course rather than Jefferson's. The immediate war proposed by Hamilton surely involved considerable risks. The seizure of New Orleans and the Floridas would likely have resulted in naval hostilities with France and Spain, hostilities for which the country was unprepared. Nor could the United States count on the help of England; short of war between France and England, such assistance was always uncertain. At the same time, the risks entailed by Hamilton's course did not include a French military force in New Orleans, a risk that Jefferson's course could not avoid taking. Moreover, Hamilton might well have calculated, in February 1803, that Great Britain and France were bound to go to war in the reasonably near future. If this calculation is not seen as unjustified in assessing Jefferson's diplomacy, there is no apparent basis for deeming it unjustified in considering Hamilton's proposal.

At the same time, these differences between Jefferson and Hamilton reflected a deeper and more significant difference separating the two men. Hamilton's proposal stemmed from an outlook which assumed that time might work against us, that we could not entrust our fortunes to the contingencies of circumstance, and that we had to resolve immediately to take our fate into our own hands as far as this was at all possible. By contrast, Jefferson's diplomacy reflected an outlook which assumed that time was on our side, that something would turn up to favor our fortunes, and that far more harm than good would result from an impatience to bring matters to a head. The difference, of course, was one of temperament. Yet it was also one that reflected sharply different attitudes toward force and the justification for its use. In the end, Jefferson's attitude toward force, his conviction that to resort to war would jeopardize everything the Republicans had accomplished and might still accomplish, was not simply important but, in all likelihood, crucial. It is this last difference that must also largely account for the fact that Hamilton was willing openly to defy Napoleon, whereas Jefferson, the apostle of independence, was not. This unwillingness drove Jefferson to a policy of dependence on Great Britain. Ironically, by comparison, in this affair of Louisiana it is Hamilton who appears as the truer representative of a policy of independence than does Jefferson.

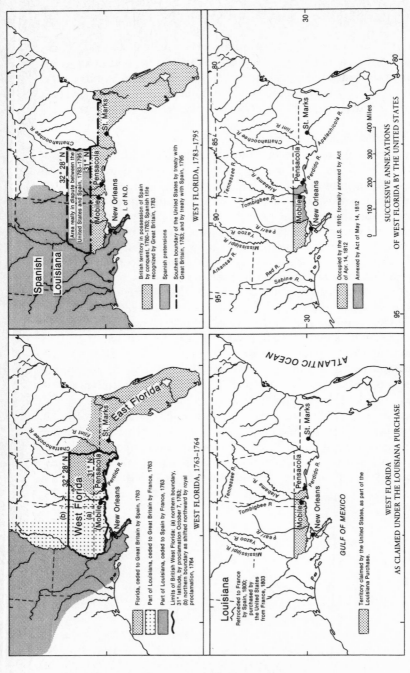

The Diplomatic Cartography of West Florida. (Map 12, "The Diplomatic Cartography of West Florida," from *A Diplomatic History of the United States*, 5th ed., by Samuel Flagg Bemis, copyright © 1965 by Holt, Rinehart and Winston, Inc., reprinted by permission of the publisher.)

14

The Gambit
for West Florida

ALTHOUGH THE ACQUISITION of Louisiana was a triumph of all but incalculable proportion, it was not without blemish. However great a bargain the purchase undoubtedly was, the wrong thing had been bought. The aim of Jefferson's diplomacy had been the acquisition of New Orleans and the Floridas. The general instructions of March 2, 1803, to Livingston and Monroe read: "The object in view is to procure by just and satisfactory arrangements a cession to the United States of New Orleans and of West and East Florida, or as much thereof as the actual proprietor can be prevailed on to part with."[109] As matters turned out, Napoleon did not have to be prevailed on to part with only some of the territory gained by the Treaty of San Ildefonso. Having resolved to abandon his plan of empire in the Western Hemisphere, he relinquished the whole of this territory to the United States. But the whole of this territory did not include the Floridas.[110]

When Jefferson learned of the retrocession of Louisiana to France, he had no expectation that Louisiana would soon come into his possession. On the contrary, he was willing to accept France in the territories on the west bank of the Mississippi if France would cede to the United States New Orleans and the Floridas. It was on the latter two objectives that his diplomacy was fixed, and the appropriation that he asked of Congress mentioned Louisiana not at all. His minister in Paris on being asked by Talleyrand whether the United States wished the whole of Louisiana had answered no and had added "that our wishes extended only to New Orleans and the Floridas." Livingston had earlier proposed that France give the United States the territory above the Arkansas River, which might serve as a buffer between the French, who would remain in the western territory to the south of that river, and the British in Canada. But Talleyrand had replied that if France once gave up New Orleans, "the rest would be of little value" and had inquired what the United States "would give for the whole." The whole, however, did not include the Floridas. Livingston had accordingly suggested to Madison that if Louisiana were purchased, "it would be good policy to exchange the west bank of the Mississippi with Spain for the Floridas, reserving New Orleans."[111]

It was Louisiana, and not the Floridas, that Jefferson received from Napoleon. This fact affected nearly all of the parties involved. Livingston was more than a little anxious over a negotiation that, in view of its objects, he had not been empowered to undertake. "This business has cost me much pains and anxiety," he wrote to Rufus King in late April 1803, "but I think our country will be essentially served"[112] The failure to obtain the Floridas—particularly West Florida—was a source of disappointment not only to Livingston and to Monroe, but to the President as well. Indeed, Jefferson very likely experienced this failure as keenly as anyone who had a role in the affair of Louisiana. It was, after all, the President who had fixed upon the twin objectives of New Orleans and the Floridas and who had insisted upon virtually tying these objectives together in the negotiations. And it was the President who was perhaps more sensitive than others in his administration to the importance placed on West Florida by the South. His disappointment, moreover, would have been far more marked had it not been for the optimistic assumptions the President appeared to have made about the ease with which the desired territory would sooner or later be obtained from Spain. The circumstances that had led France to abandon New Orleans and Louisiana would, he was persuaded, lead Spain to abandon the Floridas. Whatever the course of subsequent negotiations with Spain, we needed only to await the involvement of that country in the war that had then broken out in Europe. Our claims, Jefferson wrote, "will be a subject of negotiation with Spain, and if, as soon as she is at war, we push them strongly with one hand, holding out a price in the other, we shall certainly obtain the Floridas, and all in good time."[113]

Still, the claims to the Floridas, and particularly West Florida, had to have a plausible basis. The grounds for claiming the territory were already at hand in the treaty of purchase. That agreement did not indicate with any precision the boundaries of the territory the United States had purchased from France. Instead it incorporated the language found in the earlier Treaty of San Ildefonso, whereby Spain had agreed to retrocede Louisiana to France. Thus the treaty by which France ceded to the United States what Spain had earlier ceded to France stipulated: "the colony or province of Louisiana, with the same extent that it now has in the hands of Spain, and that it had when France possessed it; and such as it should be after the treaties subsequently entered into between Spain and other states."

The ambiguous wording incorporated into the treaty was such as to allow utterly contradictory interpretations. Indeed, an objective reading of this wording forces the conclusion that it did stipulate two very different things. The American case rested on "Louisiana, with the same extent . . . that it had when France possessed it"—that is, a Louisiana extending to the Perdido River, incorporating West Florida. Opposed to this was "Louisiana, with the same extent that it now has in the hands of Spain." For the Spanish, Louisiana and West Florida were held as two distinct territories.

Livingston had proposed that the language of the Treaty of San Ildefonso be incorporated into the treaty between France and the United States. He had initially also wanted the French to commit themselves in writing to the effect that should "florida" fall into their hands any way, they would transfer "the same" to the United States. Moreover, as further proof of friendship, France was asked to commit itself to aid the United States "in acquiring from Spain such cessions within the said province of florida as may aid the commerce of the United States."[114] But the French negotiator, Barbé-Marbois, was unwilling to make this commitment in writing, though he was quite willing to make a verbal declaration to this effect.[115]

In the concluding phase of the negotiation, Livingston appeared to have been satisfied with the verbal commitment of the French government respecting West Florida. Whether this change was due to a conviction that the already incorporated provision of the Treaty of San Ildefonso was sufficient grounds for making a claim to West Florida or simply reflected resignation over the

French position remains uncertain.[116] What is quite clear, however, is that within a matter almost of days Livingston and Monroe were certain that the treaty of cession gave the United States a claim to West Florida.

Livingston was the first not only to express this certitude but to formulate the main outline of the argument on which the claim to West Florida would be based by the Jefferson administration. Eager, in all likelihood, to steal a march on Monroe, he wrote to Madison on May 20, 1803, only eighteen days after the conclusion of the treaty, that "the moment I saw the words of the treaty of Madrid [San Ildefonso], I had no doubt but it included all the country that France possessed by the name of Louisiana, previous to their cession to Spain, except what had been conveyed by subsequent treaties." Barbé-Marbois, Livingston went on to report, "neither asserted nor denied, but said that all they received from Spain was intended to be conveyed to us." But Livingston was sure that his construction was correct and found confirmation of this in the statement of the Spanish minister, Cevallos, to Pinckney, "that France had recovered Louisiana as it formerly belonged to her, saving the rights of other parties." It was "well known that Louisiana, as possessed by France, was bounded by the river Perdido, and that Mobile was the metropolis." Livingston concluded with these words: "Now, sir, the sum of this business is, to recommend to you, in the strongest terms, after having obtained the possession, that the French commissary will give you, to insist upon this as a part of your right: and to take possession, at all events, to the river Perdido. I pledge myself that your right is good; and, after the explanations that have been given here, you need apprehend nothing from a decisive measure."[117]

Livingston's letter may be regarded as the opening salvo in the West Florida controversy.[118] In the convoluted and arcane issues it raised, in the energy it required to be thoroughly mastered, and in the handful of persons who did in fact master all of the historical details, the West Florida controversy is the American analogue of the Schleswig-Holstein question. It contained just enough built-in ambiguity to give at least some plausibility to the most contradictory positions. The French took full advantage of this feature in their dealings with the American negotiators. It was a simple matter for them to lead the Americans to believe they supported the United States's position and would continue to do so. But then

Livingston and Monroe, as well as their superiors in Washington, badly wanted to believe that France would support the United States rather than Spain.

Despite the later complications the parties gave to it, the case initially made by Livingston was simple enough. Monroe expressed it with economy in a letter written to Livingston a few days after the latter had communicated his views on the West Florida claim to Madison: "We perfectly agree in the opinion that Louisiana, as it was in the hands of France prior to the year 1763, extended to the River Perdigo, and that it was restored to her by Spain, in the treaty of Il Dephonso, precisely in the same extent." [119]

Subsequent expositions of the American claim simply built on this premise. While in possession of Louisiana, France had never dismembered it. On the same day, November 3, 1762, France had conveyed New Orleans and the territory west of the Mississippi to Spain while giving up the Floridas to Great Britain. Later, in 1783, Spain reunited West Florida to Louisiana, and thereby recreated the province as France had earlier possessed it, save for those territories now controlled by the United States. It was this same province that Spain had ceded to France at San Ildefonso and that the United States had purchased from France.

This was the burden of Monroe's opinion of June, written, it would appear, to establish his own independent standing as an initiator of the claim to West Florida. That claim, Monroe confided to Madison, was "too clear to admit of a doubt." Indeed, so compelling was the case for the American claim that Monroe believed the government of Spain must entertain the "same opinion on that point." Nor did he doubt that if the United States were to take possession of West Florida as a part of Louisiana, "that the measure will be acquiesced in by that govt., or at least that it will not be taken ill by it." [120]

In Madison and Jefferson, the two ministers had a receptive audience. At the outset, it is true, both the Secretary of State and the President expressed uncertainty over the boundaries of Louisiana. Those boundaries, Madison wrote to Livingston at the end of July, "seem to be imperfectly understood, and are of so much importance, that the President wishes them to be investigated whenever information is likely to be obtained." [121] But this initial hesitation and uncertainty soon changed. By October, it seemed "undeniable" to Madison that the eastern boundary of Louisiana extended as far as the Perdido River.[122] By the following winter, the Secretary of State had prepared a case for the American claim to West Florida that followed in its principal features the case earlier made by Livingston and Monroe.[123]

It was Jefferson's conversion, however, that was essential. Although the President initially indicated that he had an open mind on the boundaries of Louisiana, it was in fact open only to evidence and arguments that supported American claims to West Florida. In mid-July, Jefferson had left Washington for Monticello, where, as he wrote to Gallatin in late August, with the aid of his books he devoted himself to investigating the question of the boundaries of Louisiana. The results of his investigation satisfied him that the claim to the Perdido was "solid."[124] He was now persuaded that Monroe and Livingston had clearly been justified in claiming our right to West Florida as far as the Perdido. Soon thereafter Jefferson outlined in some detail his position in a paper entitled "The Limits and Bounds of Louisiana."[125] The study Jefferson made bore all the marks of a directed verdict. It was one that the President endowed with ever greater conviction despite his failure to find confirmation of the administration's position in expert opinion.[126] The disconcerting replies to his queries did not appear to affect Jefferson's determination to claim the Perdido as the eastern boundary or, for that matter, his growing conviction that the claim was founded in right. What were merely "claims" in August 1803 had become our "just limits" by January 1804.[127]

It was in this manner that the controversy over West Florida arose and became an issue to which the administration eventually mortgaged so much of its diplomatic capital. The West Florida claim was a classic case of desire pointing the way to conviction. The United States had, of course, long desired the Floridas—particularly the territory bounded by the Mississippi and Perdido rivers. But prior to the Louisiana Purchase, it had not been contended that West Florida formed a part of Louisiana. Livingston, the original champion of the West Florida claim, had argued the contrary position prior to May 1803. The abstract merits of the case apart, straightforward considerations of interest had prompted him to do so. If France were to acquire Louisiana, it was in the American interest to interpret the boundaries of Louisiana as restrictively as possible.[128] Livingston's conversion and the case subsequently made by him and Monroe for claiming West Florida depended almost entirely on the tendentious reading of the earlier treaty of cession between Spain and France.

It was a reading that the original parties to the San Ildefonso treaty did not share. Certainly Spain did not share it. A consistent negotiating position with respect to French claims to the Floridas testified to this. Madrid had rejected these claims throughout, insisting that the Floridas formed no part of the province of Louisiana. But the French had not argued the contrary. However much Napoleon may have wanted the Floridas, his envoys had not contended that the Floridas—or even West Florida—were theirs by virtue of the 1800 treaty of cession. Instead, their instructions had been to the effect that the treaty of 1763 defined the eastern limits of Louisiana and that these limits were the Mississippi (and Iberville). The secret instructions issued to General Victor, the military commander destined for Louisiana, were explicit on the point that the Mississippi formed the eastern boundary of Louisiana. The same is true of the other documents, both Spanish and French, dealing with the transfer from Spain to France and, ultimately, to the United States.[129]

The Spanish–French consensus on the meaning of the Treaty of San Ildefonso proved to be critical. It was not easy to argue that these two states did not correctly understand the terms of the agreement they had made with each other. But this was, in effect, what the American position came to. It prompted the ironic comment of Henry Adams, made with respect to Livingston (but applicable as well to Monroe, Madison, and Jefferson), that "he was forced at last to maintain that Spain had retroceded West Florida to France without knowing it, that France had sold it to the United States without suspecting it, that the United States had bought it without paying for it, and that neither France nor Spain, although the original contracting parties, were competent to decide the meaning of their own contract."[130]

In selling Louisiana to the United States, it is true, the French had encouraged the American negotiators in the belief that Paris would support the aspirations of the United States to obtain the Floridas. At the time, the French Foreign Minister had responded to American entreaties for support of their Floridas claim by stating: "You have made a noble bargain for yourselves, and I suppose you will make the most of it."[131] The first great step toward this end should have been the support of American claims to the territory between the Mississippi and the Perdido; not only was this strip the most desirable part of the Floridas, but it was also the territory that made up the eastern limits of Louisiana when, prior to 1763, the French had possessed the province. Until December 1804, Paris remained silent on the merits of the American claim to West Florida. Then, to the surprise and chagrin of the American negotiators, Talleyrand took the side of Spain in the controversy. "All the territory lying to the eastward of the Mississippi and the river Iberville, and south of the 32d degree of north latitude," he wrote on December 21 to the American minister, "bears the name of Florida. It has been constantly designated in that way during the time that Spain held it; it bears the same name in the treaties of limits between Spain and the United States; and, in different notes of Mr. Livingston of a later date than the treaty of retrocession, in which the name of Louisiana is given to the territory on the west side of the Mississippi, of Florida to that on the east of it." Spain had retroceded to France what France had earlier ceded to Spain. Spain had received no territory east of the Mississippi and could therefore give back none to France.[132] The French decision to support Spain's position in the West Florida controversy dealt a near-fatal blow to the standing of the American claim. To continue to insist upon interpreting a treaty in a manner that was plainly at odds with the interpretation given by the parties to that treaty bordered on the absurd. Yet this remained the position of the United States until the Madison administration finally occupied the territory and incorporated it into the Union.

15

Lessons of the Louisiana Purchase

ONE CONSEQUENCE OF THE Louisiana purchase was a determination to possess the Floridas. A related consequence of
the cession was the conviction that possession of the Floridas would
prove to be a simple matter. From the outset, Jefferson was confident that the addition of the Floridas to the Union was both
inevitable and imminent. It was inevitable because of our growing
strength and Spain's growing weakness. Spain could not effectively defend the Floridas against the mounting pressure of American settlers. It was imminent because Spain's intrinsic weakness
would be fully revealed once it was involved in the European war.
If France could not expect to secure its position on this continent
once engaged in war in Europe, what hope might Spain reasonably entertain of doing so?

The optimism that marked the administration's Florida diplomacy was pervasive and its root cause was apparent. The diplomacy that was seen as having been so successful in the case of
Louisiana could scarcely be less successful in the case of the Floridas. The lessons of Louisiana followed quite simply from this success. What were some of those lessons? Perhaps the most important was that the easy way is the only way. Louisiana had shown
that there was no need to arm and to make sacrifices when vital
interests are at stake. Instead, it was sufficient to hold up the prospect of alliances that would probably never be made and to threaten
wars that one had no intention of waging. Another lesson was the

wisdom of conducting negotiations in which one asks for much while offering little. In the affair of Louisiana there had been few quid pro quos in the American diplomatic armory. The conviction that there was no need to accept the principle of reciprocity in negotiations was related to still another lesson of Louisiana: that the United States held a position of central importance in the scales of European diplomacy. By virtue of our trade alone, American support of or hostility to one or the other of the great European powers was likely to prove decisive in any conflict between them.

On the morrow of the Louisiana triumph, these were the lessons that were to be applied to the Floridas. In the Florida diplomacy, France was seen from the outset as the key. In the negotiations attending the Louisiana Purchase, the French had made a verbal promise to help the administration obtain the Floridas. Monroe and Livingston were not alone in taking this promise at face value. Madison and Jefferson did so as well. They did so because they were persuaded that it was in France's interest to support American aspirations to possess the Floridas. In the dispute between the United States and Spain over the construction of the treaties of cession, Madison expressed the prevailing view in writing to Livingston that France was "bound no less by sound policy, than by a regard to right" to support our position, for "a transfer from Spain to the United States of the territory claimed by the latter, or rather of the whole of the Floridas on reasonable conditions, is in fact, nothing more than a sequel and completion of the policy which led France into her own treaty of Cession; and her discernment and her consistency are both pledges that she will view the subject in this light."[133] On this view, France had ceded Louisiana to the United States to prevent the latter from forming an alliance with Great Britain, an alliance that presumably would have resulted, on the outbreak of war, in the certain loss of Louisiana by France. A transfer of the Floridas from Spain to the United States would express France's continued interest in preventing an alliance between Great Britain and this country as well as in preventing the Spanish possessions from falling into the hands of Great Britain.

That France was bound to help the United States acquire the Floridas—at least West Florida—became an article of faith for the Jefferson administration. The belief persisted despite growing signs that Paris might not have the same appreciation of its interests the administration had. In the period immediately following the cession, Spain's indignation over and opposition to the arrangement brought it into conflict not only with this country but with France as well. It was, after all, France that had betrayed Spain, not the United States. But in January 1804, Madrid declared that it no longer opposed the cession. This change in policy resulted in the marked improvement of Spain's relations with France. The key to the administration's diplomacy then failed to turn the lock of the door to the Floridas. Throughout the rest of 1804, France stalled when appealed to by the Americans.

If France could only be importuned, Spain might be threatened. A diplomacy of intimidation quickly became the norm in dealing with Madrid. When Spain protested the Louisiana cession on the grounds that France was bound not to alienate the territory and that the title was not in any event good, France having failed to fulfill the terms of its undertaking with Spain, the American government responded by dismissing the protest. It could be directed only to France, the reply read, not to this country.[134] When Spain nevertheless persisted in protesting the cession, this opposition was interpreted by the President and his principal advisors as foreshadowing the attempt to prevent by force the French from provisionally occupying New Orleans, preparatory to the American occupation. Were the Spanish Governor of Louisiana to fail to deliver that province to the person duly authorized for that transfer, the Secretary of State, speaking on behalf of the President, warned Spain's minister to Washington, force would be employed to ensure effective possession. And if required to use force to take possession of New Orleans, the Spanish government was advised that the Floridas would also be attacked and held as compensation for the expense Spain would have made necessary by its resistance.[135]

This was only the first of a number of occasions on which Spain was threatened with force. At the same time, this behavior did not in fact reflect a deeper disposition to employ force as a means of policy when expedient. If Spain was made the object of forcible threats, this reflected the assumptions that the threats would not have to be carried out, that Spain was in no position to accept war as the outcome of its differences with the United States, and that in the end Spain would have to yield to such pressures as the United States might bring to bear.

These optimistic assumptions, together with the expectation that the French would make good on their promise to support American aspirations respecting the Floridas, formed the basis for the confidence entertained over negotiating prospects with Spain. The negotiations over the Floridas were seen as a much easier replay of the successful negotiations over Louisiana. France had ceded Louisiana to the United States rather than risk losing it in war to Great Britain. Spain, confronted by the prospect of war with Great Britain, would do the same with the Floridas. In his initial instructions to Monroe on the negotiating posture to be taken toward Spain, Madison emphasized the use Monroe was to make of Britain's supposed intention to seize the Floridas once Spain entered the war on France's side (or even while Spain was still at peace). "In every view," Madison concluded, "it will be better for Spain, that the Floridas should be in the hands of the United States, than of Great Britain; and equally so, that they should be ceded to us on beneficial terms by herself, than that they should find their way to us thro' the hands of Great Britain." [136]

This was, of course, a page taken directly from the book of Louisiana. Spain would only be acting in its own best interests in ceding the Floridas to the United States. This was so even though Spain might expect little in return for the cession save the good will of this country. [137] But the good will of this country, Madison pointed out in a subsequent letter of instructions to the American minister in Madrid, was no small matter. After all, he asked, "What is it that Spain dreads?" The Secretary of State's answer spoke volumes.

> She dreads, it is presumed, the growing power of this country, and the direction of it against her possessions within its reach. Can she annihilate this power? No.—Can she sensibly retard its growth? No.—Does not common prudence then advise her, to conciliate by every proof of friendship and confidence the good will of a nation whose power is formidable to her; instead of yielding to the impulses of jealousy, and adopting obnoxious precautions, which can have no other effect than to bring on prematurely the whole weight of the calamity which she fears. Reflections, such as these may perhaps enter with some advantage into your communications with the Spanish government, and as far as they may be invited by favorable occasions, you will make that use of them. [138]

The occasion of Madison's reflections to Pinckney was Spain's opposition in the autumn of 1803 to the cession of Louisiana. But the reflections were equally appropriate to other differences that would arise with Spain over its possessions. Spain could only lose by attempting to constrain the growing power of this country. This being the case, Madison concluded, prudence dictated that Spain should make every effort to conciliate and to appease the United States. It was an interesting commentary considering that it came from a government that prided itself on having turned its back on the old diplomacy and embraced new and more enlightened principles. Madison's reflections were worthy of a Talleyrand, just as Jefferson's threat to deprive Spain of the Floridas if it actively opposed the cession of Louisiana was worthy of a Napoleon. The day would soon come when the Spanish state secretary would act on the logic of Madison's reflections and tell the American minister to choose peace or war, the consequences being the same for Spain.[139]

The essential terms asked of Spain for a settlement of differences were first laid out by Madison to Monroe and Pinckney in the spring of 1804.[140] These terms were further refined and summarized by the President in a letter of July 5 to his Secretary of State. The sense of the cabinet, Jefferson wrote to Madison, now appeared to be

> 1. That an acknowledgment of our right to the Perdido, is a *sine qua non,* and no price to be given for it. 2. No absolute and perpetual relinquishment of right is to be made of the country East of the Rio Bravo del Norte even in exchange for Florida. . . . 3. That a country may be laid off within which no further settlement shall be made by either party for a given time, say thirty years. This country to be from the North river eastwardly towards the Rio Colorado, or even to, but not beyond the Mexican or Sabine river. . . . This relinquishment, and 2 millions of Dollars, to be the price of all the Floridas East of the Perdido, or to be apportioned to whatever part they will cede.[141]

Spain was expected to give up the Floridas, West and East. For the former, Spain would receive nothing; for the latter, two million dollars. In the West, although Texas was to be claimed by right, a borderland was to be laid off between the Rio Grande and the Colorado River which was to be free of settlement for thirty years. For Spain to have agreed to these terms would have required nothing less than a capitulation on its part, for they stipulated the surrender of territory without receiving in turn any real equivalent.

The expectation that Spain would nevertheless accept terms that were so one-sided may be accounted for by assumptions to which attention has already been drawn. Certainly, that expectation was not supported by Spain's recent behavior. In the early winter of 1804, Pinckney had initiated negotiations in Madrid on claims that had arisen from the quasi war of the late 1790s. These claims on behalf of American commerce were comprised of two categories. The first resulted from captures made by Spanish warships; the second resulted from captures made by French ships in Spanish waters and condemned by French consuls in Spanish ports. In earlier negotiations over these claims, the Spanish government had refused to entertain claims arising out of French actions, insisting that it bore no responsibility for those actions. Instead, it offered to conclude a convention dealing with its own actions, while reserving the disputed French captures for later discussion. In August 1802, a convention to this effect was concluded. Sent to the Senate in early 1803, the convention was narrowly defeated in March of that year.

In the summer of 1803, the claims against Spain were revived and seen as a means for purchasing the Floridas. Monroe, who was to have proceeded from Paris to Madrid, was instructed by Madison to "hold a strong language" on the claims, including those "not yet admitted by the Spanish government."[142] But Monroe subsequently went to London, where he replaced Rufus King, and during 1804 Pinckney alone conducted the negotiations in Madrid. From the start, they went badly. Although the Spanish government had decided in January 1804 to abandon its opposition to the cession of Louisiana, it was no more inclined to accommodate the American insistence on pressing claims resulting from French captures in Spanish waters than it had been two years earlier. Moreover, in the course of these negotiations, the Spanish government learned of the passage of the "Mobile Act" by Congress. Instigated by the administration, the Mobile Act was intended to express legislative support and approval of the claim that West Florida was part of Louisiana and, as such, included in the territory purchased by the United States. The Spanish reaction was unexpectedly strong and, as a result, Jefferson so interpreted the legislation as to render it innocuous.[143] In Madrid, however, the aroused Spanish government refused to conclude any claims convention with the United States unless the Mobile Act was first revoked. It also insisted that the claims for French spoliations be simply abandoned.

The record of the negotiations with Spain had shown—or should have shown—that Madrid would not be easily frightened into submitting to American proposals on the disputed territories. Indeed, Pinckney had terminated the negotiations in July by virtually threatening the Spanish government with war because of its refusal to ratify the claims convention. But the Spanish minister, Cevallos, although apprehensive over the prospect, had refused to back down before Pinckney's threats.[144]

This resistance might have served to warn the President and the Secretary of State against persisting in their assumptions respecting Spanish policy and behavior. There was no sign that it did. That Spain's position dictated an accommodating response to this government's terms for settling territorial differences remained an unshaken premise. Nor was this premise affected by growing evidence that French support of American aspirations to possess the Floridas could not be counted on. When, in the autumn of 1804, Monroe arrived in Paris on his way to Madrid, he had refused to credit this evidence, just as he had refused to entertain the prospect of paying for West Florida. Instead, he wrote to Talleyrand reminding him of his earlier promise to aid this country in securing the Floridas, and in response to French hints that if America wanted West Florida, it must be prepared to pay for it, Monroe replied that having paid for it once, the United States would not pay for it a second time.[145]

In December, the event occurred that the administration had so heavily counted on for the success of its Spanish diplomacy. Spain entered the war on the side of France. But the consequence of entering the war was not to make Spain more yielding to American demands but, if anything, less yielding, for Madrid now had France as an ally and guarantor. In early January 1805, Spain and France concluded a treaty in which France promised the return of colonies seized from Spain in the course of the war.

In these circumstances, Monroe and Pinckney might have been expected to retreat, if not from the substance of their position, then at least in the manner of presenting it. Instead, the American negotiators appeared almost unaffected by developments that all but destroyed their position. In part, this was simply the result of ignorance, for word of France's repudiation of the American claim did not reach Madrid until early in February.[146] In part, though, this was the result of the conviction that a true regard for its interests could only lead France to conciliate the United States. Monroe in particular had long cherished this conviction and was loath to accept that it had little if any foundation. American trade, he believed, was much too important to France for it to side with Spain against the United States. Then, too, a close understanding between the United States and England was something Monroe was certain that France would go to great lengths to prevent.[147]

This conviction entertained of France was put to a test in March by the American minister to Paris, in response to the promptings of Monroe. General John Armstrong, upon sounding out official opinion, wrote to his colleague in Madrid that there was no support for the American positions, whether on claims or on boundaries. To leave no doubt in Monroe's mind over the meaning of his message, Armstrong concluded by noting "to the question, what would be the course of this Government in the event of a rupture between us and Spain? They answered, we can neither doubt nor hesitate; we must take part with Spain."[148]

The report of Armstrong marked a turning point in the protracted negotiations with Spain. From this moment on the negotiations were taken up by exchanges between the two sides that were intended to register positions rather than settle differences. In effect, the negotiations were at an end. Pursued over the better part of two years, Jefferson's Spanish diplomacy proved utterly barren of positive results. All that it appeared to have accomplished was the humiliation of the American negotiators and, beyond them, of the government they represented. The moral indignation with which this defeat was greeted served only to underline the setback the Jefferson administration had suffered. After grasping at the Floridas for almost two years, Jefferson could blandly write in the spring of 1805: "We want nothing of hers. . . . But she has met our advances with jealousy, secret malice and ill-faith."[149] And the President's negotiators could find their failure in the "very different spirit [that] animated this government in every respect," a spirit that allowed neither "candor nor conciliation in the management of the business."[150]

What had led to a result that stood in such startling contrast to the outcome of the Louisiana affair? Groping for an answer, the participants sought one largely in the unexpected and, in their view, almost inexplicable behavior of France. Monroe was only the most unguarded in his faith that France would support the United States at the expense of Spain. The President and the Secretary of State also believed this and were almost equally at a loss to account for the French support of Spain. In a letter to Armstrong, Madison acknowledged his perplexity over French behavior. "The part which France takes in our controversies with Spain," he wrote,

> is not a little extraordinary. That she should wish well to her ally, and even lean towards her, in the terms of an adjustment with the United States, was perhaps to be expected. But that she should take side wholly with Spain, and stimulate pretensions, which threatening the peace of the two countries might end in placing the United States on the side of Great Britain, with resentments turned against France as the real source of their disappointment, this is more than was to be expected, and more than can easily be explained.[151]

Madison's explanation of French behavior was that it derived either from a very sanguine estimate of France's prospects at the time in Europe or (as Armstrong thought) from the desire "to convert the negotiations with Spain into a pecuniary job for France and her agents." Having been persuaded for so long that French support for the transfer from Spain to the United States of the territory claimed by the latter was, as he had once expressed it, "nothing more than a sequel and completion of the policy which led France into her own treaty of cession," Madison could attribute France's unexpected course only to overblown estimates of her power or to venality.[152]

What were the interests that Napoleon presumably neglected and that should have guided his policy? At a low point in the final round of negotiations with Spain, Monroe summarized these interests in a letter to Armstrong that expressed his stubborn conviction that a rupture with the United States was an event France must seek to avoid, even at the risk of its relationship with Spain. France, he wrote, "is not prepared to quarrel with us, for many reasons . . . she relies altogether for supplies on our flag, and on our Merchants and People for many other friendly offices in the way of trade, which none others can render . . . our government pursues, by prohibiting our trade with St. Domingo, and in many other respects, a system of the most friendly accommodation to the interests of France." [153]

If these were the compelling interests that should have led France to support American aims, why had they failed to do so? The explanation was not difficult and it did not consist in Napoleon's failure to look beyond the momentary and immediate. It was quite true that France depended on American shipping to carry cargoes from her colonies to ports in France. It was also true that the carrying trade had become an important American interest. The trade was undertaken not out of friendship for France but because it was very profitable. There was no reason to suppose that Napoleon failed to appreciate this or to understand that a threat on our part to shut it off would be the merest bluff. Nor was there reason to suppose that Napoleon credited the threat of a close relationship between the United States and Great Britain. Certainly, he appreciated that any intimacy could materialize only on the basis of a resolution of the maritime issues separating the two countries and that such settlement was unlikely. Even then the French might have been expected to have learned something from the Louisiana affair. One lesson was that the United States would prefer by far the continued presence of Spain in the Floridas to that of Great Britain. Another was that the periodic threat of an Anglo-American alliance was not to be taken too seriously.

These considerations apart, there remains Madison's point that the Floridas were simply the sequel and completion of Louisiana and that the same considerations that had led France into its own treaty of cession could be expected to lead Spain into ceding the Floridas. And if not Spain, then surely the power that dominated Spain. But the parallel Madison drew between the two cases was a false one. Napoleon had ceded Louisiana on the eve of a renewal of the great contest in Europe. Faced with a choice between selling the vast territory to the United States and having it taken from him by Great Britain, he chose the former and thereby helped to finance the war on which he was determined. In selling Louisiana, Napoleon offended the Spanish government. But that government—Godoy and Cevallos above all—had gone out of its way to delay and even to oppose his plan of empire in the West. Having done so, it was entitled to no special consideration from him.

In the case of the Floridas, the governing circumstances were very different. To Spain, the Floridas formed a vital line of defense against American domination of the Gulf of Mexico.[154] Nor did it much matter to Spain whether the Floridas, if they were to be lost, were lost to the United States or to Great Britain. The administration's argument that only the transfer of these possessions to the United States could secure against their falling into the hands of Great Britain might weigh with France but it could have little weight with Spain. If a choice had to be made, Spain might well prefer losing the colonies to Great Britain rather than to the United States. For the English might restore the Floridas in a peace settlement, whereas the Americans, once in possession of the territories, would never do so.

These considerations might have been of little moment had France enjoyed a complete control over Spain and Spanish policy. This was indeed one of the principal assumptions on which the administration based its Florida diplomacy. But there were limits to the control Napoleon could safely exercise over Madrid. It was one thing to cede to the United States territory that France had a claim to, even if the claim was imperfect and even if the cession violated a promise made to Spain. It was quite another thing to "deliver" to the United States territory that France had no real claim to and had never seriously pretended otherwise (only the United States had done so).

In December 1804, Spain became an ally of France in the war. Not only did it provide a substantial subsidy to France; it also contributed to French naval power. It may be argued that Napoleon was not one to hesitate in rewarding a faithful though weaker ally with his ingratitude. Still, he might be expected to do so only if the temptation were of appropriate magnitude. Yet until the autumn of 1805, Jefferson did not hold out any real temptation, unless the offer to purchase East Florida for two million dollars is so considered. When the offer did go substantially higher, and did eventually arouse Napoleon's interest, it had become apparent to Paris just how important the Floridas were to Jefferson. It was with this realization that Napoleon, feeling no great urgency, began to squeeze the United States to pay an even higher price.

Did Napoleon, by taking Spain's side in the West Florida dispute, squander the opportunity of creating a community of interest with the United States, one that might have been of great benefit to him by committing the United States to a more vigorous opposition to English violations of American neutral rights? Even more, might a more intimate relationship with France have led to an Anglo-American war before 1812, a war that could only have been to Napoleon's advantage?[155] These questions raise the reverse side of the issue already taken up in considering Jefferson's Louisiana diplomacy. There the question arose whether the threat to form an alliance with Great Britain influenced Napoleon's decision to sell Louisiana. The same threat was made in the diplomacy over the Floridas, though with markedly less emphasis. Not having been very plausible in 1802-3, it was still less plausible in 1804-5. But if the plausibility of the threat of an alliance with Britain was low in the case of the Floridas, the same must be said of the prospect of a Franco-American community of interest. Napoleon could have had few illusions on this score. If American gratitude over Louisiana had narrow limits, there was no reason to think that its gratitude over the Floridas would prove any greater. There is no evidence that Napoleon seriously doubted Jefferson's commitment to a policy of neutrality, and this despite his subsequent position in 1808 that the price of the Floridas was America's entrance into the war against Great Britain.

The circumstances that had made possible the success of Louisiana were no longer present in the case of the Floridas. Although the diplomacy of the administration remained substantially the same in the two instances, the results were strikingly different. After two years of diplomatic effort, the Floridas diplomacy had issued not only in failure but in humiliation as well. Spain had been presented with more than one ultimatum and made the object of more than one threat. In each instance, the ultimatum had been allowed to lapse without consequence and the threat to remain without effect. Still, the President would not give up his quest for the Floridas. If anything, time—and failure—had only deepened desire and strengthened determination.

16

The Empire of Liberty:
The Conflict between
Means and Ends

WHY WAS THE PRESIDENT so persistent over the possession of the Floridas? Why did he make the issue of West Florida so central for his administration that he was prepared to subordinate most other concerns to it? There is no clear answer to these questions. Certainly, a search of Jefferson's utterances—public and private—on the Floridas affords no clear answer. In striking contrast with his analysis of the interests at stake in the Louisiana crisis, in the case of the Floridas he is curiously silent. Only his determination to possess them is clear.

Nor have historians thrown much light on the matter. Henry Adams is clear enough about the importance the Floridas had for Jefferson. West Florida, Adams wrote, was his "overmastering passion."[156] But why it was, Adams does not say with anything approaching his usual persuasiveness. His explanation for what amounted to an obsession was that "among the varied forms of Southern ambition, none was so constant in influence as the wish to obtain the Floridas." A government "Southern in character and purpose," Adams concluded, responded to the desires of its principal constituency.[157]

These desires of the South, however, were not shared by the rest of the country. The North, as Adams noted, "turned impatiently away whenever the Floridas were discussed." Whereas the affair of Louisiana had engrossed virtually the entire country, the same could not be said of the Floridas. And whereas a successful outcome of Louisiana promised to secure the triumph of Jefferson's party even in New England, success in the case of the Floridas did not hold out any comparable result. Indeed, it is not even clear that in pursuing his Florida diplomacy Jefferson was responding to southern pressures.[158]

Jefferson's determination to secure the Floridas has also been attributed to a concern for security. A historian of the Floridas diplomacy has recently written that by 1804 both Jefferson and Madison "were convinced not only that the United States was sufficiently strong to incorporate the area into the American Union, but that the national security demanded expansion at Spain's expense lest a third power intervene."[159] It was this threat of intervention that presumably altered the traditional view, one that Jefferson fully shared, which looked with favor on Spanish possession of territory adjacent to the United States until such time as this country might effortlessly become Spain's successor.

Clearly, a successful French or British intervention in the Floridas would have threatened the nation's security interests. The occupation of West Florida by either power would have put to risk American control of the lower Mississippi and threatened access to the Gulf by way of the river and port of Mobile. There is no evidence, though, that Great Britain ever contemplated this step once the United States had taken possession of New Orleans and Louisiana. A British occupation of the Floridas was a prospect the administration held out to Spain (and to France) as an incentive to accept this country's negotiating position. In fact, a British seizure and occupation of the Floridas would have met with the strongest opposition by the Jefferson administration. Great Britain had no sufficient reason for taking a step that was certain to incur American hostility. Even the reason for considering the occupation of New Orleans in the spring of 1803—to block what was believed to be at the time an imminent threat of French occupation—was no longer apparent.

While the prospect of a French intervention in the Floridas could not be dismissed, it was not an immediate danger. It could not be undertaken as long as France remained at war with Great Britain. Even more, it was scarcely a serious prospect unless France emerged victorious from the war in progress.[160] Then, and really only then, might an American government have had reason to fear an occupation of the Floridas. But then, as well, the Floridas might have been among the least of Jefferson's worries, for in the event of victory Napoleon's grander scheme for empire in this hemisphere was quite likely to be revived.

Nearer the mark, it would seem, in accounting for Jefferson's Florida obsession was a sense of frustration and disappointment at having failed to achieve the goals he had initially set out. This sense might be partially compensated by the triumph of Louisiana, but it could not be entirely removed. Besides, there was the feeling that having gone so far in expanding the Empire of Liberty, a further effort to round it out was in order. The acquisition of Louisiana had not sated Jefferson's appetite for expansion. Instead, his appetite appeared only to have grown with the eating. And it had grown with the eating because expansion represented something close to an end in itself for Jefferson.

The issue that is here raised goes beyond the indisputable. That Jefferson was not only a committed expansionist but among the generation of Founding Fathers perhaps the greatest expansionist is generally acknowledged. Was his commitment to expansion, however, as compelling as his commitment to liberty? In the Empire of Liberty, was empire as significant as liberty? The response of historians to this question has generally been negative. While Jefferson is seen as having done more than his share to consolidate the Union, his devotion to the Union is carefully distinguished from his devotion to liberty. Dumas Malone has written that while Jefferson was the great architect of orderly expansion, "to him the Union was always a means, not an end in itself."[161] Merrill Peterson has written in the same vein: "Liberty was the ultimate value, Union the means, to be cherished only so long as it furthered the end of its being. Such was Jefferson's philosophical view of the matter."[162] Julian Boyd, in his now classic essay, "Thomas Jefferson's 'Empire of Liberty,'" went still further in declaring: "Though he called it an Empire of Liberty, it was to be neither an isolated political entity nor an imperialistic force for compulsory extension of ideals of liberty: its domain and compulsions would be in the realm of the mind and spirit of man, freely and inexorably transcending political boundaries, incapable of being restrained, and holding imperial sway not by arms or political power but by the sheer majesty of ideas and ideals."[163]

In the prevailing view, then, the commitment to empire was a commitment to means, never to be confused with the commitment to the end of liberty, a commitment that had no political and territorial boundaries. That Jefferson left little doubt about the nature of his hierarchy of commitment is presumably borne out by his willingness to countenance "sister republics" rather than to insist on an ever greater and expanding union. He expressed this willingness, it is often noted, at the time of the Louisiana cession, when critics of the purchase voiced the fear that so vast an enlargement of territory was sure to increase the danger of disunion. The prospect did not appear to alarm Jefferson, however, for he had never shared the belief that the stability and cohesion of republics required that they remain small. On more than one occasion, he had noted that this "doctrine of Montesquieu," that a republic can be preserved only in a small territory, was false and that the reverse was true.[164] But even if the acquisition of Louisiana were to lead to the eventual separation of the western lands from the Union, Jefferson expressed himself ready to accept this outcome. In a letter written to John Breckenridge in the summer of 1803, he asked why the Atlantic states should dread the possibility that the acquisition of Louisiana might lead to "the formation of a new confederacy, embracing all the waters of the Mississippi, on both sides of it, and a separation of its Eastern waters from us." Jefferson's answer was this:

> The future inhabitants of the Atlantic and Mississippi states will be our sons. We leave them in distinct but bordering establishments. We think we see their happiness in their union, and we wish it. Events may prove it otherwise; and if they see their interest in separation, why should we take side with our Atlantic rather than our Mississippi descendants? It is the elder and the younger son differing. God bless them both, and keep them in union, if it be for their good, but separate them, if it be better.[165]

Although union was desirable, only liberty was necessary. The Empire of Liberty need not consist of one, and only one, sovereign government. Its indispensable requirements were otherwise. "However our present interests may restrain us within our own limits," Jefferson noted to Monroe during the first year of his presidency, "it is impossible not to look forward to distant times, when our rapid multiplication will expand itself beyond those limits, and cover the whole northern, if not the southern continent, with a people speaking the same language, governed in similar forms, and by similar laws; nor can we contemplate with satisfaction either blot or mixture on that surface."[166]

The Empire of Liberty was to be made up of one people, dedicated to liberty under republican institutions. There was to be no place here for subjects, only for citizens. This was why, in principle, Negroes could have no permanent position within the palladium of freedom and why, in practice, Indians as well had to be excluded from it. Although confined to this hemisphere, there were no other discernible limits placed on the physical extent of the empire. Occasionally, the attempt has been made to find such limits in the physical barriers of nature—whether mountains, deserts, or rivers.[167] But these barriers, along with that of distance in general, had a significance that was bound to change with improvement in communications. It is the political and social barriers to expansion that are relevent here—or rather that are largely irrelevant given Jefferson's virtually limitless vision of the Empire of Liberty.

The question remains whether Jefferson was as prepared as he professed himself to be to see this vision eventuate in sister republics rather than to remain united under one government. In practice, the issue never had to be squarely faced. His statements apart, however, there is no evidence that he was so prepared. What the record does indicate is that Jefferson's position would have required showing that the path of political fragmentation—that is, of sister republics—was as effective a means of maintaining or promoting the greater Empire of Liberty as one government coincidental with this empire. A demonstration to this effect would not have been easy in the best of circumstances. Whatever theory might decree, practice could be relied on to support the path of union, indeed, for all practical purposes, to equate empire with union. This being the case, how meaningful is it to insist that union was only a means to the end of liberty, if it constituted the most effective means by far of achieving and maintaining this end. Viewed from this perspective, the empire (union) might just as well be considered an end and one that in practice is as significant as liberty itself.[168]

What may be said of union may also be said in large measure of the extent of territory within which union, and certainly a truly republican union, was to be maintained. Jefferson's commitment to expansion was one that was strongly rooted in the equation of expansion with the security and well-being of a republican America. For it was through expansion, he believed, that the republican character of the Union would be preserved. Expansion would remove the presence of dangerous neighbors, hence the prospect of wars that must result in the imposition of unbearable burdens on society. Even in the absence of war, such neighbors must threaten republican institutions simply by requiring the creation and retention of a large military establishment. Expansion would also ensure that the predominantly agricultural political economy considered synonymous with the republic of virtue could be sustained despite a steady increase of population. Clearly, the Louisiana Purchase represented a giant step toward the fulfillment of the promise held out by expansion and was so seen. But the Floridas as well were a step in the same direction, though a more modest one.[169]

To be sure, the time might come when expansion would no longer promote the security and well-being of America, when it might even threaten the Union and liberty. That time did come for another great expansionist, John Quincy Adams, when he had once concluded that territorial expansion could not be separated from the expansion of slavery. But for Jefferson that time never came (though on more than one occasion he had sought to place constraints on the territorial expansion of slavery). Throughout, he saw expansion as the indispensable concomitant of a stable, secure, and prosperous Empire of Liberty.[170]

Expansion, then, was at the center of Jefferson's reason of state. It constituted the great necessity of his statecraft, one that was no less imperious because he was ultimately unwilling to employ force in its pursuit. His unwillingness to do so testified to his conviction that war must imperil the ends for which expansion was undertaken and was therefore likely to prove self-defeating. This conviction represented the most significant limit to Jefferson's expansionism. There were, of course, other limits. They can be summed up, for the most part, in the commitment to make citizens, not subjects, for it is this commitment that largely explains the insistence on similarity in language, religion, customs, and laws. But how great a constraint on expansion they represented remains unclear. Jefferson's aspirations to Canada and Cuba alone reflected their potential elasticity.

It is against the background of Jefferson's reason of state that his pursuit of the Floridas may be fully understood, just as it is against the same background that his treatment of the constitutional issues raised by the Louisiana Purchase may best be seen. In the case of the Floridas, Jefferson was tempted—indeed, driven— by his reason of state to confound the necessary with the desirable. No compelling interest required wagering the fortunes of his administration on the possession of West Florida. Yet Jefferson was insistent on doing so. Even with the advantage of hindsight, he saw in this wager a pure case of *salus populi* and, accordingly, one in which a duty to follow a higher law took precedence over the written law. With the Floridas serving as one case to exemplify the principle that the laws of necessity form the higher obligation of the statesman, Jefferson wrote in 1810 that "it is incumbent on those only who accept of great charges, to risk themselves on great occasions, when the safety of the nation, or some of its very high interests, are at stake." [171]

In the case of Louisiana, the temptation differed. Here Jefferson yielded to his reason of state by abandoning a long and deeply held view of the Constitution and of the powers of the federal government. By contrast with the Floridas, a truly compelling interest together with the fear that it might be lost led him to do so. The prospect that Napoleon might have a change of mind helped move Jefferson to accept a view of the Constitution that he had before bitterly opposed and that he would again oppose.

The Louisiana cession raised these questions: Was there a constitutional power to acquire territory? If such power existed, what was the status of the territory acquired? The first question arose by virtue of the cession itself. The second question arose because Article 3 of the treaty of cession provided that

> The inhabitants of the ceded territory shall be incorporated in the Union of the United States, and admitted as soon as possible, according to the principles of the Federal Constitution, to the enjoyment of the rights, advantages and immunities of citizens of the United States; and in the meantime they shall be maintained and protected in the free enjoyment of their liberty, property, and the religion which they profess.

In the winter of 1803, anticipating the day when New Orleans and the Floridas might be purchased, Jefferson had expressed his

constitutional scruples over the possible action to his cabinet officers. His Attorney General, Levi Lincoln, had outlined a proposal
for avoiding constitutional difficulties that would entail, with French
consent, the extension of the boundaries of Georgia and the Mississippi Territory. But Gallatin had poured cold water on this plan,
noting that if it was unlawful for the federal government to acquire territory, it was equally unlawful for a state to do so. Instead, Gallatin advised the President that the United States "as a
nation have an inherent right to acquire territory," that when acquisition was by treaty "the same constituted authorities in whom
the treaty-making power is vested have a constitutional right to
sanction the acquisition," and that when territory has been acquired "Congress have the power either of admitting into the Union
as a new state, or of annexing to a State with the consent of that
state, or of making regulations for the government of such territory."[172] Gallatin's view, a remarkable summary of what was to
become accepted constitutional doctrine on the matter, elicited the
response from Jefferson that "there is no constitutional difficulty
as to the acquisition of territory, and whether, when acquired, it
may be taken into the Union by the constitution as it now stands,
will become a question of expediency. I think it will be safer not
to permit the enlargement of the Union but by amendment of the
Constitution."[173]

There the matter rested until shortly after the news of the cession had reached this country. On August 9, 1803, Jefferson wrote
to John Dickinson that there was "a difficulty in this acquisition
which presents a handle to the malcontents among us, though they
have not yet discovered it. Our confederation is certainly confined
to the limits established by the revolution. The general government has no powers but such as the constitution has given it; and
it has not given it a power of holding foreign territory, and still
less of incorporating it into the Union. An amendment of the Constitution seems necessary for this."[174]

These are the words of the true believer in strict construction,
a principle that until this time had commanded Jefferson's complete allegiance. As applied to the Louisiana Purchase, however,
an unyielding faithfulness to strict construction might well give
rise to no little inconvenience. At worst, it might even jeopardize
interests considered vital. Whereas Jefferson's letter to Dickinson
stated the need of a constitutional amendment to authorize even

the acquisition of Louisiana, thus reversing the view earlier expressed to Gallatin, the backsliding to a more flexible position had already begun. It can be detected in Jefferson's letter of August 12 to John Breckenridge, in which he speaks of Congress appealing "to *the nation* for an additional article to the Constitution, approving and confirming an act which the nation had not previously authorized."[175] This was the language of centralization, only now it was being spoken by Republicans and for a purpose they deemed absolutely vital. In the same letter to Breckenridge, Jefferson invoked the plea of necessity. The executive, he wrote, "have done an act beyond the Constitution." The legislature "must ratify and pay for it, and throw themselves on their country for doing for them unauthorized what we know they would have done for themselves had they been in a situation to do it."

Since Jefferson believed that he had gone "beyond the Constitution" in purchasing Louisiana, he continued to insist on the need of an amendment to legitimize the action. By mid- to late August a note of urgency marked his communications; this has been attributed to a fear that Napoleon might change his mind. In a letter of August 23 to Gallatin he expressed growing anxiety that the agreement with France might fall apart and stressed the need for quick action by Congress. The same letter contained a draft of Jefferson's proposed amendment (an earlier version having been discarded) which provided that Louisiana "is made a part of the United States" and that "its white inhabitants shall be citizens, and stand, as to their rights and obligations, on the same footing with other citizens of the United States in analogous situations." The portion of Louisiana north of the Arkansas River was reserved to the Indians. Finally, Jefferson added the proviso that Florida "whenever it may be rightfully obtained, shall become a part of the U.S."[176]

Jefferson's growing anxiety that the agreement with France might fall apart was prompted largely by his ministers in Paris, particularly Livingston. On June 7, Livingston and Monroe wrote Madison of Napoleon's growing dissatisfaction with the treaty and of his fear that he might never receive payment since this "was made dependent on the delivery of possession of the country to our commissary [and], might, by accident or other causes, become nugatory; the Spaniards might not surrender it at once, the British might take it, etc." The ministers warned that the treaty must be ratified promptly and without making any changes that might give Napoleon an excuse to cancel it.[177] On June 25, Livingston was still more emphatic, writing to Madison that "I hope that nothing will prevent your immediate ratification, without altering a syllable of the terms. If you wish anything changed, ratify unconditionally and set on foot a new negotiation. Be persuaded that France is sick of the bargain; that Spain is much dissatisfied, and that the slightest pretence will lose you the treaty." [178]

Although Jefferson's apprehensions had increased as a result of these letters, they do not appear to have had a similar effect on his two principal advisors. Both Madison and Gallatin rejected the notion that France desired to get out of the bargain. In reply to Jefferson's letter of August 23, Gallatin declared: "I feel . . . any apprehensions that France intends seriously to raise objections to the execution of the treaty; unless intoxicated by the hope of laying England prostrate, or allured by some offer from Spain to give a better price for Louisiana than we have done, it is impossible that Bonaparte should not consider his bargain as so much obtained for nothing; for, however valuable to us, it must be evident to him that pending the war he could not occupy Louisiana, and that the war would place it very soon in other hands." [179] Gallatin suspected that the warnings the administration had received were due in part to Livingston's "natural desire to persuade us that he has made an excellent bargain."

What effect Gallatin's and Madison's views had on Jefferson is not clear, but they cannot have been very marked. For the President's anxiety over the treaty and its prompt ratification appeared, if anything, to have grown in this period. Insistent, nevertheless, on settling by formal amendment the constitutional issues raised by the treaty of cession, Jefferson found himself at odds with a number of his cabinet officers and leading members of his party. The denouement of this estrangement came in Jefferson's exchange with Wilson Cary Nicholas, Senator from Virginia and political ally. In a letter to Jefferson of September 3, Nicholas, after expressing his differences with Jefferson on the constitutional power of the federal government to acquire territory and to admit new states into the Union, a power which Nicholas found "as broad as it could well be made," went on to warn the President that if he gave an opinion on the competence of the treaty-making power before the Senate acted on the Louisiana treaty, the consequences would prove very unfortunate:

By giving an opinion before the Senate act upon it, you would take the whole responsibility of that opinion upon yourself in the public estimation, whereas if the Senate act before your opinion is known they will at least divide the responsibility with you. I should think it very probable if the treaty should be by you declared to exceed the constitutional authority of the treaty making power, that it would be rejected by the Senate, and if that should not happen, that great use would be made with the people, of a wilful breach of the constitution.[180]

In replying to Nicholas, Jefferson urged that "whatever Congress shall think it necessary to do, should be done with as little debate as possible, and particularly so far as respects the constitutional difficulty." Having said this, he reaffirmed his unshaken belief in strict construction of the Constitution. "Our peculiar security is in possession of a written Constitution. Let us not make it a blank paper by construction. I say the same as to the opinion of those who consider the grant of the treaty making power as boundless. If it is, then we have no Constitution." But while Jefferson confessed that he thought it important "to set an example against broad construction, by appealing for new power to the people," he was nevertheless prepared to concede. "If . . . our friends shall think differently, certainly I shall acquiesce with satisfaction; confiding, that the good sense of our county will correct the evil of construction when it shall produce ill effects."[181]

This was the end of the matter as far as any utterances by Jefferson were concerned, for he never again pronounced himself on the subject. Why had he given in as he did? Dumas Malone has written that Jefferson had to make a choice between proposing a constitutional amendment and not doing so, that the former course was the more dangerous in terms of the country's welfare and Jefferson's continued leadership, that Jefferson "was characteristically undogmatic about means," and that it was inherent in his style of leadership not to press it unduly and to accept counsel. And Malone concluded: "To have expressed his constitutional scruples publicly would have endangered an agreement that he deemed essential to national security, while putting weapons into the hands of his political enemies; but he wanted his intimate friends to know that these scruples were still present in his mind."[182]

This explanation, coming as it does from Jefferson's foremost biographer, is startling. It asserts that in making the choice he did, Jefferson departed from *his* constitutional scruples in order to serve the public's welfare (national security). He relented on the "means" he would rather have employed because his political allies insisted that his chosen means might jeopardize the end sought and thereby give an advantage to his political enemies. Malone's explanation would be unexceptionable were it not for the fact that the principle of strict construction was, as Henry Adams long ago insisted, "the breath of his political life. The Pope could as safely trifle with the doctrine of apostolic succession as Jefferson with the limits of Executive power." [183] Jefferson did not think of strict construction as something that was *his* in the sense that it represented his preference—even his strong preference. Instead, it was seen as the way of truth in matters political. Yet Jefferson had departed from this way. Why?

One explanation on which both Adams and Malone place emphasis has to do with Jefferson's style of leadership. Adams put it this way:

> The leadership he sought was one of sympathy and love, not of command; and there was never a time when he thought that resistance to the will of his party would serve the great ends he had in view. The evils which he foresaw were remote: in the hands of true Republicans the constitution, even though violated, was on the whole safe; the precedent, though alarming, was exceptional.[184]

Of course, as matters turned out, the precedent was significant precisely because it was exceptional. Indeed, so exceptional was it that, as Adams emphasized, it dealt "a fatal wound to 'strict construction.' " One does not have to agree with John Quincy Adams's judgment that the manner in which the purchase was conducted subverted every principle of Jefferson's party to find in this precedent a great turning point, which, once passed, resulted in changing the nature of the Union. That change, considered no less than revolutionary by historians, was effected by a broad construction of the Constitution. It is scarcely surprising that in later years, Jefferson could neither repudiate nor defend his silence over the constitutional difficulties when Congress met to consider the treaty on October 17, 1803. Moreover, despite the hope he expressed at the time about the limited effects of his silence, the very importance of the occasion had to have seriously raised the question for him whether this had been a lapse of faith that could never quite be atoned for.

It is certainly the case that his trusted and intimate friends were increasingly insistent that Jefferson take the path they saw as the safer and easier one and that Jefferson's leadership was not of the kind to impose his will on dissident followers. Dumas Malone is surely right in saying that the great success of Jefferson's political leadership "was that he forbore to press it unduly, that he was a good party man, that he did take counsel." Still, this was not an ordinary party issue; it was one that touched the most sensitive of Jefferson's political convictions. Those near to him—certainly Gallatin and even Madison—were not similarly affected. Their concern was to get the treaty expeditiously ratified. If this could be done without running much risk, they were relatively indifferent to the means.

In the end, then, the main reason—and defense—that must be given for Jefferson's silence is that he feared a constitutional amendment could be introduced only by risking the treaty. How great was the risk? Historians have generally agreed that it was substantial. Even Henry Adams, after making so much of Jefferson's abandonment of strict construction in September 1803, concluded that "there were good reasons for silence." Not only were Livingston's letters alarming; Spain had vigorously protested the sale of Louisiana. "A probable war with Spain stared Jefferson in the face," Adams wrote, "even if Bonaparte should raise no new difficulties."[185]

The Spanish danger, however, was ultimately the French danger, and not because Spain was the entire creature of France, but because Spanish resistance posed no real danger unless it was supported by France. That the Jefferson administration appreciated this may be seen in the firm and even aggressive response it made to Spanish complaints over the treaty. Nor is it reasonable to suppose that the American response would have been as firm with respect to Spain had it not been thought that Paris would provide no support for Spanish complaints. The danger, if danger existed, was from France and France alone. But if the reasoning of Gallatin is accepted—and it is difficult not to accept it—that danger could have been no more than marginal. An examination of the French sources bears this out; there is no real support for the view that Napoleon was seriously thinking of canceling the treaty of cession.

There remains the risk that Nicholas raised—that the proposal of an amendment would result either in the Senate's rejection of the treaty or the charge by the Federalists of a willful breach of the Constitution. It is not apparent, though, why the Senate would have rejected the treaty for this reason. The Republicans enjoyed a great advantage in the Congress: in the Senate 25 to 9 and in the House 102 to 39. Why would the introduction of an amendment have jeopardized the treaty? The Federalists would have opposed in any event, and the majority of Republicans were clearly inclined to a broader construction of the Constitution than was Jefferson. They might have thought an amendment entirely unnecessary—as Nicholas had—but they would scarcely have rejected the treaty for this reason. Everett Brown, whose inquiry into the constitutional history of the Louisiana Purchase still is without peer, wrote: "Doubtless such an amendment as Jefferson desired could have been carried without great difficulty, but it was not proposed, and an important precedent for future action in regard to the acquisition of territory was established." [186]

At the time, this had been the view as well of John Quincy Adams, the Senator from Massachusetts. Adams was virtually alone in his defense of strict construction in the Louisiana affair. His efforts to propose an amendment met with rejection on all sides. Adams thought, as he later wrote in his memoirs, that the purchase of Louisiana entailed "an assumption of implied power greater in itself and in its consequences, than all the assumptions of implied power in the twelve years of the Washington and Adams Administrations put together." [187] There was no power conferred by the Constitution to bring new lands into the Union. Nor, he insisted, was there a power that allowed either Congress or the Executive to govern a people without their consent. Adams's position was that the treaty could be ratified but not carried into effect without an amendment. Retrospectively, he justified his position by estimating that once through Congress, the amendment process could have been completed by the states within a period of a week. [188] If this estimate appears overly sanguine, it should be recalled that the treaty had very considerable support in the country and the administration was well aware of this.

Even if these remarks are considered well taken, it does not follow that *no risk* was involved in introducing an amendment. Clearly, there was a risk in doing so. But what follows from recognition of this? Is it not this: that in order not to run *any* risk Jefferson put aside his constitutional scruples in the case of Louisiana? And if this conclusion seems reasonable, what does it say about Jefferson's devotion to constitutional principle when it appeared to pose even a modest risk to the imperious demands of reason of state?

IV

THE MARITIME CRISIS
(1805–9)

17

The Nature of
Jefferson's Failure

A<small>T THE OUTSET OF HIS</small> second administration, Jefferson stood
at the height of his popularity, rivaling Washington as an object of popular adoration. The triumphs of his first term had isolated the Federalists and disproved on every tangible measure their
forecasts of inevitable ruin. The Republicans, by contrast, stood
forth as a national party, barely disturbed by factional infighting.
The purchase and incorporation of Louisiana, though it subjected
the President to the reproach of having violated the Constitution,
had been heartily welcomed by the people as a whole, who cared
little for constitutional subtleties. This great triumph in foreign
policy had not required the President to abandon most of the "sound
principles" with which he had entered office. His refusal to tax
"the industry of our fellow citizens to accumulate treasure for wars
to happen we know not when, and which might not perhaps happen but from the temptations offered by that treasure" had seemingly been rewarded.[1] It had yielded not increased danger from
foreign nations but an addition to the national domain that placed
the security of the country on a firm foundation and would in any
case soon pay for itself. Though internal taxes had been nearly
abolished, customs revenues had increased substantially. With the
debt on its way to extinction, the President even broached the idea
in his Second Inaugural of what might be done with the surpluses
that would accumulate in the future, and he laid out, subject to

constitutional amendment, a broad program of internal improvements.

The course of Jefferson's second administration would soon disrupt, and then utterly destroy, this happy picture. In the following four years, Jefferson found himself embarrassed by a schism in his own party, thwarted by a national judiciary still in control of the Federalists, and badly compromised in the area of foreign affairs. It was, indeed, in foreign policy that the ruin of his second administration was most complete. His various projects to secure the Floridas from Spain would end in failure; so, too, would his attempt to gain from England a satisfactory arrangement on the high seas. Worse, the means Jefferson chose to meet the growing crisis on the high seas—an embargo on American shipping—recoiled upon him in the end and led to a host of unforeseen and pernicious consequences. His retreat to Monticello in 1809 was a release from a score of accumulated troubles, which broke his health and shattered the people's confidence. The diplomatic method by which he had won the triumphs of his first term gave him no victories in the second. On the contrary, it progressively reduced his freedom of action and left him at the end immobilized.

Wherein lay the nature of Jefferson's failure? One view is that Jefferson was the victim of circumstances beyond his control. Most of Jefferson's biographers, though not uncritical of his various lapses in judgment, revert to this idea in explaining the predicament in which he found himself. The embargo, on this view, was the inevitable result of the vise in which the belligerents progressively placed the United States. France's victories on land and Britain's at sea left the two great antagonists without a ready means of striking at each other, and in resorting to commercial sanctions they sucked the one great remaining neutral into a vortex that even the wisest American statesman could not have avoided. In these circumstances, submission was out of the question; but so, too, was war. The people were not psychologically ready for the test of arms, and though Jefferson urged military preparations as early as late 1805, the Republican majority in Congress—reflecting the sentiments of the people—could not be roused to any substantial effort. The embargo, in Jefferson's words, was a "lesser evil" than war and deserved a fair trial. Had it not been undermined by politically inspired opposition at home, it might yet have worked its will on the belligerents. Jefferson's domestic opponents—whether dissident Republicans like John Randolph or Federalists like Timothy Pickering—had no better course to suggest. The opposition of the one to the administration reflected no consistent principle; the opposition of the other was tainted by disloyalty to the Union.[2]

The interpretation of Henry Adams does not differ in all respects from that presented above—Adams, too, was merciless toward Randolph and Pickering—but its central proposition is far different. For Adams, it was Jefferson's political vision, and not simply the exacting circumstances in which he found himself, that explained the President's progressive immobility and ultimate helplessness. Jefferson's problem, Adams held, was that he would not compromise his principles, but he would not fight for them, a contradiction that inevitably reduced him, as Josiah Quincy put it, to "a dish of skim milk curdling at the head of our nation." Jefferson's dread of war was well known in London and Paris, and it served to invite assaults that might otherwise not have been undertaken. For Adams, Jefferson's second administration was thus a study in political pathology, which consisted, above all, in the weakness with which the Republicans met English aggression, and of which the embargo was merely the culminating episode. War, not economic coercion, represented in Adams's eyes the true path of statesmanship once the English began their hostile proceedings in 1805; and for him the question was not simply, or even primarily, one of policy. Rather, it was one pregnant with historical significance, turning on the question of the kind of country America might become, whether it would be, as Jefferson had imagined, a country unlike any other, exceptional not only in its political institutions but in the relations it maintained with foreign powers. For Adams, the failure of embargo meant the repudiation not merely of Jefferson's theory of foreign relations but of the whole notion of American exceptionalism: "America began slowly to struggle, under the consciousness of pain, toward a conviction that she must bear the common burdens of humanity, and fight with the weapons of other races in the same bloody arena; that she could not much longer delude herself with hopes of evading laws of Nature and instincts of life; and that her new statesmanship which made peace a passion could lead to no better result than had been reached by the barbarous system which made war a duty."[3]

The interpretation presented here differs materially from each of these rival explanations. The defectiveness of Jefferson's foreign policy cannot be ascribed merely to exiguous circumstances. His political vision was central to his failure. Yet the fatal weakness of his vision consisted not of his aversion to war—an aversion that reflected a sound appraisal of the interests of the country—but of the outlook that made him incapable of reaching an accommodation with England on impressment and neutral rights, the two great issues that arose between the two states. Such an accommodation was desirable mainly because the United States had a vital interest in the maintenance of the British navy as a shield against Napoleon, whose ambition to universal empire constituted a menace not only to American security but also to international order. Jefferson magnified the threat represented by England, depreciated that represented by France, and in the end gave his allegiance to neutral rights rather than the balance of power. This was his cardinal error, and though it did not prove fatal to American security (because Napoleon was ultimately defeated), it wedded American foreign policy to a course of action that badly injured American commerce and ultimately led to war.[4]

Jefferson regarded an accommodation with England as nothing less than submission, and indeed once the United States had adopted an uncompromising attitude toward impressment and neutral rights it would certainly have been humiliating to accept England's terms. Yet it may be doubted that the position the administration adopted on neutral rights was consistent with the maintenance of either security or commerce, even though both were invoked on its behalf. In the initial form in which this struggle took place, the administration's attitude toward a solution of impressment was defective, and its defense of the carrying trade—which it saw as embodying the dictates of both law and morality—was in fact questionable on those very grounds. Only with the enforcement of the Berlin Decrees against American commerce and the concomitant Orders in Council of late 1807 did the belligerent edicts infringe on an extremely important interest: the right of American vessels to trade freely, in American goods, with both England and France. This was, in normal circumstances, an entirely reasonable and justifiable claim. Nevertheless, Jefferson's response would have been more consistent with the maintenance of both security and commerce had it given sufficient weight to the importance of the British Navy as a bulwark against Napoleon, and had it appreciated that neither war nor embargo might secure its favored objects.

These traditional calculations of security and interest were not unknown to the President. He had previously expressed the understanding that American security depended on withholding the empire of the sea from the land power of France; as early as the 1780s he had seen that the advantages that European war gave American commerce were attended by the liability that too active a commerce might breed war and was not indispensable to domestic security or prosperity. As Jefferson approached his great crisis with England, however, these cautionary admonitions were forgotten. The President threw in his lot with France, and then, in a great test to secure the neutral rights that shielded commerce, he brought to a standstill the commercial life of the country. That he came to see positive virtue in this violent action on the economic life of the country and turned with a vengeance on the sordid motives of commercial men was the final irony of the embargo. His readiness to do so showed not only the instability of his conception of commercial interest but also, and most importantly, that his deepest tendency was to convert questions of interest into matters of right and wrong, which then assumed a kind of independent character and became inseparably annexed to the honor and independence of the country. His tendency to do so is the measure of his moralism; his moralism, in turn, not only constituted the central aspect of his diplomatic outlook but is also identifiable as the primary corrupting factor within it.

18

Jefferson's
Diplomatic Design

THE FIRST YEAR OF THE second administration—1805—is significant because the administration then fixed upon a diplomatic design that, in its essential features, remained intact until the end of its term. The design consisted of two great desiderata: a peaceful strategy of gaining the Floridas under the auspices of France, and a determination to reaffirm the uncompromising stance (for which the Republicans had always stood) toward British violations of American rights on the high seas. The fall of 1805 has often been regarded as a "turning point," because Jefferson then flirted with a different policy—war with Spain, alliance with England. Had he embraced the policy at this time much trouble would later have been avoided. But because it was merely a flirtation, its status as a turning point may be doubted. It rather constituted a reaffirmation of a set of attitudes—represented in the cabinet discussions by both Gallatin and Madison—that was central to the Republicans' public philosophy. One of these was that war would subvert the promises made at the outset of the administration, with its deep aversion to debt and taxes. The other was that the compromises on neutral rights required if an alliance were to be made with England would be profoundly degrading and incompatible with the maintenance of American independence.[5]

The idea that the way to the Floridas lay through France had informed the American position since the first difficulties arose over Louisiana and had indeed been encouraged by French officials at the time. But the first bout of diplomacy in Paris and Madrid was barren of result, as Talleyrand repudiated the American contention that West Florida lay within the limits of the Louisiana Purchase. In the absence of French pressure, Spain had no motive to acquiesce in a further dismantling of its empire, and the President still rejected the idea that the United States might pay money to secure what by right belonged to the American government. The Floridas, Madison had held, were not to be converted into a "French job." "The United States owe it to the world as well as to themselves," the Secretary of State said, "to let the example of one government at least, protest against the corruption which prevails."[6] And though he insisted that the destinies of no country could be injured by "adherence to the maxims of virtue," virtue had as yet not been rewarded. From the American diplomats at Madrid and Paris—Monroe, Pinckney, and Armstrong—came recommendations for vigorous action: the seizure of the Floridas and the occupation of the territory between the Colorado River and the Rio Grande.[7] The idea had a certain appeal to Jefferson, whose determination to acquire the Floridas remained fixed. Of the justice of the American claim to West Florida he had by this time fully persuaded himself—that claim reflected, as Madison said, "a principal of universal justice, no less than of municipal law."[8]

War with Spain meant the defiance of Napoleon, and that was not a step to be lightly hazarded, especially with the possibility that the expected peace in Europe might free France and Spain from their engagements elsewhere. If American claims were not to be abandoned, safety could be found only in an alliance with England, which Jefferson recommended for the consideration of his cabinet in the late summer of 1805. The eventual treaty of alliance he then proposed would come into force whenever the United States went to war with Spain or France but did not require the United States to join Britain's war against Napoleon. Madison immediately noted the inequality of sacrifice inherent in Jefferson's terms. He did not see "the least chance of laying [Great Britain] under obligations to be called into force at our will without corresponding obligations on our part," obligations which Madison was decidedly unwilling to undertake. Gallatin's objections were even more pointed. He thought that neither of the two points asked by the United States of Spain—a settlement of the boundaries of Louisiana and the payment by Spain of spoliation claims arising from

Spanish depredations against American commerce—constituted a just cause of war; the policy of such a war he doubted even more. War would depress the revenue to around eleven million dollars, which was adequate only for paying the debt and maintaining current expenditures: "All the expenses of the war," therefore, "must be supported by loans or new taxes." Though he thought that the United States, with only a little addition to its military resources, might take possession of the Floridas and alarm the outposts of Mexico in the Southwest, he worried over the attitude that Napoleon would take. "If Bonaparte, haughty and obstinate as he is, shall think proper to persevere, notwithstanding our taking Florida, then our fate becomes linked to that of England, and the conditions of our peace will depend on the general result of the European war. And this is one of the worst evils which the United States could encounter; for an entangling alliance, undefined debts and taxes, and in fine a subversion of all our hopes, must be the natural consequence."[9]

Yet Jefferson persisted. Once the possibility of war with Spain and France had suggested to him the necessity of an English alliance, the possibility of an English alliance soon convinced him that war with neither power would be necessary: "It being generally known to France & Spain that we had entered into treaty with England, would probably ensure us a peaceable & immediate settlement" of American claims against Spain. He clung to this idea until October, when the latest European news convinced him that war would continue for at least another two years, thus radically lessening the prospect that he might have to face Spain and France alone. The news that Britain had hardened its attitude toward neutral rights came at the same time, and the administration soon returned to the policy that best suited its instincts: It would, as John Randolph later charged, seek to bully England and truckle to France.[10] It would renew (while refining) the 1804 instructions to James Monroe, U.S. Minister to England, calling for an end to impressments and to English attempts to restrict America's neutral trade, all the while holding out the possibility of a resort to economic sanctions if these objects were not secured. And it would revive the negotiation over the Floridas by dangling before Napoleon a monetary incentive to bring Spain to terms.

These were fateful decisions. The instructions to Monroe placed the administration on a collision course with England, and the attempt to secure the Floridas through the good offices of France made the administration dependent on the fleeting moods and unknown purposes of Bonaparte. Jefferson had no intimation at this moment that his freedom of action had been reduced, but in fact it had been considerably narrowed. He persisted in his faith that peaceable methods could secure both his principal objects—that England, dependent on American commerce, would not dare push matters to extremes with his administration, and that France, eager to make a good bargain with the United States, more solicitous of American than of Spanish friendship, and fearful of an American entente with Great Britain, would bring the Floridas in his grasp.

The administration's strategy was presented to the Congress in two messages, one public, the other private. The public message recounted the numerous injuries of Spain and Britain against American rights and demanded redress. Though the President still professed optimism "that time and a more correct estimate of interest, as well as of character, will produce the justice we are bound to expect," he warned that if negotiations failed, "we must join in the unprofitable contest of trying which party can do the other the most harm. Some of these injuries may perhaps admit a peaceable remedy. Where that is competent it is always the most desirable. But some of them are of a nature to be met by force only, and all of them may lead to it."[11] Jefferson's warning was meant to dispel the impression that the administration was "entirely in Quaker principles." The fortification of America's coastal cities, a fleet of gunboats, the classification of the militia by age—all this, he thought, was necessary, and he even hinted that the construction of ships of war of seventy-four guns would not meet with his disapproval.

Yet the bellicose tenor of the public message was misleading. In his initial draft, Jefferson had enlarged on the peaceable remedies that were available to the United States—"We may suspend intercourse with nations which harass it by stem. We may tax the commerce of the wrong doers to relieve the individuals wronged. We may pass a navigation act, adapted to our position & circumstances, only avoiding to confound the just with the unjust"—but he struck the passage from his presentation to the Congress.[12] In truth, the passage struck was more revealing of his true thought than his public utterance: he had no intention of going to war with Great Britain at this time, and even with Spain he was confident that war would not be necessary. This he made clear in a confidential message to the Congress on the Floridas, in which he reviewed the negotiations with Spain in terms decidedly more optimistic than he had employed in his public message. His failure to request a specific appropriation in this formal communication brought forth charges from John Randolph that the administration was attempting to foist upon the Congress responsibility for measures that belonged to the Executive, but the administration's strategy for the Floridas was clear enough. The talk of war was for the purpose of keeping the pressure on Spain; the real design was to repeat the triumph of Louisiana by offering to Napoleon a monetary incentive to bring Spain to terms.

As a token of its good faith toward France, the administration also discretely supported a bill introduced by Senator George Logan of Pennsylvania which called for total nonintercourse with the black rebels on Haiti. France had demanded such a measure. If the Florida negotiation were to succeed, France's requests had to be satisfied (at least apparently). There were, to be sure, ancillary reasons for supporting the Logan bill—the main one being the fear that a successful slave uprising on Haiti would be followed by similar attempts in the United States—and no doubt the bill gained the support of southern congressmen mainly for that reason. But that danger, such as it was, had existed for a long time, and only the negotiation for the Floridas explains the administration's support for it at this juncture.[13]

The attitude the administration adopted toward England at this time was for the most part shielded from public view. The President took a strong stand in his annual message against the "new principles" the British had "interpolated into the law of nations," but he declined to specify a remedy. The administration even remained silent during the congressional debates on nonimportation, which dominated the attention of the Ninth Congress. Its reticence stemmed not from lack of faith in the coercive potential of American commerce, but from its sense that the timing of the measure might complicate the negotiation for the Floridas and its desire to preserve the Executive's freedom of action. Napoleon, the President thought, had still to fear the possibility of an Anglo-American rapprochement, and though Jefferson had himself given up on the idea, it served no purpose to advertise his reticence.

Congress was therefore left to its own devices during the 1st Session of the Ninth Congress, and the measure it finally adopted—though it earned the approbation of the administration—showed the ambivalence of the country in facing down the English challenge. Total nonimportation was defeated, largely because of the fear that it would damage the revenue, and a milder resolution was substituted in its place, which exempted those goods for which England was the sole supplier. The bill was then further weakened by delaying its effective date to the following November—which earned for it Randolph's famous characterization as a "milk and water bill—a dose of chicken broth to be taken nine months hence." For fear of embarrassing James Monroe, Jefferson had opposed the suggestion of adding another envoy to the English negotiation, but the habit of dispatching envoys extraordinary at moments of national crisis had already become the first recourse of American diplomacy, and it posed no obstacle in his English negotiation to ask for the nomination of William Pinkney, a Federalist lawyer from Baltimore, to join Monroe in London.

In truth, Jefferson had reason to be satisfied with the results of the Ninth Congress. It had appropriated the money for the Florida negotiation, made the necessary gesture to France in regard to Haiti, and brandished the commercial weapon against England. Little was done in the way of military preparations, but Jefferson thought that they were "adequate to the present state & prospect of things."[14] Publicly, the administration had committed itself to very little; as always, its methods of political control had been indirect. No specific appropriation had been requested of the Congress to buy the Floridas, and the administration gave out only the barest of hints as to its true views regarding the embargo against Haiti and nonimportation against Great Britain. Such was the President's reserve, indeed, that later historians have doubted whether the administration had any responsibility at all for the latter two measures.[15]

However much credit is given to Jefferson's peculiar manner of indirection, there was something odd in this behavior. That it was foreign to modern conceptions of the presidency goes without saying. Even at the time, however, and seen in light of Republican expectations, its peculiarity was noted by more than one observer. When John Randolph broke from the administration in 1806, he noted the failure of the President to provide a clear statement of where the administration stood. "I have before protested, and I again protest, against secret, irresponsible, overruling influence." The first question he asked on learning of the nonimportation resolutions introduced by Congressman Gregg was, "Is this a measure of the Cabinet? Not of an open declared Cabinet, but of an invisible, inscrutable, unconstitutional Cabinet, without responsibility, unknown to the Constitution." Well knowing that "whatever measures we might take they must execute them, and therefore that we should have their opinion on the subject," Randolph deplored the secret message over Spain and the failure to request a specific appropriation or even name the specific object of the negotiation the President now planned. And Randolph concentrated equal fire on the failure of the executive branch to give its own views over the appropriate policy toward England. "Let not the master and mate go below when the ship is in distress, and throw the responsibility upon the cook and the cabin-boy!" [16]

What explains the President's preternatural silence? That Jefferson entertained some secret misgiving over the direction Congress took in its 1805–6 session is an idea that continues to find favor, yet it is not persuasive. There is no evidence, in particular, that Jefferson looked upon the escalating confrontation with England in a spirit of pessimism, or that he considered American claims against Britain to be anything other than well-founded. Gregg's initial proposal for immediate and total nonimportation from Britain he doubtless considered too extreme under the circumstances, and he no doubt looked with some apprehension over the developing split within Republican ranks over the issue. Yet the main lines of policy Congress ultimately adopted were perfectly consistent with the sentiments expressed in his Fifth Annual Message, and even more so with the strong language that he and Madison used in communicating the results of the congressional session to Monroe.

The President's sensitivity to Republican ideology and the closely circumscribed role therein allotted to executive power is a more plausible explanation of his reserve. He felt unfairly maligned by Randolph's bitter attacks, and he was certain that there were many who would attack him whatever course he chose. "If we recommend measures in a public message, it may be said that members are not sent here to obey the mandates of the President, or to register the edicts of a sovereign. If we express opinions in conver-

sation, we have then our Charles Jenkinsons, and back-door counsellors. If we say nothing, 'we have no opinions, no plans, no Cabinet!' "[17] The dilemma was inescapable; Jefferson normally resolved
it by taking the second course. Randolph had not objected to this
procedure when he still enjoyed the confidence of the administration; he could not, the President thought, fairly object to it now.

This method of indirect control had an additional advantage.
His show of deference toward Congress increased his diplomatic
flexibility and allowed him a greater share of freedom in spinning
the web that would entrap his foreign adversaries. In his diplomacy with each of the European powers, he found secrecy useful
and indeed indispensable. This was certainly the case in the Spanish business, but it was present also in nonimportation against
Great Britain and embargo against Haiti. Allowing Congress to
wave the sanction against England would distance his own administration from a charge of undue pressure, while nevertheless making use of it (just as Madison would seek credit from France for
an ineffectual measure the French had no right to demand). It was
one of Jefferson's characteristic delusions that he imagined that
this exquisite moral accounting would impress his English adversaries or his French friends; still, these calculations were perfectly
consistent with the diplomatic design he was now pursuing, and
both undoubtedly occurred to him.[18]

Jefferson's sense of satisfaction with the congressional session
was buoyed still further by the death of William Pitt and Pitt's
replacement at the head of the new ministry by Charles James
Fox, whose conciliatory disposition toward the United States was
well known. "With England," Jefferson wrote, "I flatter myself
our difficulties will be dissipated by the disasters of her allies, the
change of her ministry, and the measures which Congress are likely
to adopt to furnish motives for her becoming just to us."[19] In a
letter to Monroe, he half apologized for the nonimportation bill.
"We had committed ourselves in a line of proceedings adapted to
meet Mr Pitt's policy & hostility, before we heard of his death,
which self-respect did not permit us to abandon afterwards." But
though conscious of the delicacy of threatening a friendly administration with a hostile measure, he wanted his envoys to understand that America held the cards in this conflict. Not only was
Britain dependent on American commerce and thus on American
friendship; America's growing strength upon the ocean made it a

formidable potential enemy on that element. "We have the seamen & materials for 50. ships of the line, & half that number of frigates; and were France to give us the money & England the dispositions to equip them, they would give to England serious proofs of the stock from which they are sprung, & the school in which they have been taught; and added to the efforts of the immensity of sea coast lately united under one power, would leave the state of the ocean no longer problematical."[20] This piece of bravado was not the language of a man who recoiled from the prospect of a confrontation with England. On the contrary, it reflected Jefferson's continuing faith in his ability to threaten each of the great European powers with an American combination with the other— a threat that was, after the coercive potential of American commerce, the oldest gambit in his bag of diplomatic tricks. The moment was fast approaching when these warnings would be put to the test, but Jefferson had little idea that it would come to that. His adversaries in London and Madrid, he thought, would come to see that the American position was unassailable, and with his benefactor Napoleon he believed that he had practically closed the deal.

19

The Anglo-American
Dispute: Neutral Rights
and Impressment

THE DISPUTE WITH ENGLAND on which the administration now embarked centered on two points. One arose from the British practice of impressing seamen from American merchant vessels; the other, Britain's growing restrictions on American trade. Both would bedevil Anglo-American relations until the United States attempted, in the War of 1812, to settle them by force of arms. Even then, the war between Britain and the United States did not yield a meeting of the minds between the two sides, and neither issue was mentioned in the Treaty of Ghent, which brought that conflict to an end. They died away only with the defeat of Napoleon and the end of the hurricane that had swept over the world for nearly a quarter of a century.

Of the two issues, impressment raised the thornier problems. England had long manned its fleets through impressment, by which the British claimed a right to strip from merchant vessels British nationals to serve for the duration in the Royal Navy. It was a hard practice: discipline in the Navy was severe, and wages were miserably low in comparison with the American merchant service. Britain claimed no right to impress American citizens, but the similarity in dress, manners, and language made the American virtually indistinguishable from his British counterpart. Nor were those seamen who had managed to obtain American citizenship after 1783 exempt from the press-gang. Though recognizing the American citizenship of those who had adhered to the revolutionary cause during the War of Independence, Britain still clung to the doctrine of indissoluble allegiance, which bound the subject to his sovereign from birth and conferred upon the former a status that could not be renounced. The United States did not, as is often assumed, contest the right of Great Britain to act upon this doctrine within its own jurisdiction; the American government did assert, however, that Great Britain acted in manifest violation of the law of nations when it searched American vessels on the high seas for the purpose of removing its own nationals. The contested issue involved not so much the right of expatriation as contrasting assumptions regarding territoriality at sea.

Impressment was closely related to the stupendous growth of America's neutral commerce, for the war itself created the circumstances for an enormous growth in the American merchant marine. During the war, American ships dominated not only the direct trade between England and America but the indirect traffic between Europe and the French and Spanish West Indies. The American merchant service, which had labored under numerous restrictions in the years before 1793, grew by leaps and bounds during the subsequent years of European war, as British competitors were hurt by the pressing demands of the Royal Navy for seamen and as French, Dutch, and Spanish shipping was driven from the seas by Great Britain. Registered American deep sea tonnage grew even during the years of increasing commercial restrictions by Britain and France, advancing from 558,000 tons in 1802 (a year of peace) to 981,000 at the end of 1810.[21] This growth was mostly owing to the European war; a century later the tonnage carried by American vessels still had not surpassed the level it reached in 1807.[22] Jefferson's prophecy in 1790 that a European war would allow the United States to "become the carriers for all parties as far as we can raise vessels," thus allowing the New World to "fatten on the follies of the old," had been fully realized. But the achievement was a precarious one; it depended, at least in part, on the acquiescence of the power that ruled the seas, and in 1805 the British signaled that a much harder attitude would henceforth be taken toward a commerce that notably aided Napoleonic France.

The rationale for the new British policy was set forth most fully by James Stephen, an Admiralty lawyer with close ties to the Pitt ministry. Stephen's pamphlet, *War in Disguise; or, The Frauds of the Neutral Flags,* began with a paradox: "The finances of France appear scarcely to be impaired, much less exhausted, by her enormous military establishments, and extensive enterprises, notwithstanding the ruin so long apparently imposed on her commerce. Poverty, the ordinary sedative of modern ambition, the common peace-maker between exasperated nations, seems no longer to be the growth of war." This staggering fact, which shook the "first principles" of the statesman's art, was attributed by Stephen to the fact that "the commercial and colonial interests of our enemies, are now ruined in appearance only, not in reality." Their seeming retreat from the ocean was a "mere *ruse de guerre*": in fact, the French had merely "changed their flags, chartered many vessels really neutral, and altered a little the former routes of their trade." The effect was to gravely undermine the efficacy of British naval power at the very moment when Britain's maritime superiority was most decisive.[23]

Stephen argued for the rigid application of the Rule of 1756, by which neutrals were forbidden to enter a trade that had been denied them in time of peace. The neutral, in the British doctrine, had "a right to carry on, in time of war, his accustomed trade," but had no right to enter "a trade which the neutral has never possessed, which he holds by no title of use and habit in times of peace; and which, in fact, he can obtain in war, by no other title, than by the success of the one belligerent against the other, and at the expense of that very belligerent under whose success he sets up his title."[24] The opening of the colonial trade to American vessels was "a measure not of French councils, but of British force"— it proceeded not from the enemy's "will, but his necessity." When neutrals entered such a trade they severely compromised the rights of the belligerent to profit from his conquest. "You have in effect, and by lawful means," as Sir William Scott characterized the position of the neutral,

> turned the enemy out of the possession which he had exclusively maintained against the whole world, and with which we had never presumed to interfere; but we will interpose to prevent his absolute surrender, by the means of that very opening, which the prevalence of your arms alone has effected:—supplies shall be sent, and their products shall be exported: you have lawfully destroyed his monopoly, but you shall not be permitted to possess it yourself; we insist to

share the fruits of your victories; and your blood and treasure have been expended, not for your own interest, but for
the common benefit of others.

Great Britain never abandoned the rule of 1756 in principle
but had relaxed it in practice. The *Polly* decision of 1800 conferred immunity on the carrying trade in colonial goods so long
as neutral vessels broke their voyage by stopping in a neutral port,
unloading their cargo, and paying import duties. The intention of
the ruling was to confine importations to domestic consumption,
but it was entirely frustrated by the failure of American customs
officials to insist upon a bona fide importation, and, even more,
by the payment of drawbacks on customs duties (which was granted
by congressional legislation in 1799). American reexportations had
grown to over sixty million dollars by 1805, doubling their aver-
age level during the prosperous years of 1800–1802; it was at this
moment that the admiralty courts reversed the lenient *Polly* rule
with the 1805 *Aurora* and *Essex* decisions, which revived the rule
of 1756.

Jefferson greeted the "new principles . . . interpolated into the
law of nations" with a ringing denunciation in his Fifth Annual
Message: the British doctrine was "founded neither in justice, nor
the usage or acknowledgment of nations." Reason revolted at the
inconsistency by which "a belligerent takes to itself a commerce
with it's own enemy which it denies to a neutral, on the ground
of its aiding that enemy in the war."[25] Madison devoted the fall
and early winter of 1805 to laboring on a refutation of the British
position. Running to nearly two hundred pages, it appeared in
early 1806 as *An Examination of the British Doctrine, Which
Subjects to Capture A Neutral Trade, Not Open in Time of Peace*.
Jefferson thought that Madison had "pulverized" the Rule of 1756,
"by a logic not to be controverted."[26]

Yet the pamphlet was a strange performance, and was so considered even by those who agreed with its main proposition. It
contained no comprehensive view of the great crisis unfolding in
Anglo-American relations; it never once discussed the commercial
interest underlying America's "right" in relation to the political
economy of the new nation; it ignored the effect that America's
neutral trade was having on the power of the two great belligerents; it specified no remedy. Madison did not even attempt to meet
the British position that acceptance of the American rule would

deprive Great Britain of the effective use of its naval power in the prosecution of the war. Instead, he devoted his prodigious energy to establishing, to his satisfaction at least, that America's right to carry on the trade of Britain's enemies was incontrovertible, whatever the consequences for the belligerents; that unless this right was admitted in its full extent, American's entire commerce would be placed in jeopardy and Great Britain would "renew with effect the old and exploded tyranny of interdicting *all neutral commerce* whatever with her enemies."[27]

Yet the issue for which Great Britain was then contending did not reach American's accustomed commerce, the trade that brought, in American ships, the manufactures of Europe and the products of the West Indies to the United States in exchange for the rich harvest of the seas and the diverse productions of the American soil. Instead, it centered on a commerce opened entirely as a consequence of the European war, and that would largely come to an end with the termination of the conflict. As John Randolph truly said, in his scathing indictment of the administration in the Congress, "the question in dispute" did not consist of "the fair, the honest, and the useful trade that is engaged in carrying our own productions to foreign markets, and bringing back their productions in exchange," but rather "that carrying trade which covers enemy's property, and carries the coffee, the sugar, and other West India products, to the mother country." It was a "mushroom," a "fungus of war," "a trade which, as soon as the nations of Europe are at peace, will no longer exist."[28] The whole premise on which the administration rested its interpretation of neutral rights—that the belligerents in war had no right to interfere with the commerce of the neutral, "that the war among others shall be, for [the neutral], as if it did not exist"—was a fiction belied by the plain facts of the case. For it was only the war that swept the merchant marines of France, Spain, and Holland from the high seas and gave the United States a virtual monopoly of the carrying trade. Only the war, and British naval supremacy, opened this trade to American shipping and allowed the United States to fatten on the follies of Europe. These are facts that cannot be ignored in any assessment of the legality and morality of the American claim.

There was no more common idea in the Republican confrontation with the external world than the contrast between might and right, between the rapacity and iniquity of the European powers and the virtue and innocence of the United States. It was largely over the question of neutral rights and impressment, and largely in the context of the dispute with England, that this theme was endlessly elaborated. Yet the legal and moral rights to which the United States laid claim are by no means self-evident. Even with respect to impressment, the American position is not easy to defend. Britain's seizure of American citizens, it is true, constituted violations of the person before which no American government could remain silent. The practice, as was said in Congress, was "an outrage which no nation but Great Britain practices, and to which no nation but America submits." But the whole question cannot be seen apart from the extraordinary growth of the American merchant marine, for this growth would not have been possible without the contribution of the able-bodied seamen to which, even on the American definition of citizenship, Great Britain might lay legitimate claim. As the American government itself would come to realize, there were far more British sailors employed on American ships than American citizens impressed into the Royal Navy. It was no doubt true that "a wound inflicted by a British cat-of-nine-tails is not less severely felt than if it had proceeded from the lash of an Algerine," and that Britain, in its impressment of American seamen, compelled them "to fight her battles against a Power between whom and [their] own Government there exists no difference."[29] It was no less true, however, that the vast expansion of American shipping had raised up "a crowd of dangerous rivals for the seduction of [Britain's] sailors, and put bribes in their hands for the purpose"; and that Britain, though dominant on the oceans, was placed in the mortifying situation of seeing the trade of its enemies "carried on by the mariners of Great Britain." The carrying trade between the Napoleonic empire and the West Indies—which in its very nature made a substantial contribution to French economic and military power—was in fact, as Stephen argued, "nourished with the lifeblood" of the British Navy.[30]

A genuine settlement over impressment required the satisfaction of two great points: Britain had to be able to man its navy and American citizens had to be protected from arbitrary seizure. In theory, there was no disagreement between the two sides on either issue; in practice, however, there was a set of formidable obstacles to a satisfactory arrangement. For the Jefferson administration, the sine qua non of any agreement was that Britain renounce its proclaimed right to search and remove British subjects from vessels flying the American flag on the high seas—a renunciation that would still have allowed for search and removal in British ports. It was not unthinkable that Britain might renounce this practice; it had almost done so in 1803, but the United States refused the proffered agreement when the Addington ministry reserved a right of search and removal in the "narrow seas" surrounding Great Britain. The striking thing about the subsequent negotiations over impressment—culminating in 1806—is that the United States was unwilling to make any effort to return British deserters, or to prohibit their employment on American vessels, unless Britain first agreed to renounce its own capacity to ensure the effectiveness of whatever municipal regulations the United States might make. The American demand was too extreme, for it provided no guarantee that the problem of desertion or of the extensive employment of British sailors on American vessels would be adequately resolved.

In truth, the solution to the manifold aggravations over impressment lay mainly in American hands. Had the United States made efforts to remove the large number of British sailors in the employ of American shippers, it is quite likely that the British would have reduced substantially, and perhaps even eliminated, the oppressive practice of which the United States complained. American municipal regulations, enforced by the national government, that imposed penalties on shippers who employed British subjects, alongside a system for providing papers of American citizenship with stringent precautions against forgery, which the existing system of "protections" clearly lacked, could have virtually eliminated the problem. Solutions to the crisis were not wanting, so long as the United States was willing to limit its appetite for Britain's sailors. Such a self-denying ordinance, however, held no appeal for the administration, and no effective offer was ever made.

Impressment would not have arisen as a dangerous and intractable issue between Britain and the United States in the absence of the huge expansion of the American carrying trade, and the legal and moral issues that divided the two sides must therefore largely turn on the right of the neutral to undertake this trade at the expense of the belligerent. Jefferson and Madison located this right in both reason and usage, but in neither case was their argument persuasive. The practices of European states at war had merely established that there was a great and overriding conflict of interest between the pretensions of the power superior at sea and those states—neutral and belligerent—comparatively weak in naval armament. That Britain had given up the principle for which it now contended in particular treaties and had relaxed its enforcement through unilateral administrative and judicial decisions does not show that Britain had abandoned the claim to make its naval power effective in the prosecution of war. From the age of Elizabeth to the Napoleonic wars, naval supremacy was always seen to be inseparable from the interdiction and harassment of the enemy's commerce and the invigoration of Britain's own. And though the rule of 1756 was of comparatively recent origin, having first been elaborated in the Seven Years' War, it was only in that war that neutrals began to enter in substantial numbers a trade forbidden them in time of peace. It was not a new principle for which the British contended, but the application of an old principle to new circumstances. The appeal to usage—to, as Madison put it, "the law of nations, as an established code, as an actual rule of conduct among nations . . . founded in consent, either tacit or express"[31]—could fairly yield only the conclusion that conflict, not consensus, characterized the law of neutrality on this issue.

Once international law is freed from the test of usage and the appeal is made to reason, the distinction between legal and moral judgment can become a very fine one, and may indeed disappear altogether. Yet the appeal to reason is no more favorable to the American claim. The test of reasonableness cannot be seen apart from the circumstances of the conflict, and, most fundamentally, the relative vitality of the interests at stake for the two sides. Britain's claim to restrict a neutral trade forbidden to the neutral in time of peace was based on the contention that only by doing so might it effectively bring the war to Napoleon. Britain's armies were small in number; its main contribution to the defeat of the Napoleonic empire, to say nothing of its security against conquest, necessarily rested on its supremacy at sea. The American claim, if recognized, would have severely weakened Britain's principal weapon and sharply compromised its rights as a belligerent. The interest for which the United States contended, by contrast, was of far lesser weight. The carrying trade, to be sure, was immensely profitable, and it suited well the talents of a maritime people who had "derived cargoes from the depths of the ocean, and laid the cod, the seal, and the whale, under contribution."[32] But the pursuit of this trade reflected no compelling necessity; the happiness and prosperity of the Union, though they assuredly depended on commerce in some measure, did not rest on Americans becoming the carriers for all the world, and it would be absurd to pretend otherwise. The contrary was closer to the truth, as Jefferson himself had once recognized (and as he would recognize again near the end of his presidency, when the commerce of the country and his passion for peace lay in ruins).[33]

In seeking to understand the motivation that led the President and his Secretary of State into embracing the position they did, it may be argued that the considerations talked of most frequently—the vindication of right—were of secondary importance, and that those talked of least frequently—traditional calculations of interest—were primary in their thinking. Of the latter, three may be identified. One stemmed from the needs of the Republican party, which Jefferson dearly wanted to make into a national party, and which therefore required the embrace of the doctrine dearest to the hearts of northern Republicans (many of whom, like Crowninshield of Massachusetts, were deeply involved in the carrying trade to France). Another lay in the conviction that France, even under Napoleon, would remain a necessary counterweight to British power. And the third was the sense that American commerce was all of a piece, that to cut off one branch of the carrying trade would jeopardize the whole, either directly by complicating the voyages of American vessels or indirectly by its tendency to whet Britain's appetite through concessions. All of these considerations were important; all served to reinforce Jefferson and Madison's chosen course. But in the end it is misleading to see these calculations of interest as determinative; the question of right had an independent standing. Their equation of it with national honor and independence set it apart from considerations of political power and economic interest, and indeed invested it with all the moral drama that had characterized the struggle for independence in 1776.

20

The Abortive
Peace Settlement

THE ENVOYS CHARGED WITH NEGOTIATING the issues dividing England and the United States—James Monroe and William Pinkney—faced an enormously difficult task. Madison framed his instructions in such a manner as to preclude—if not with the intention, then certainly the effect—the possibility of a successful negotiation. A satisfactory treaty would have to include an English pledge to refrain from all impressments on the high seas; and, at the least, it would have to restore the favorable conditions for the reexport trade laid down in the 1800 *Polly* rule. Innumerable other items—relating to blockade, indemnification for past seizures, contraband, immunity of neutral waters—were pressed on the negotiators in the strongest terms. But impressments and the *Polly* rule were sine qua non; if left unsatisfied, Madison warned, the nonimportation bill would go into effect.

Monroe and Pinkney could not share the optimism of their government. Fox's ministry was weak, vulnerable to charges from the Tories that it would knuckle under to American demands. Though convinced of the rectitude of their government's position, the negotiators saw that there were some issues on which Britain could not be made to yield: on the main points of the negotiation, "interests of the most vital character were involved. . . . At a time too when the very existence of the country depended on an adherence to its maritime pretensions."[34] Recognizing the imbalance of interest that characterized the Anglo-American conflict, they departed from their instructions and put their signatures to an agreement quite similar in character to the Jay Treaty of a decade before. The treaty itself was silent on impressment, though the British pledged, in a separate note, to exercise "the greatest caution in the impressing of British seamen." A substantial British concession was made on the reexport trade, which legalized this trade but required that duties equivalent to 2 percent ad valorem be paid on West Indian produce reexported to Europe, and 1 percent on European merchandise reexported to the West Indies.[35]

When Jefferson and Madison first heard, in January 1807, that impressment would not be treated in the desired manner, they resolved to oppose the treaty. On its arrival in Washington, Jefferson summarily rejected it, refusing even to submit it to the Senate for consideration. This initial reaction surprised and angered many congressmen, who thought it smacked too much of executive arrogance. The treaty met neither of the administration's ultimata, and it was accompanied by a reservation that stung the government and quickened its anger. The British held that, unless the United States resolved to oppose Napoleon's Berlin decree, they would not consider themselves bound by the treaty. Napoleon's famous decree, which he delivered November 21, 1806—together with the British Order in Council of January 7, 1807—advanced the crisis to a new stage, raising issues that the Monroe–Pinkney treaty did not address. Napoleon, in an action that established the outlines of what would later come to be known as the Continental System, declared England in a state of blockade, prohibited all intercourse with it, made a prisoner of war of every Englishman found within French authority, and, most importantly for neutrals, made all English merchandise, as well as any ship coming directly from England or its colonies, a prize of war. The British Order in Council retaliated by declaring subject to seizure vessels engaged in the coastal trade between any French (or French-controlled) port in Europe; its object was "to restrain the violence of the enemy and to retort upon him the evils of his own injustice."[36]

Much obscurity surrounded these decrees at the outset. Armstrong, the American minister in Paris, received guarded assurances that the Berlin decree was not meant to be applied to American commerce, though in fact it would be so applied in September 1807. The British Order in Council was equally puzzling. If it was an act of retaliation, as it was represented in Parliament, it might be lifted as soon as it was established that the Berlin decree exempted American commerce. But it might equally be justified as an application of the rule of 1756, which prohibited to neutrals a trade normally closed to them in time of peace. The coasting trade in Europe was certainly of this character, and the Whigs would later defend the measure in these terms. Initially, Madison gave it a still different interpretation. "No fair objection can lie against it," he told the British minister in Washington, "provided it be founded on, and enforced by, actual blockades, as authorized by the law of nations."[37] The more he reflected on it, however, "the more unjustifiable it appears in its principle, the more comprehensive in its terms, and the more mischievous in its operation."[38] He thought it merely a pretext "eagerly seized for unloosing a spirit, impatient under the restraint of neutral rights, and panting for the spoils of neutral trade."

The rejection of the treaty left the President in a peculiar situation. In the spring of 1807, he persisted in the policy on which he had set the country in the winter of 1805-6, though it was now visibly in disarray. Neither of his objects—concession by Great Britain in the maritime conflict and the acquisition of the Floridas under the auspices of Napoleon—had yet been achieved. Napoleon, it appeared, set no such high price on American friendship as Jefferson imagined, and the President was not even close to agreement with the traditional "friends of America" among the Whigs. The administration boiled at the thought that it was not being taken seriously, that its estimate of American power was not shared by others, and it brought out, as it were, the big guns.

In his summary rejection of the treaty, communicated to the American envoys in May 1807, Madison warned that if an agreeable treaty was not made, America had the power to deliver a fatal blow to Great Britain. "The necessaries of life and of cultivation, can be furnished to [the British West Indies] from no other source than the United States," he said; "immediate ruin would ensue if this source were shut up." The United States might cut off the supply of naval stores and grain, on which Britain would likely become increasingly dependent. Britain could retaliate with no commercial injury worth mentioning, and a resort to war would be a losing proposition.

> To say nothing of the hostile use that might be made against Great Britain of 50,000 seamen, . . . nor of her vulnerable possessions in our neighborhood, which tho' little desired by

the United States, are highly prized by her, nor of the general tendency of adding the United States to the mass of nations already in arms against her; it is enough to observe, that a war with the United States involves a complete loss of the principal remaining market for her manufactures, and of the principal, perhaps the sole, remaining source of supplies without which all her faculties must wither.

And Britain would lose "all the advantages which she now derives from the neutrality of our flag, and of our ports." [39]

Jefferson, on some occasions, still professed optimism that the British might be brought to terms. To his envoy in Spain, however, he confessed that he had "but little expectation that the British government will retire from their habitual wrongs in the impressment of our seamen," and he thought it likely that a "war of commercial restrictions" might ensue. He had suspended the non-importation act to show "the sincerity of our desire for conciliation," but he betrayed some anxiety that his action might be misunderstood by Bonaparte: "Every effort," he told Bowdoin, should be "used on your part to accommodate our differences with Spain, under the auspices of France, with whom it is all important that we should stand in terms of the strictest cordiality. In fact we are to depend on her & Russia for the establishment of Neutral rights by the treaty of peace, among which should be that of taking no person by a belligerent out of a Neutral ship, unless they be *soldiers* of an enemy." Jefferson's hope that Napoleon and Alexander would put an end to the impressment of American seamen was destined to be disappointed, for the Franco-Russian agreement at Tilsit did not awe the English into a settlement along such lines. Surprisingly, Jefferson then immediately vented his spleen against Spain: "Never did a nation act towards another with more perfidy and injustice than Spain has constantly practiced against us. And if we have kept our hands off her till now, it has been purely out of respect for France, & from the value we set on the friendship of France. We expect therefore from the friendship of the emperor that he will either compel Spain to do us justice, or abandon her to us. We ask but one month to be in possession of the city of Mexico." [40]

This was quite a vision, this picture of a string of conquests extending from the arctics to the tropics, of privateers swarming the seas, of England prostrate at America's feet. Whether the two letters, taken together, are more notable for their gross overestimation of American military power—which was at the moment almost nonexistent—or for the covert alliance with Napoleon that underlay them both, or for the absurd charges against Spain, which was holding on for dear life against Napoleon's deadly embrace, does not permit an easy answer. One fact is clear. The administration's sense of American power was so extensive and its conception of American rights and interests so unbending that no negotiated settlement with England was possible, and even had the treaty been more favorable than it was, it would likely have met with rejection from the President and the Secretary of State.

The instincts of the President and his Secretary of State are perhaps no better revealed than in their reaction to Gallatin's startling finding that the number of British seamen employed on American vessels was far larger than was previously thought. Though Gallatin's estimates were rough, he concluded that around half of the total number of able seamen involved in the foreign trade were English, even on the American definition of nationality.[41] Jefferson did not make this finding, which badly undercut the American case against Great Britain, the occasion for a reassessment of American policy. New propositions that would pledge the United States to return British sailors in American employ were not forthcoming, even though it was now apparent that a failure to offer such a concession would make a deal with England impossible and would leave the issue as a standing impediment to satisfactory relations. Jefferson's reaction was the precise opposite: "Mr. Gallatin's estimate of the number of foreign seamen in our employ," he wrote Madison, "renders it prudent, I think, to suspend all propositions respecting our nonemploiment of them."[42]

The rejection of the treaty was a fateful step. Had it been accepted at this moment, much danger and difficulty might have been avoided—even though new hurdles would have had to be crossed with the crisis brought on by the further tightening of belligerent restrictions. The commercial prohibitions that would destroy American commerce in the attempt to save it, and, ultimately, the war that would confirm the Hamiltonian institutions most dreaded by Jefferson (and that he would loathe in his old age), were the natural outcomes of this defiant stance, though of course Jefferson was blind to the nature of the choice he had made. Most of these consequences lay in the still distant future; of more immediate import was the relationship between the rejection of the treaty and an incident that would gravely complicate Anglo-American relations. In an effort to reclaim four sailors who had deserted from a British man-of-war and reenlisted on the American warship *Chesapeake*—one of whom taunted British officers on the streets of Norfolk and whose return to the Royal Navy was refused by American authorities—the British warship *Leopard* hailed the American frigate on June 22, 1807, and demanded the peaceful return of the four men. This being refused, the British captain ordered a ten-minute cannonade, killing three and wounding eighteen Americans. This action yielded four men (one British, three Americans); it also produced an explosion in American opinion.[43]

The attack on the *Chesapeake* crystallized the growing resentment in the country against Great Britain; and though the wave of popular indignation and the cries of war it provoked soon dissipated, it constituted a running sore in Anglo-American relations for the next few years. It fixed both sides to the positions they had adopted previously and gave a powerful impetus to the growing storm in Anglo-American relations. Jefferson responded with an executive proclamation ordering all British warships out of the limits of the United States, and tied the British reparation for the *Chesapeake* attack to a settlement, on American terms, of impressment. Congress, he thought, would then be free to decide in the fall "whether war is the most efficacious mode of redress in our case, or whether, having taught so many other useful lessons to Europe, we may not add that of showing them that there are peaceable means of repressing injustice, by making it the interest of the aggressor to do what is just, and abstain from future wrong."[44] An embargo on American shipping and exports would be imposed six months later, in December 1807, and would be enforced—at least on paper—with ever greater severity for another fifteen months, until Jefferson's final days in office. Yet the road to this destination was not a straight one, and Jefferson sometimes wavered from this mild sentiment of early summer. Not the least mystery of the period following Jefferson's threat of peaceable coercion in June 1807 was the depth of his conviction thereto.

21

Jefferson and
the Embargo

THE EMBARGO IS NORMALLY referred to as a thing unto itself, but it was in fact composed of a multitude of measures whose successive manifestations in law, administration, and social thought were complex. The original measure, adopted by Congress in hurried and secret sessions of which no good records remain, was signed by the President on December 22, 1807. It prohibited all American vessels, save those under the immediate direction of the President, from sailing to foreign ports. Foreign-owned vessels might depart in ballast or with cargoes taken on before word of the embargo reached them; otherwise, all ship-bound exports were forbidden. The initial measure failed to provide penalties for violations and also excluded vessels engaged in the coasting trade. Subsequent acts—of January 9, March 12, and April 25, 1808— attempted to close the loopholes existing in the original measure, made the penalties for evasion progressively more onerous, prohibited exports by land, and vested an extraordinary degree of power in the Executive and his agents. The capstone bill—the so-called Fifth Embargo Act—was signed on January 9, 1809. Under its terms, Leonard Levy has written, "the privilege against self-incrimination was rendered meaningless; the right to trial by jury made a farce; the protection against property being taken without due process of law ignored; and the freedom from unreasonable searches and seizures abolished." It remains to this day, Levy ar-

gues, "the most repressive and unconstitutional legislation ever enacted by Congress in time of peace."[45]

Imports were not forbidden under the embargo. The weakened
nonimportation bill Congress had passed in 1806, however, did
go into effect on December 14, 1807. Troubled by the difficulties
of enforcing the measure, Gallatin had recommended that Congress delay its implementation until modified, but Congress took
no action at the time. When it did so, on February 27, 1808, the
measure was not strengthened, and numerous British goods made
their way into the United States throughout 1808, amounting to
around 50 percent of the previous year's importation.[46] Since nonimportation received an extraordinary amount of attention in the
1806 congressional session, the sudden move to embargo and nonexportation was peculiar. It reversed the sequence of the Continental Association during the preliminaries to the American Revolution, when a ban on British imports in 1774 preceded the total
ban on commercial intercourse in 1775. If seen as an instrument
of economic coercion and in light of traditional Republican strategy for securing commercial redress from Great Britain, the embargo of 1807–9 had a strange design, ill-calculated in many respects to achieve its object.

Jefferson's relationship to the protean measure of embargo is
not simple and direct. At the end the President considered the embargo to be an honorable attempt to find a substitute for war in
settling disputes between nations; a positive blessing by virtue of
its reordering of the American economy; and above all, a test of
republican virtue against the "parricides" and "unprincipled adventurers" who sought to evade it. These themes, however, did
not emerge with real clarity in Jefferson's correspondence until the
summer of 1808, nearly six months into the embargo's life, and
they came into public view only with his Eighth Annual Message
in November. In its adolescence (March–April 1808), the embargo was allied with a diplomatic strategy in which the United
States sought to make clear that it would go to war against whichever belligerent failed to lift its offending restrictions on American
commerce. Considered in this light, it appeared as a temporary
measure, to be in all likelihood followed in the fall by war. At
birth, the emphasis was still different. In his message to the Congress, Jefferson spoke of no coercive purpose and placed exclusive
emphasis on the precautionary motive that underlay the em-

bargo—the need, in view of "the great and increasing dangers with which our vessels, our seamen, and merchandise, are threatened on the high seas," "to keep in safety these essential resources." At the same moment, but in private, he spoke of the embargo's value in terms of the time it provided and not of the coercive powers it possessed. In embryo, the whole question appeared still differently. Though Jefferson raised the prospect of peaceable coercion in the immediate aftermath of the *Chesapeake* incident, he also gave considerable thought to the prospect of war with England. At various moments in the late summer and fall of 1807, he appeared to consider it, on balance, a path superior to a trial at economic coercion.

In view of these changes in Jefferson's outlook, the meaning of the protean measure for the President—indeed, the whole state of his mind in the final eighteen months of his term—has always been one of the great riddles in Jeffersonian historiography. What the President wanted, and when he wanted it, was equally a puzzle at the time, for Jefferson did not enlighten the public as to his real purposes. He took upon himself the entire responsibility for the measure, but he provided the public with no reasoned defense of it. The sacrifices he asked of the people were nearly as great as those to which they had been called a generation earlier, in 1776. In December 1807, however, there was no appeal to a "candid world." As violations of the embargo mounted, he responded not with argument but with ever more draconian measures of enforcement.[47]

The President's silence can be understood only against the background of the events of the preceding summer and fall. The attack on the *Chesapeake* stirred in him all the emotions of 1776. Never since Lexington had the country been so united. Though he refused to call Congress into immediate session, his reasons for doing so did not reflect a desire to dampen the war fever sweeping the country; on the contrary, he hoped that it would maintain itself at that point until Congress met in the fall. Whatever the merits of war, during the summer the timing was not right. There was an extraordinary amount of mercantile property on the ocean (amounting to perhaps $100,000,000), much of it uninsured. An immediate declaration of war would leave it vulnerable to British seizure. Congress, in any case, could not choose between war and economic coercion until the British had time to respond to the

American demand, agreed upon by Jefferson and Madison in the summer, to suspend all impressments on the high seas. By tying the requirement for avoiding a rupture to this demand—which even a Whig ministry had been unwilling to concede—Jefferson knew that his demands were highly unlikely to be met. He did not care; he was psychologically ready for the test of strength. He believed then, as he was to say in 1812, that the acquisition of Canada would be a mere matter of marching, and he thought that this was time to settle matters with Spain as well. "Our southern defensive force can take the Floridas, volunteers for a Mexican army will flock to our standard, and rich pabulum will be offered to our privateers in the plunder of their commerce & coasts. Probably Cuba would add itself to our confederation."[48] Circumstances could not be better. With the French defeat of Russia at Friedland and the subsequent agreement between Napoleon and Alexander at Tilsit, Napoleon now commanded the whole of the continent and would likely renew his attempt at an invasion of England, which would leave Canada defenseless and Britain ever more dependent on American naval stores (brought on by the closure of the Baltic to English shipping). The fall, he thought, would be the time to strike.

Jefferson's newfound enthusiasm for military conquest has normally been regarded with skepticism by historians. It is too contrary to his inveterate tendency to "palliate and endure," too inconsistent with his faith in peaceable coercion, too antagonistic to the dread of war that had long formed the foremost maxim of Republican statecraft.[49] Yet as the congressional session approached, he if anything became more bellicose. Gallatin thought the first draft of the President's Seventh Annual Message appeared "in the shape of a manifesto issued against Great Britain on the eve of a war," and Jefferson himself ridiculed the idea that nonimportation would secure a redress against Great Britain. It would end in war, he told his son-in-law, and would simply give Britain "the choice of the moment of declaring it."[50]

It is probably true that Jefferson, in his own mind, had not fully resolved upon war, and that he stated the case for it while secretly holding in reserve all the objections that might be raised against so perilous a course. He approached, but did not cross, the Rubicon. He was, in any case, never a man to fly in the face of popular sentiment, and by the time the Tenth Congress con-

vened he could read the shape of the popular will as well as anybody. This was decidedly against war. Nor could he ignore the
incomplete state of military preparations. Both considerations were
pointed out by Gallatin in his review of the Annual Message.

The Secretary of the Treasury was the bearer of a host of un-
pleasant facts, which Jefferson had allowed himself to overlook
but which he was now forced to acknowledge. If the justification
for war, Gallatin argued, "be England's refusal to disavow or to
make satisfaction for the outrage on the Chesapeake," universal
domestic support would be forthcoming. But he was confident "that
we will meet with a most formidable opposition should England
do justice on that point, and we should still declare war because
she refuses to make the proposed arrangement respecting seamen."
Nothing would be gained by attacking Canada immediately, and
the United States remained acutely vulnerable to British attack.
The "essential preparations" for American military actions "are in
some points hardly commenced, in every respect incomplete." On
the sea, military action by the United States would be confined, as
Jefferson had always urged, to privateers, yet America's China and
East India trade was still on the seas "to an immense amount."
Nothing had been done in the way of making the militia fit for
war; American seaports still remained unfortified. If Congress were
to declare war in November, Great Britain might lay "New York
under contribution before winter"—which in turn would deal a
devastating blow to American finances. "Great would be the disgrace attaching to such a disaster; the Executive would be particularly liable to censure for having urged immediate war, whilst so
unprepared against attack."[51] The popularity of the administration might be gravely compromised, perhaps irreparably.

Jefferson bowed to the realities that Gallatin had set forth. The
Annual Message was toned down; war was for the moment rejected. The President nevertheless took it for granted that some
positive step was required to meet British aggression; "submission" continued to be unthinkable. In December, the latest news
from Europe reinforced the need for action. From France came
word that the Berlin decree was now being applied against American vessels, in direct violation of the 1800 Franco-American treaty.
From Britain came the news that George III, in an October proclamation, had called for a return of all British sailors to the Royal
Navy—which signaled a tightening rather than a loosening of Brit-

ish policy on impressment—and that the British government would soon announce sweeping measures of reprisal against France and neutral commerce. "The whole world," Jefferson thought, "is thus laid under interdict by these two nations, and our vessels, their cargoes and crews, are to be taken by the one or the other, for whatever place they may be destined, out of our limits." Since American vessels, cargoes, and seamen were certain to be lost if they left port, Jefferson thought it best to keep them at home.[52]

On December 18, Jefferson recommended to Congress an embargo on American shipping, and he hoped that Congress would "also see the necessity of making every preparation for whatever events may grow out of the present crisis." This enigmatic statement presumably referred to military preparations. No coercive purpose was attached to the embargo in this message; he spoke only of "the great and increasing dangers with which our vessels, our seamen, and merchandise, are threatened on the high seas and elsewhere" and the "great importance to keep in safety these essential resources." This precautionary purpose was of great importance to the President, and it continued to form his chief justification of the embargo for some months hence. Its great advantage was that it gave time and might alternatively serve the purposes of either war or economic coercion. It served the former by withdrawing American ships and seamen from the danger of English seizure and prepared for the moment when they might venture forth as privateers. It served the latter by withholding from England and its colonies supplies that Jefferson had always believed they could not do without.

The embargo was thus not undertaken solely as an instrument of economic coercion. The precautionary motives were real and pressing, and the possibility of war had not been foreclosed. Yet the coercive purposes were always there from the start. The very issue had been raised in debate within the cabinet and centered on whether the embargo would be one of indefinite duration. Gallatin strenuously objected to the "permanent embargo" that was urged by Madison. "In every point of view, privations, sufferings, revenue, effect on the enemy, politics at home, &c., I prefer war to permanent embargo." He thought "entirely groundless" the hope that the embargo might have a productive effect on the negotiations with Britain over reparations for the *Chesapeake*; it would not, he thought, "induce England to treat us better."[53] Gallatin's advice, however, was rejected. This crucial decision must in the end be attributed to Jefferson's hope that the embargo might serve a coercive purpose, for had his purposes been wholly precautionary, a temporary embargo would have amply suited his purposes. Though he bent to Gallatin's advice in some respects, however, he rejected it on this crucial point.

As he had done so many times previously, Jefferson took the advice of his Secretary of State, whose anonymous contribution to *The National Intelligencer*—the administration's chief press organ—laid out the blend of precautionary and coercive purposes that inspired the measure. Because of the rival restrictions of the belligerent powers, the ocean presented a field "where no harvest is to be reaped but that of danger, of spoliation and of disgrace." There was no alternative but "a dignified retirement within ourselves; a watchful preservation of our resources; and a demonstration to the world, that we possess a virtue and a patriotism which can take any shape that will best suit the occasion." Since the embargo was "a measure of precaution, not of aggression," it could afford neither Britain nor France "the slightest ground of complaint." At the same time, the embargo would make it "the interest of all nations to change the system which has driven our commerce from the ocean." Great Britain would feel it "in her manufactures, in the loss of naval stores, and above all in the supplies essential to her colonies," while France would lose "those colonial luxuries, which she has hitherto received through our neutral commerce," and France's colonies, like those of Britain and Spain, would wither once "cut off from the sale of their productions, and the source of their supplies."[54]

Save in this indirect way, the administration thus did not advertise the coercive character of the embargo. It believed that to do so might invite danger, and neither Madison nor Jefferson wished to give British Foreign Secretary Canning a handle to condemn the measure and perhaps even to make it the basis of a bold reprisal. Madison's insistence that no nation could find in it "the slightest ground of complaint" served a diplomatic purpose. So, too, did the administration's insistence that the embargo was meant to favor neither belligerent but was a measure of strict neutrality. Britain was supposed to feel the embargo's deadly effects but not to take offense. Like the 1806 nonimportation act, which now also went into effect, the embargo was "at once a domestic and a foreign regulation, a pacific and a hostile act, a measure with which England had no right to be angry, and one which was calculated to anger her,—strictly amicable and at the same time sharply coercive."[55]

The two diplomatic fictions with which the administration embarked on the embargo—that it was wholly precautionary and that it favored neither belligerent—were soon given up. "The embargo which appears to hit France and Britain equally," Jefferson told the French minister Turreau, "is for a fact more prejudicial to the latter than the other by reason of a greater number of colonies which England possesses and their inferiority in local resources."[56] The bias toward France was shown even more in the conditions the administration set, early in 1808, for lifting it. Jefferson believed, in March, that there would come a time "when our interests will render war preferable to a continuance of the embargo." When that time arrived—which Jefferson normally placed at the beginning of the congressional session in the fall— he suggested to Madison that the United States ought then to declare war on whichever power had not withdrawn its hostile decrees.[57] There existed the possibility that the decrees of both powers might remain in effect, in which case the United States would have to "take our choice of enemies between them." But the administration fully expected that France would go along with its plan. The Continental System, as such, was not made a ground of American complaint. The municipal regulations that prohibited British ships and merchandise from landing in Europe constituted, according to the United States, no violation of international law. All France need do was to lift those portions of the decrees that were enforced upon the ocean. Such a concession, as Madison correctly saw, would "immaterially diminish" the operations of the Continental System "against the British commerce, that operation being so completely in the power of France on land, and so little in her power on the high seas."[58]

Napoleon, as it happened, did not rise to the bait. With the Bayonne decree, April 17, 1808, he sequestered American vessels and made their release dependent on an American declaration of war against Britain. He then made the American claim to the Floridas dependent on the same decision. Jefferson squirmed under his inability to make Napoleon yield to his designs; but by the summer of 1808, when Bonaparte's recalcitrance had become crystal clear, it had all ceased to matter. Jefferson's dread of war had by this time resurfaced and was again his ruling passion. Had Napoleon played his expected part, the administration would not have gone to war against England. For the more the embargo wore on, the more its meaning assumed a new shape in the President's hands. He still talked of the possibility of war and recognized that the embargo could not last forever; but as public opposition mounted, he grew steadily more determined that the experiment in peaceable coercion should have a fair trial. As the end of his administration approached and as he reflected on the meaning of his presidency, the prospect of war became increasingly distasteful. "I think one war enough for the life of one man," he told John Langdon in early August;[59] war, he reflected, could not be his final legacy to the nation. As this thought took hold, so, too, did an even more striking idea. In its final months, the embargo, which had been undertaken in part to free commerce and to overthrow the belligerent restrictions, now became a device to insulate America by sharply curtailing its commerce with the rest of the world.

This transformation of the embargo's purpose was the ultimate irony of the measure. At the end, Jefferson came full circle.[60] The Republican dream of overthrowing European barriers against free trade—which Jefferson had embraced in the late 1780s and pursued throughout his public life—now yielded to Jefferson's still older conviction that commerce was an instrument of corruption by virtue of its tendency to involve the country in Europe's endless wars. His draft of the Eighth Annual Message in November 1808 celebrated the economic changes wrought by the embargo and considered it in signal respects a positive blessing. Gallatin considered these passages to be "the most objectionable in the message" and "little less than a denunciation of commerce." If left to stand, they would serve to confirm all the charges of a hostility to commerce that Federalists had hurled at him throughout the preceding year. The avowal, Gallatin warned, "that a positive benefit is derived from the introduction of manufactures caused by the annihilation of commerce . . . will produce a pernicious effect and furnish a powerful weapon to the disaffected in the seaports and in all the eastern states."[61] Jefferson toned down the message but was otherwise unrepentant. His growing frustration over violations of the embargo became fixed on the low motives of commercial men, and he felt betrayed that his own administration had somehow been lured into defending them. It occurred to him that the neutral rights on whose vindication his administration had staked its whole diplomacy were contrary to the interest of the United States to exercise. "As to the rights of the United States as a neutral power," he told one correspondent,

> our opinions are very different, mine being that when two nations go to war, it does not abridge the rights of neutral nations but in the two articles of blockade and contraband of war. . . . With respects to the interests of the United States in this exuberant commerce which is now bringing war on us, we concur perfectly. It brings us into collision with other powers in every sea, and will force us into every war of the European powers. The converting this great agricultural country into a city of Amsterdam,—a mere headquarters for carrying on the commerce of all nations, is too absurd.[62]

This sober recognition came too late in the day to have any bearing on policy. It mocked the principles for which the administration had stood in its contest with England, for it made incomprehensible the attitude Jefferson had adopted in his English diplomacy since the *Essex* decision; indeed, it vindicated the conception of an "American system" that Hamilton had glimpsed a generation earlier, when he set forth his vision of an economy not critically dependent "on the combinations, right or wrong, of foreign policy."[63] Jefferson had gotten himself into this fix because he had converted the whole question of neutral rights into a great moral drama, separate from the commercial interest of the country and indeed greatly transcending it. But in the end his moralism betrayed him, for it prevented him from seeing that no great injury would be done to the country if it compromised its far-reaching claims on the high seas. It equally prevented him from understanding how far-reaching the claims of the United States really were; paradoxically, it therefore paved the way for a policy of commercial isolation just as extreme as the policy of commercial reformation that was now abandoned.

22

Neutral Rights versus
the Balance of Power

THE MARITIME ISSUES dividing the United States and Great Britain were often seen then, and are frequently seen now, apart from the course and stake of the European war. Yet the outcome of this conflict, which was of unprecedented scope and violence, was of great importance for the future of American security. Had Napoleon been victorious and actually reduced Britain to terms—as Jefferson thought likely—the security of the United States would have been gravely endangered. Jefferson understood, in some dim way, the dangers of his chosen course, but he thought that the immediate perils presented by England were of far greater moment than the future hypothetical dangers represented by France. The embargo, as it evolved, was thus strongly weighted toward France, and represented, in Jefferson's mind, a potent supplement to Napoleon's Continental System. That the embargo failed and that Napoleon was ultimately defeated has tended to deflect historical attention from the consequences of success. Yet the peculiarity of the embargo is such that success would likely have been far more detrimental to American security than failure. On this final episode of Jeffersonian statecraft the verdict must be, with Themistocles, that "we should have been ruined, if we had not been undone."

There is no question that the British Orders in Council of November 1807 bore heavily on the United States. By these orders, all American commerce with Europe had to pass through the Brit-

ish entrepôt. Colonial productions from the West Indies were subjected to an equalizing duty whose intent was to prevent France
from obtaining produce from its own colonies "at a cheaper rate
than that of the colonies of Great Britain." (This adjustment was
similar to that effected in the Monroe–Pinkney treaty, with the
signal difference that voyages were now to be broken in England
rather than America, and with the taxes falling to the English treasury.) Ships carrying American produce—such as flour, grain, tobacco, and rice—were required to obtain licenses in Britain (thus
subjecting them to French seizure) but were otherwise free to
smuggle goods onto the continent. A prohibitory duty, however,
was placed on American cotton at the outset and subsequently its
export to Europe was forbidden altogether.[64] These were humiliating regulations. They placed the United States, from a commercial
point of view, in a situation perilously close to that which had
prevailed before the War of American Independence. Though justified by Britain as measures of strict reprisal that would be lifted
as soon as Napoleon repealed the Berlin decree, the orders could
not fail to recall the issues for which the Revolution had been
fought. They were seen in America as evidence of a boundless
spirit of commercial monopoly at best, and of an insidious plan
to reduce America again to the condition of colonies at worst.[65]

The Berlin decree was also menacing to American commerce.
Just as England tried to ensure that American trade would pass
through the British entrepôt, the Continental System aimed at excluding all British commerce with Europe and thus destroying the
basis of English financial power; it was made possible by the enormous string of French victories on the continent, which lay prostrate at Napoleon's feet in the summer of 1807. First Austria (in
1805), then Prussia (in 1806–7), and finally Russia (in the summer of 1807) were reduced to terms. A prime condition of the
pacification was, in each case, participation in the commercial war
against England. Napoleon, with scarcely a ship on the ocean, intended to bring Britain to its knees, by excluding the manufactures
that were its lifeblood. The policy that Britain adopted was the
perfect inverse of this system; it aimed at turning "the provisions
of the French decree against themselves; and as they have said that
no British goods should sail freely on the seas, you might say that
no goods should be carried to France except they first touched at
an English port. They might be forced to be entered at the custom-

house, and a certain entry imposed, which would contribute to advance the price and give a better sale in the foreign market to your own commodities."⁶⁶ The issue that this would pose for the United States, as the leading neutral, was as simple as it was painful: would it cooperate with Napoleon in upholding the Continental System or with Britain in undermining it?⁶⁷

Jefferson's choice was to adopt the former course, although the administration tried mightily to preserve the fiction that it was following a policy of strict neutrality. Even from the perspective of neutral rights, the choice was open to question. The danger presented by France was in no way mitigated by the fact that the Continental System was largely enforced on land and was not in the main in technical violation of neutral rights; from a commercial point of view, Napoleon's unification of the continent posed a serious future danger to the principle of freedom of commerce, for it is very unlikely that Napoleon's pose as a champion of neutral rights would have survived the circumstances—the need to defeat England—that gave rise to it. The necessity of "crushing the Tyrant of the Sea," as Hamilton put in 1798 (and which was no less true in 1807), "has been trumpeted as a motive to other powers to acquiesce in the execution of a plan, by which France endeavors to become the Tyrant both of Sea and Land."⁶⁸

The central issue raised by the rival belligerent decrees in 1807 was the importance of the balance of power as an objective of American policy. That American security depended on the maintenance of a balance of power was often urged upon the administration by its domestic opponents. John Randolph noted it in his initial break from the administration in 1806. He questioned then the wisdom of the United States throwing its weight "into the scale of France at this moment, from whatever motive—to aid the views of her gigantic ambition—to make her mistress of the sea and land—to jeopardize the liberties of mankind. Sir, you may help to crush Great Britain, you may assist in breaking down her naval dominion, but you cannot succeed to it. The iron scepter of the ocean will pass into his hands who wears the iron crown of the land. You may then expect a new code of maritime law."⁶⁹ The same point was the ritual refrain of Federalists such as Fisher Ames. A war that gave Napoleon the possession of the British Navy, "or a peace, like the last, that should humble England, and withdraw her navy from any further opposition to his arms, would give the civilized world a master," fixing upon mankind "the weight and ignominy of a new Roman dominion." Indeed, the fate of the modern world would be worse than the ancient one, "for Rome preserved her morals till she had achieved her conquests; France begins her career as deeply corrupt as Rome ended it."[70]

The importance of the balance of power as an objective of American diplomacy had not escaped Jefferson's attention. Since the 1780s, the rivalries of the European powers were always of the first importance in the shaping of his diplomatic outlook. Though he normally saw France as the more useful counter to Britain, he prided himself on his skill in playing both sides of the street: he had waved the threat of a British alliance before Napoleon in the negotiations over Louisiana, and he had again entertained the idea of a British alliance in 1805. Each of the two great European powers, he thought, was a "necessary instrument to hold in check the disposition of the other to tyrannize over other nations."[71] Sometimes he sought advantage, at other times merely safety, in this competition among the great. It often occurred to him to play the jackal and to fatten on the spoils of European rivalry; from the 1780s on, his commercial and territorial objectives were always reflected through this prism. At the same time (and often in the same breath), he was fond of comparing the United States to a "common herd of cattle"—fat and innocent—looking upon a mighty "battle of lions and tigers."[72]

Insofar as Jefferson saw the balance of power as an objective of American statecraft (as opposed to simply an existing constellation of forces to be manipulated), he saw it as a function of the balance of sea and land power. It is misleading to see in his thought a concern for the continental balance of power as such. He tended rather to look favorably on France's triumphs in Europe. These victories presented no immediate danger to the United States; they were over monarchies with which (save for the Tsar) he had little sympathy; and they were useful in restraining England and would induce the British, he thought, to respect American "rights." Though he thereby tended to sharply depreciate the hostile significance of French victories on the continent, he had in the past recognized that it would be very dangerous to American security if France could join the empire of the seas to its prodigious power on land. "Our wish," he said after Trafalgar and Austerlitz, "ought to be that he who has armies may not have the Dominion of the sea, and that he who has Dominion of the sea may be one who has no armies. In this way we may be quiet; at home at least."[73]

The "wish" that Jefferson expressed in 1806 did not inform the policy he pursued in 1807 and 1808. So enraged was he over English transgressions against neutral rights that he no longer troubled himself over the danger that "he who has armies" might gain "the Dominion of the sea." He acknowledged his change of attitude by confessing that it was "mortifying that we should be forced to wish success to Bonaparte, and to look to his victories as our salvation."[74] He had never expected that he would be placed in this position. "But the English being equally tyrannical at sea as he is on land, & that tyranny bearing on us in every point of either honor or interest, I say, 'down with England' and as for what Buonaparte is then to do with us, let us trust to the chapter of accidents. I cannot, with the Anglomen, prefer a certain present evil to a future hypothetical one."[75] Jefferson here expressed, in a nutshell, the central problem of American diplomacy during the Wars of the French Revolution and Napoleon. Which was more important: neutral rights or the balance of power, the "present evil" represented by Britain, or the future danger held out by France? It was on this question, more than any other, that Republicans and Federalists divided over foreign policy from the 1790s on. The resolution Jefferson gave to it in 1807 constituted his most serious error in foreign policy.[76]

The danger represented by Napoleon cannot be known with certainty, any more than can most other attempts to unify the continent since the Renaissance. All have ultimately been beaten back, so the consequences of failing to do so cannot be readily glimpsed. Yet there can be little question that Napoleon entertained the ambition of universal empire. His defeat was a close-run thing; and in the long history of war from 1793 to 1815 his victory over Great Britain never appeared more likely than in late 1807. Napoleon was, after all, a man of prodigious talents and fanatical determination. Penned in on the continent by the British Navy, he had unleashed his restless and unquenchable energy on the peoples of continental Europe. He had achieved, by 1807, the reduction not only of the three other great powers in Europe—Austria, Prussia, and Russia—but of a host of neutrals as well. Had he once succeeded in wresting from England the command of the seas, there is no telling where his ambition would have taken him. India, South America, Canada, Louisiana—the world would have danced in his head, and the world would have been open to him. His intervention in Iberia, which appears in retrospect as the beginning of the end, might well have gone differently had the British not lent material assistance to the resistance of Spain and Portugal. He might even have avoided his greatest blunder, the invasion of Russia, had maritime empire truly beckoned. The defeat of England would have given him, at a stroke, a concentration of power unprecedented in the world since Rome; that such a power might seriously endanger the security of the United States was the only safe calculation.

That the balance of power ultimately prevailed and was the ordering principle of the settlements that ended the Napoleonic wars thus does not show that his defeat was inevitable.[77] And if one sees the manner in which he achieved his victories, the danger that he represented becomes all the more apparent. He was a genius not merely as a military commander but as a statesman. By playing a policy of *divide et impera* to a kind of sublime perfection, he broke up the numerous coalitions raised against him and managed to seduce nearly all the neutral powers. His method of doing so consisted, at bottom, of his ability to raise up a delicate mixture of hope and fear in his adversaries. Neutrals were kept on a leash that gave them the illusion of freedom; he never forgot to make himself in some way indispensable to them. At the same time, he knew how to unleash the sword in devastating fashion, and in each of his relationships with foreign states he cultivated the fear of the other party. By 1807, every state in Europe had proved amenable to this treatment save England; here he had a determined and remorseless enemy.

Napoleon's diplomacy toward the Floridas was the merest sideshow for him, but it is interesting as a demonstration of his general method. He saw at a glance that this territory, so important to Jefferson, might be held out as bait in a dalliance whose consummation would be indefinitely protracted. The diplomatic game over the Floridas, as Henry Adams observed, "was one which Napoleon played with every victim he wished to ensnare, and the victim never showed enough force of character to resist temptation. German, Italian, Russian, Spaniard, American, had all been lured by this decoy; one after another had been caught and devoured, but the next victim never saw the trap, or profited by the cries of the last unfortunate."[78] Part of the Emperor's purposes had been divined by Armstrong, the American minister to France, as early as September 1805, when he said that "Spain and America will in fact be a couple of oranges in [France's] hands which she will squeeze at pleasure, and against each other, and that which yields the most will be the best served or rather the best injured."[79] Jefferson understood the nature of the transaction in which he was engaged; thinking that France set a high price on American friendship, he played along even after being twice deluded by France—first in 1804, when the French refused to support American claims at Madrid, then in 1806, when Napoleon brusquely repudiated Talleyrand's efforts to settle the matter. Initially, the administration was firmly set against the payment of any money; then it consented to pay a bribe to France; but the price kept rising. Ultimately, after Napoleon had set upon Spain, the price for the Floridas became participation in the war against England. Only then did Jefferson balk.

Napoleon had no particular grievance against the United States or Thomas Jefferson. As Madame de Staël once said, "He regarded a human being as a fact or a theory, and not as a fellow creature. He did not hate anymore than he loved, there was only himself for himself." Had he subdued Great Britain, the danger to the United States would not necessarily have been immediate, and Federalists exaggerated their case when they said that it was.[80] But danger there nevertheless would have been. Both Canada and Louisiana offered natural French bastions in North America, and it appears highly likely that at some point he would have sought to exploit them. He would have enjoyed the ability, if displeased with the United States, of sending a powerful army of invasion against American territory, before which the United States would have been virtually helpless. In point of numbers, this capability to introduce an army of fifty thousand men in North America would have constituted a danger quite different from that traditionally posed by Great Britain, whose land forces were small. Such a vitally important new factor in the strategic equation would have been serious enough; it would have been complemented, moreover, by a commercial power capable of denying access to the European market to the United States. The Continental System that he erected against Great Britain might easily have been turned against the United States as well.

It is certainly no reproach to Jefferson that he had a conception of international order in which law and morality played a signal part. His understanding of what these required, however, was defective. Invariably he invested the narrowest of American interests with a status so exalted that they could not be compromised; hence his moralism constituted a formidable obstacle to an accommodation with England. Worse, his conception of international order was one that gave to neutral rights a status far higher than that accorded the principle of the balance of power; this ordering of priorities can be defended from the perspective of neither international order nor American security. Together, they afford a profound illustration of how the insistence on "reason and morality" without reference to "power and expediency" can subtly derange a statecraft, subvert the modest tasks of diplomacy, and end by betraying both physical security and economic interest.[81]

23

Embargo and War

HISTORIANS HAVE NORMALLY been more censorious toward Jefferson over the means he saw fit to employ rather than the ends that he embraced. The embargo has been criticized as ineffective and for having led to a host of pernicious consequences that Jefferson might have foreseen. The President has had few defenders, even among those—such as Dumas Malone and Merrill Peterson—who are otherwise sympathetic to the predicament in which he found himself. The most damning and memorable portrait of the embargo's failure remains that of Henry Adams, who saw it as the culminating episode in Jeffersonian statecraft and a revealing commentary on the theory of foreign affairs to which Jefferson had adhered throughout his public life: "Financially," wrote Adams,

> it emptied the Treasury, bankrupted the mercantile and agricultural class, and ground the poor beyond endurance. Constitutionally, it overrode every specified limit on arbitrary power and made Congress despotic, while it left no bounds to the authority which might be vested by Congress in the President. Morally, it sapped the nation's vital force, lowering its courage, paralyzing its energy, corrupting its principles, and arraying all the active elements of society in factious opposition to government or in secret paths of treason. Politically, it cost Jefferson the fruits of eight years painful labor for popularity, and brought the Union to the edge of a precipice.[82]

It did all this without achieving its purpose of coercing the belligerents to lift their various restrictions on American commerce—a failure that in retrospect appears virtually inevitable. It was not to be expected that the British political nation, having steeled itself against the hardships of war for many years, would concede the issue in dispute merely from fear of economic distress. It had endured far worse at the hands of an enemy far more powerful, yet its will had not been broken; it stretched probability to believe that the slow pressure of embargo might yield what the terror of Napoleon had failed to produce.

The apparent plausibility of the scheme rested on the threat, in conjunction with Napoleon's Continental System, to deprive British manufacturers of their two most important markets and British producers of indispensable supplies. A chain of bankruptcies, imperiling the English financial system, would, it was thought, inevitably follow. The great manufacturers of cloth would expire for want of American cotton; the West Indies would rapidly face starvation. Yet none of these results ensued. As it happened, a substantial number of British manufactures were imported during 1808; the West Indies were partly fed by the great diversion of produce and lumber that opened up along the Canadian border and on the sea route between New England and Nova Scotia—which Jefferson, despite ever more draconian measures of enforcement, was powerless to stop. Though prices rose sharply in the cotton industry, reflecting increased speculation and hoarding, they began to break in 1809 when smuggled American cotton began getting through in substantial amounts to British manufacturers. Worse, the flight of the Prince Regent of Portugal to Brazil and the uprising of the Spanish nation against Napoleon opened the markets of the Latin empires to British manufacturers and doomed all visions of imminent bankruptcy.

It is idle to ask whether the embargo, had it been more stringently enforced, would have worked, for it soon became clear that it could not be enforced save by a veritable war against the violators of the embargo at home. A "little army" along the Canadian border, a naval blockade on the New England coast, "arbitrary powers" of seizure "equally dangerous and odious" (and in clear violation of the Fourth Amendment to the Constitution) were thought necessary to carry the embargo into effect by Gallatin, to whom fell the thankless task of administering the measure. Gallatin's presentation of the harsh alternatives was meant to shock, but Jefferson was not dismayed. Revolted by the widespread violations, he nevertheless resolved to push ahead. "I did not expect a crop of so sudden & rank growth of fraud & open opposition by force could have grown up in the U.S.," he told Gallatin. But he believed "that if orders & decrees are not repealed, and a continuance of the embargo is preferred to war (which sentiment is universal here), Congress must legalize all *means* which may be necessary to obtain it's *end*." [83]

Congress did just that in the session that gathered in November 1808. Its choices were the continuance of the embargo, war, or submission; though the President affected to make no choice that would bind his successor, it was well understood that he wished for neither war nor submission and was resolved to "hug the Embargo and die in its embrace." The final revision of the embargo law in January 1809 appeared to honor his wishes, for it was the most draconian yet. But its very harshness paved the way for its abandonment in the spring, as the northern Republicans who had been most zealous on behalf of the vindication of neutral rights came ultimately to appreciate that the choice lay between embargo and Union. Their decision was to protect the latter by abandoning the former.

Most historians share the view, which then began to take hold in the country, that war was the only honorable alternative left at this point. That eight years of Jeffersonian parsimony in military expenditures had left the nation woefully unprepared for a trial of arms confirms for them the bankruptcy of Jefferson's theory of foreign relations. If seen from the standpoint of simple consistency, the charge is not without merit. Since the administration's foreign policy pointed toward war, it was indeed chargeable—on its own premises—with a blameable inattention to the necessary means. Because the administration had long deprecated the need for military establishments of any kind, the movement of the Congress and the public to a different policy was necessarily slow, and Jefferson put very little of his prestige behind it. Worse, the preparations that were made reflected, for the most part, Jefferson's own visionary conception of military policy: he threw his support behind a fleet of gunboats (which proved utterly inadequate in the War of 1812), and he urged on Congress a scheme to classify the militia that was ill-suited for expeditionary purposes (and whose performance in the War of 1812 was equally poor). The ships of the line and the standing army that Republican ideology had always deprecated were, in a military point of view, much superior to the forces recommended by the President, whose support for even these was too little and too late.[84]

Yet though it may be acknowledged that there was a need, on general purposes of precaution, for military establishments more effective than those possessed by the United States, it by no means follows that war would have been consistent with the interests of the United States. In the event, war, too, bankrupted the Treasury, left the nation vulnerable to predatory raids on its capital, and brought the commercial life of the country to a virtual standstill. War, too, stretched the limits of constitutional power (at least on the old Republican theory) and brought the Union to the edge of a precipice. The latter danger was potentially the most serious of all, though its scope and character are not readily calculable. The only real point of vulnerability that Britain had was in Canada, and though the American invasion of Canada failed in 1812, the consequences of a successful conquest there might have proved quite dangerous.[85] As it was, the acquisition of Louisiana had provoked rumblings of disunion; and it was not then (nor is it now) easy to know what retroactive effects the conquest of Canada might have had on the cohesion of the Union. The relationship among expansion, sectionalism, and disunion in American history is an intricate one—there always existed the danger that the Union might become too big for itself and would splinter at the moment of its greatest power. (Something of this sort had happened as a consequence of the Great War for the Empire [1754–63], which led directly to the American Revolution; a comparable dynamic was at work, a century later, in the relationship between the Mexican War and the War Between the States.)[86] Jefferson may well have been right in thinking that no Constitution was so well fitted as America's for extensive empire and self-government, but the conquest of Canada would have violated the "federative principle" and greatly disturbed the existing balance among the sections. The old idea which saw the Union as one based on the consent of the governed had barely survived the incorporation of Louisiana; it would have been gravely endangered by a Canadian conquest.[87]

The benefits of exploiting the one great vulnerability that Britain presented to the United States were thus of an uncertain character. That the United States presented many vulnerabilities to Great Britain, however, was as clear as could be. Along the vast American seacoast lay a string of unprotected cities that were exposed to British sea power. The mobility of sea power made the task of defending the American coastline virtually insuperable, and even had the will been there to do so—which it was not—the task would in all probability have eluded even an energetic government. However this question of coastal defense presented itself, and whether the answer to it was sought in a seagoing navy or a system of fortifications and gunboats (which were the two leading alternatives), the enterprise had about it a paradoxical character. Acting alone, the United States could not hope to raise naval forces sufficient to contend with the Leviathan of the Deep. Only if the American Navy were employed in conjunction with French forces might it be possible to "render the state of the ocean no longer problematical," but in that case France would in all likelihood succeed to maritime dominion—a result, as Jefferson himself had once acknowledged, that would have increased the danger to American security.

Ultimately, these considerations were of little weight with men who had come to regard the rights of neutrality as inseparable from the maintenance of national honor and independence, and who saw a readiness to compromise on such issues as "submission." That neither war nor embargo was likely to attain the object at which they aimed and that the slim chance of success for either measure rested on tacit collaboration with Napoleon were dismissed when held up alongside the need to vindicate neutral rights. The issue of submission was a weighty one. More than any other consideration, it animated the War Hawks, who brought war to the country in 1812. In and of itself, it was an admirable sentiment: no nation, if it wishes to preserve its honor or self-respect, can abandon a position once it has identified the sacred rights and independence of the country with its vindication. To do so cannot fail to invite the contempt of other powers and the loss of national self-confidence. This truism, however, affords no guidance to the wisdom of the policy to which these intangible considerations of pride are attached, and it is there that one must find the real deformity of Jefferson's foreign policy.

For it is apparent that a different policy—in its essential features, one long recommended by Jefferson's domestic enemies—

would have produced a course of action that would not have entailed an abridgment of national independence or an insult to national honor. Had the administration publicly recognized that England was in truth engaged in a contest for public liberty and international order, and that by virtue of its own stance against Napoleon Britain protected the United States from the peculiar menace that Bonaparte embodied, a different attitude toward neutral rights would inevitably have followed. Jefferson would not say this because he did not believe it. A readiness to show restraint in the employment of British seamen on American warships and merchant vessels could have reduced substantially the difficulties over impressment; and a recognition that both neutral rights and the balance of power were better served by breaking down Napoleon and his Continental System than by coercion against Britain would have created the conditions for a satisfactory resolution of the crisis brought on by the 1807 Orders in Council—and probably would have avoided that crisis entirely. Cooperation with Britain on the high seas, of course, would inevitably have brought forth the withering opposition of Napoleon, and might have led to war with France. If war had come, however, it would have represented no threat to America's physical security: with the Atlantic intervening, and with the Royal Navy barring a direct Napoleonic strike at the United States, no such invasion was possible. The United States might have taken as much or as little of the war as it pleased, allowing shippers and merchants to calculate the balance of risk and reward raised by attempts to smuggle goods onto the continent.

Such a course of action represented the road not taken: the whole trend and design of Jeffersonian foreign policy was against it. Paradoxically, however, it was far more compatible with Jefferson's domestic vision and even with many of his cherished objects in foreign policy than the course of action that he did pursue. For it would have allowed the steady retirement of the national debt and been consistent with the deep aversion of the people to large military establishments. It would have provided for America's physical security, and would have served the commercial prosperity of the country much more than the course of action he did follow. It posed no obstacles to the Floridas; and it did not threaten the working out of the great republican experiment at home. Its only real drawback was that it tied the United States, as Gallatin had warned, to "the general result of the European war," but the truth is that the structure of American security—resting as it did on the need to deny to Napoleon the "empire of the seas"—was tied to that result in any case. In most other respects it would have constituted not the subversion but the fulfillment of Republican hopes.

It was not to be. War inevitably came, as did the consolidation of national institutions from which Jefferson recoiled in his old age. As a final irony, the course of action he pursued set him against the cause of liberty both at home and abroad. In national memory, no association comes more readily to mind than that between Jefferson and liberty; yet from the vantage point of 1808, nothing seems more absurd. For then a foreign policy dedicated to the vindication of neutral rights ranged the United States against the Spanish patriots fighting for their freedom against Napoleon, as it would later, under Madison's administration, place the United States against the whole European movement to throw off Napoleon's domination; at home it produced violations of civil liberties on a scale that would not be equaled in American history until the Civil War.

V

THE JEFFERSONIAN
LEGACY

24

The Role of a Democratic
Foreign Policy

THE JEFFERSON LEGACY in foreign policy has always been identified with a distinctive conception of the role that external
affairs ought to play in the American scheme of things. In this
view, the purposes and objectives of foreign policy may be properly understood only as a means to the end of protecting and promoting individual freedom and well-being. No end of foreign policy can be morally autonomous, self-justifying, an end in itself.
Instead, all the ends of foreign policy must be seen as means to
the ends of society, which are in turn ultimately the ends of individuals.

In this respect, as in so many others, the outlook of Jefferson
has been regularly contrasted with that of Alexander Hamilton.
Yet the principle that subordinated foreign to domestic policy was
not only characteristic of Jefferson and of those sharing his position. It did not as such distinguish Republican thought and conviction. Federalists, too, believed that the ends of foreign policy
were not morally autonomous or self-justifying but had to be seen
as means to the ends of society, that is, the ends of individuals.
Certainly Hamilton believed this. As much as Jefferson, he belonged to those who, in George Kennan's telling division, wish to
conduct foreign policy in order to live rather than live in order to
conduct foreign policy. When Hamilton replied to Charles Pinckney of South Carolina, in the Federal Convention, that the distinction Pinckney had sought to draw between a government capable

of making its citizens "happy at home" and one that made them "respectable abroad" was an "ideal distinction," he said nothing that supported the principle endorsing the primacy of foreign policy. The distinction was ideal, Hamilton argued, for the reason that "tranquility and happiness at home" depended on "sufficient stability and strength to make us respectable abroad." [1]

Although the ends of foreign policy must be viewed as means to the ends of society, circumstances may require giving primacy to these means. So long as the security and independence of the state are the indispensable means to the protection and promotion of individual and societal values, so long as the state remains the indispensable condition of value, there is necessarily a point at which foreign policy has primacy over domestic policy. "Safety from external danger," Hamilton observed in *The Federalist,* "is the most powerful director of national conduct. Even the ardent love of liberty will, after a time, give way to its dictates." [2] When the safety of the state is threatened, individual rights may have to be curtailed. In such moments, Hamilton insisted, external necessity may dictate that republics "run the risk of being less free."

Jefferson did not deny this assertion of the primacy of foreign policy after his experience as President, an experience that led him to alter in some measure the rigid views he had earlier entertained. Now he was prepared to acknowledge what he had once resisted, that circumstances which "constituted a law of necessity and self-preservation" might render the "*salus populi* supreme over the written law." [3] The public welfare might dictate measures curtailing individual rights. Unavoidably, he insisted, it was the responsible statesmen who had to determine when the need for such measures arose. "It is incumbent on those only who accept of great charges, to risk themselves on great occasions, when the safety of the nation, or some of its very high interests are at stake." In his several attempts to enforce the embargo, Jefferson acted in circumstances which he was persuaded left him no choice but to take measures that went far in suppressing personal liberties.

The difference in this critical respect between Hamilton and Jefferson did not arise over the desirability, in principle, of subordinating foreign to domestic policy but over the prospects for doing so in practice. Jefferson saw the circumstances marking the early life of the republic in a different light than did Hamilton, largely because of their contrasting views of the value of American commerce to the European powers. This conviction underlay the strategy of peaceable coercion Jefferson followed in his second administration, a strategy that, in the form of the embargo, led to such disastrous results. Had he been right about the nature of the American position, the new republic clearly would have been in a far stronger position in relation to the participants in the armed struggle then raging than proved to be the case. In the event, the embargo demonstrated that the leverage Jefferson had expected to enjoy from commerce was misplaced and that the attempt to exploit such leverage as the nation did possess worked a harder necessity than did war itself.

In undertaking the embargo, Jefferson had aspired to transcend the imperatives of a traditional reason of state. Indeed, the embargo was ultimately rooted in what was more than mere aspiration, for Jefferson came very close to indulging the pretension of actually having banished the old necessity. Yet it was the external face of reason of state that was banished and then only provisionally since war eventually was chosen. The internal face of reason of state was soon apparent in the form of the embargo's increasingly harsh domestic effects. Thus the pretense of transcending the old reason of state resulted in a more terrifying expression of this reason than in all likelihood would have been the case had he never entertained the pretension. For the logic of the embargo was to give a primacy to foreign policy that has remained to this day as onerous as any the nation has experienced. The measures taken to implement the embargo brought much of the economy to a virtual standstill while the measures taken to enforce it brought Massachusetts and Connecticut to the verge of rebellion. Yet great as the financial, political, and moral costs of the embargo were, it proved a failure. Taken to escape the alternatives of national humiliation or war, it led first to humiliation and ultimately to war. The system of war that Jefferson had hoped after hope to reform by the embargo was not reformed, and this despite his commitment to take any and all measures necessary to give peaceable coercion a fair trial. That in the end Jefferson came close to embracing measures which called into question almost every principle of government he professed to believe was the ironic though inevitable result of attempting to defy the limits imposed on the statesman.

In part, of course, this ironic outcome must be traced to events that were not of Jefferson's doing. In greater part, however, the dominant role of foreign policy during his presidency followed from the domestic vision that Jefferson and Republicans generally entertained, a vision that dictated an active and assertive foreign policy. Although Jefferson had a profoundly isolationist outlook, one that might have been expected to lead to passivity and avoidance in foreign policy, his vision of domestic society was contingent on the fulfillment of expansive goals. Until those goals were realized, he was driven to pursue a foreign policy that was anything but isolationist. By contrast, his great Federalist adversary, Alexander Hamilton, entertained a view of the nation's internal development that was far better suited to a policy of isolation. In the circumstances attending the American position at the time, it was the Republicans rather than the Federalists who had, by virtue of their domestic vision, a greater dependence on foreign policy and, accordingly, a greater need to accept the primacy of foreign policy. For it was the Republicans who defined domestic welfare and happiness in such a way as to make the realization of these ends— ends that were equated with the very continuity of the nation— dependent largely upon external change.

Did the primacy of foreign policy nevertheless have a meaning in Jefferson's statecraft that was essentially different from the meaning given it in the statecraft of the *ancien régime*? In its classic meaning, the necessities imposed by the vital interests of the state overrode all other values. In Jefferson's hands, the primacy of foreign policy took on much the same meaning. What he regarded as the necessities of the state and nation overrode the principles that appeared to jeopardize these necessities, including principles that had otherwise commanded his undeviating allegiance.

Thus it was that as President, Jefferson abandoned his constitutional scruples rather than run the risk that Louisiana, once acquired, might be lost as a consequence of adhering to them. An interpretation of the Constitution that he had before resolutely opposed and that he would again oppose was simply dropped in silence, never to be mentioned again, in the face of apparent threats to the treaty ceding Louisiana to the United States. Constitutional principle was abandoned even though the treaty of cession does not appear to have been seriously endangered. Napoleon was not yearning to tear up the treaty. The amendment Jefferson had initially sought to satisfy his grave doubts about the constitutionality of incorporating new territory into the Union, let alone in creating additional states out of such territory, did not pose a great obstacle for the Congress. That he nevertheless abandoned his commitment to strict construction rather than hazard even a modest risk of seeing the treaty fail is eloquent testimony to the dominance territorial expansion enjoyed over even the most strongly held principles.

The abandonment of constitutional principle in the case of Louisiana was but the most dramatic illustration of how Jefferson resolved conflicts between the commitment to territorial expansion and principles that stood in the way of this vital interest. The diplomacy pursued over the Floridas was perhaps less striking, though it was no less instructive in this respect. The claims made to West Florida demonstrated the ingenuity and single-mindedness of Jefferson and his administration in the cause of expanding the nation's territory. They scarcely demonstrated devotion to principle, particularly such principle as might operate to inhibit expansion. Spain, possessor of the Floridas, was both cajoled and threatened. The most tenuous of claims were insistently pressed on Madrid. The despotic ruler of France, who had only recently so threatened the security and well-being of the United States that Jefferson had contemplated tying the nation's fate to that of Great Britain, was now repeatedly appealed to for assistance in depriving Spain of its coveted possessions.

In his determination to acquire the Floridas, Jefferson also showed not the slightest hesitation in siding with Napoleon against the freedom of those who were resisting the great Corsican's efforts to reduce them to submission. Although the black people of Santo Domingo had by their desperate struggle successfully resisted Napoleon's plan to enslave them anew, and by so doing had prevented the French from undertaking at a critical point the military occupation in force of New Orleans, Jefferson showed no sympathy for their plight. Instead, he sought to appease Napoleon and to enlist support for American claims to West Florida by appearing to embargo American trade with the island. Again, when the Spanish nation in May 1808 rose up in resistance to Napoleon, Jefferson, while voicing in private his sympathy for Spain's struggle, did not seriously consider deviating from a policy that, in effect, supported the French effort. In this instance as well, a response that appeared indifferent to the cause of a nation's freedom was motivated in large part by the hope that Napoleon might at last favor the American aspiration to the Floridas.[4]

The diplomacy of Louisiana and the Floridas not only illustrates the extent to which expansion was Jefferson's reason of state;

it also illustrates that in pursuing expansion the primacy of foreign policy had a meaning and led to behavior distinguishable from European statecraft only in its reluctance to use force. The disjunction that emerged between his dread of force and his territorial and commercial ambitions was undoubtedly the most curious feature of Jeffersonian diplomacy. This disjunction notwithstanding, what distinguished the nation's subsequent territorial expansion was the astonishing good fortune that marked its course. An easy necessity prompted the belief then, as well as later, that there was no necessity at all. Such a belief came readily to the statesman who believed that "our interests soundly calculated will ever be found inseparable from our moral duties." For on this view, interests "soundly calculated" were not only harmonious but good. Evil was but mistaken calculation. There were, then, no true dilemmas in statecraft, in the sense of choices that required the serious sacrifice of value. Instead, what were viewed as dilemmas were simply the result of wrong reasoning over interests.

The persistence of this view, despite its evident shortcomings, is a testimony to its great attractiveness. Foreign policy is above all the making of difficult choices. For the possible ends of foreign policy are unlimited, whereas the power at the disposal of the statesman—even the statesman of a great power— is always limited. If statecraft did not impose difficult and painful choices, there would be no apparent limit to the ends that might plausibly be entertained. Was this not, however, Jefferson's message to a receptive public: that ends need not be limited by means and that painful choices need not be made? In the period leading up to the embargo, never once did Jefferson tell the nation that some of its interests would have to be sacrificed (at least, for the time being), that they were simply excessive and could not be satisfied, and that a choice would have to be made between preparing for war to defend at least some of these interests or substantially modifying them. Whether out of expediency or conviction, Jefferson was not prepared to urge a consistent policy on the nation. Although he refused to prepare seriously for war, he also refused to modify a position that could only have war as its outcome. In so acting, he faithfully reflected public opinion.

In challenging the belligerents' measures for controlling neutral commerce, Jefferson was ultimately driven to intervene in the great contest between France and England. That he did not do so as a belligerent does not alter the interventionist character of the course that was taken. Intended to serve as an alternative to war, the embargo, had it only been successful, would have just as surely affected the outcome of the war as the choice of belligerency. It failed and the system of war that Jefferson had hoped after hope to reform by the embargo, that "great engine for national purposes," was not reformed. Instead, a remorseless necessity drove him from one repressive measure to the next in his effort to escape from the terrible dilemma that was, in part, of his own making.

It was only later in the nineteenth century, when the objectives of Jefferson's policies on land and at sea were finally realized, that a policy of isolation could be given effective expression and the primacy of domestic policy realized both in principle and in practice. The external world could be ignored only after expansion over the continent had been completed and an open trading system at sea had at last been achieved. Even then, the complete subordination of foreign policy rested as much on a favorable balance of power as it did on the fortunate circumstances of the American position. A seemingly effortless policy of isolation was the result of exceptionally benign circumstances, the persistence of which over a long period persuaded so many that what was, after all, a transient state of affairs represented a permanent condition.

What does indeed appear to be an enduring trait, however, is Jefferson's conception of the role that foreign policy ought to play in the American scheme of things. The view that the purposes and justification of foreign policy are the promotion of domestic well-being and happiness remains a matter of general agreement today, as it was in Jefferson's time. It does so despite the great significance of foreign policy in the nation's life since World War II. But if a consensus persists on the primacy that in principle ought to be accorded domestic policy, differences also persist over the circumstances in which, as a practical matter, domestic concerns may be accorded primacy over foreign policy. Now as in Jefferson's day, these differences continue not only because of disagreement over the nature and immediacy of external threats to American security and well-being. They persist as well because of the aspirations entertained by the nation and its leaders. In the course of the postwar era, administrations have entertained interests that were certainly no more a matter of necessity than Jefferson's need to expand territorially and to reform the international commercial order. These interests—most of which went well beyond maintaining a balance of power—may be seen as the functional equivalents of Jefferson's interest in expansion. In practice, their pursuit often resulted in the subordination of domestic to foreign policy. In one way or another, they were virtually all expressions of the American purpose. That purpose, articulated by Jefferson at the outset of the republic's history, is to promote the cause of freedom in the world.

It has been in pursuit of that purpose almost as much as it has been in pursuit of a conventional security interest that domestic policy has in practice been subordinated to foreign policy. So long as security is conceived in a traditional and restricted manner, as largely a function of the balance of power, the ultimate primacy of foreign over domestic policy need not detract significantly from the normal order of things in which domestic happiness and welfare are primary. This is particularly so for states which enjoy, by virtue of relative power and geography, a highly favorable measure of security. Even for such states, however, the commitment in principle to the primacy of domestic policy may mean little in practice if security is interpreted as a function both of a favorable balance of power between states and of the internal order maintained by states, that is, as a favorable ideological balance. Given that commitment, foreign policy may come to dominate over domestic policy in a way that is all-pervasive.

While this condition has never yet been experienced, even in the years of the classic Cold War, on more than one occasion in recent decades it has been approximated. It was perhaps most closely approximated in the 1960s, during the years of the war in Vietnam, when the disposition to define American security in terms of a purpose that went well beyond conventional security requirements achieved its fullest expression. But if this disposition has waxed and waned, it has also persisted, for the belief from which it springs—that until the world is made safe for the institutions of freedom, that is, until the world is enabled to emulate the American example, these institutions cannot be entirely safe at home— is deeply rooted. Its most significant early expression must be found in the thought of Jefferson, who is the *locus classicus* in the American tradition both of the principle subordinating foreign to domestic policy and of the commitment that has posed the greatest domestic challenge to this principle.

25

The Isolationist
Impulse

FOR MOST OF HIS LIFE, Jefferson championed a policy of isolation for the new nation. In doing so, he joined his voice to
those of many others. The desire to pursue a political destiny separate from Europe enjoyed a virtual consensus among the founding generation of American statesmen. It was given repeated
expression in the period immediately preceding the War of Independence. In *Common Sense*, Thomas Paine held out the promise
of a destiny separate from Europe and Europe's wars as one of
the great attractions of independence. In 1783, at the end of the
war, in rejecting negotiations over American entry into the League
of Armed Neutrality, the Congress declared that the true interest
of the thirteen states required "that they should be as little as possible entangled in the politics and controversies of European nations."[5] The admonition was frequently repeated in the years between the close of the war and the constitutional convention. It
was put forth on behalf of the cause of union. The absence of
union, Hamilton argued in *The Federalist,* left the Old World in
the position of dictating the terms of the connection with the new.
But let the states unite, he urged, and the situation would be reversed. Then Americans might "disdain to be the instruments of
European greatness."[6]

The classic expression of the desirability of pursuing a policy
of isolation from Europe is of course Washington's Farewell Ad-

dress. "The great rule of conduct for us in regard to foreign nations," Washington declared,

> is, in extending our commercial relations to have with them as little *political* connection as possible. . . . Europe has a set of primary interests which to us have none or a very remote relation. Hence she must be engaged in frequent controversies, the causes of which are essentially foreign to our concerns. Hence, therefore, it must be unwise in us to implicate ourselves by artificial ties in the ordinary vicissitudes of her politics, or the ordinary combinations and collisions of her friendships or enmities. Our detached and distant situation invites and enables us to pursue a different course.

The different course advised was "to steer clear of permanent alliances with any portion of the foreign world, so far . . . as we are now at liberty to do it" and to "safely trust to temporary alliances for extraordinary emergencies."[7]

Drafted by Hamilton, Washington's advice was cast in qualified and tentative terms. The "great rule of conduct" was not set forth as an absolute principle that represented, whatever the circumstances, the timeless interests of the nation, but as a policy for a state of only modest power that was consolidating a newly won independence and a still precarious security. It expressed what Washington and Hamilton considered the true interest of the nation in the given circumstances ("our detached and distant situation"). It reflected no expectation that the system of European politics might soon be changed. Instead, it intimated that this system would persist and that this country should take such advantages as it could from "so peculiar a situation."[8]

Was the outlook embodied in the Farewell Address the outlook of Jefferson as well? Certainly, Jefferson may reasonably be cited to the same effect with respect to policy. "I am for free commerce with all nations," he wrote in early 1799; "political connection with none; and little or no diplomatic establishments. And I am not for linking ourselves by new treaties with the quarrels of Europe."[9] Later, as President, he expressed himself in terms that bear a distinct similarity to Washington's Farewell Address:

> Separated by a wide ocean from the nations of Europe, and from the political interests which entangle them together, with productions and wants which render our com-

merce and friendship useful to them and theirs to us, it cannot be the interest of any to assail us, nor ours to disturb them. We should be most unwise, indeed, were we to cast away the singular blessings of the position in which nature has placed us, the opportunity she has endowed us with of pursuing, at a distance from foreign contentions, the paths of industry, peace, and happiness; of cultivating general friendship, rather than of force.[10]

Jefferson acknowledged as well Washington's distinction between permanent and temporary alliances. He recognized, if only by his actions, that to abjure any and all political connections might prove very difficult, even impossible, and that a strict policy of isolation represented as much aspiration as it did reality. So long as the new American state did not enjoy security of its frontiers, so long as it had not obtained possession of the Mississippi valley and of the great waterway providing the outlet to the Gulf, it could not rule out the prospect of an alliance with a European power. Jefferson could not do so during the Louisiana crisis, though the seriousness with which he contemplated an alliance with Great Britain remains an open question. (Without doubt, he seriously contemplated an alliance that "entangled" Great Britain by obligating it to a certain course of action. Whether he ever reconciled himself to an alliance that entangled America is another matter.)

Even after having secured Louisiana, Jefferson appeared momentarily to consider the possibility of an alliance with Great Britain. He did so in 1805 when his quest for the Floridas was resisted by the joint power of Spain and France, only to abandon the prospect when the price he might have to pay for English assistance became apparent. As in the Louisiana crisis, his alliance policy in 1805 was simplicity itself: Though the British were to pursue a policy of alliance toward us, we were to pursue a policy of isolation toward them.

Nearly fifteen years after his retirement from the presidency, Jefferson had a final occasion to consider a political connection with Great Britain. Toward the end of 1823, he was informed by his Republican successor James Monroe that the American government had received from Great Britain a proposal to join in a common declaration warning the European powers of the Holy Alliance not to attempt the reconquest of Spanish America or the transfer of any part of it from Spain to another power. The danger of European intervention appeared substantial. Monroe, though conscious that a joint declaration was contrary to the avoidance of foreign entanglements, was inclined to cooperate with Great Britain. Before embarking on such a novel course, however, he wanted the advice of his Republican predecessors, Jefferson and Madison.

In his reply to Monroe, Jefferson observed that the course to be taken by the nation ought to be consistent with two fundamental principles—the first, "never to entangle ourselves in the broils of Europe"; the second, "never to suffer Europe to intermeddle with Cis-Atlantic affairs." Nevertheless, he believed, with Monroe, that the United States ought to accept the British proposal: "The war in which the present proposition might engage us, should that be its consequence, is not her war, but ours. Its object is to introduce and establish the American system, of keeping out of our land all foreign powers, of never permitting those of Europe to intermeddle with the affairs of our nation." [11] As it turned out, no such common declaration was made. By the time Jefferson wrote his famous letter to Monroe, George Canning, the British Foreign Minister, had discontinued his conversations with Richard Rush, the American minister to London, and reached an understanding with France by which the latter disclaimed the ambitions against which the joint declaration would have been directed. The American government nevertheless pressed forward and decided to issue a unilateral declaration in President Monroe's annual message to Congress in December. That declaration, known to history as the Monroe Doctrine, bore unmistakable marks of Jefferson's influence.

In this context, what is of interest in Jefferson's response is his frank acknowledgment of the nation's dependence on British sea power for its security, yet his insistence that no price be paid for this dependence. "Great Britain is the nation which can do us the most harm of anyone, or all on earth," Jefferson acknowledged, "and with her on our side we need not fear the whole world." The reason was simple enough: "With Great Britain withdrawn from their scale and shifted into that of our two continents, all Europe combined would not undertake such a war. For how would they propose to get at either enemy without superior fleets?" Still, Jefferson was adamant about the price he was willing to pay—or, rather, not to pay—for the support of British sea power. Although

the war in which Great Britain and the United States might once
again fight "side by side, in the same cause" was "not her war,
but ours," to reciprocate for British support "at the price of taking
part in her wars" was the last thing he was willing to consider.

The significance accorded British sea power affords one basis
for distinguishing Jefferson's isolationism from that of Hamilton
and Washington, though Federalists invariably preferred informal
cooperation to formal alliance with Great Britain. Another and
more significant basis for distinguishing Jefferson's policy of isolation from that reflected in the Farewell Address was the central
importance given to the avoidance of war. In Republican thought,
war was the great evil to be feared above all. It was, as Madison
had once pointed out, the enemy of public liberty the most to be
dreaded. From it, most other evils could be traced. "No nation,"
Madison had warned, "could preserve its freedom in the midst of
continual warfare." [12] These were, for Republicans, self-evident
truths; Jefferson believed in them no less fervently than did Madison. Indians and Barbary powers apart, it was the idea that peace
depended in the last analysis on a willingness to use force that
Jefferson could never quite bring himself to accept. The aversion
to war might still be overcome, but only in the most compelling
of circumstances. The desirability of a policy of isolation owed its
strength, above all, to this profound aversion, for alliances were
one of the great means and avenues by which war was entered
into.

Jefferson also feared the infection expected to result from the
abandonment of a policy of isolation, though this fear was not
peculiar to him alone. His political adversaries also believed that
one danger of foreign entanglement was the corrupting effect it
might have on American institutions and society. For Hamilton
and the bulk of Federalists, this fear was entertained toward any
relationship of intimacy with revolutionary France; for Jefferson
and the Republicans the dread of infection was aroused by the
prospect of closer ties with Great Britain. At least this was true of
Jefferson until the late 1790s. By then he had come to conclude
that any political connection held out the threat of infection and
was to be avoided though the cause of liberty itself be at stake. In
March of 1799 he wrote to a friend: "I sincerely join you in abjuring all political connection with every foreign power; and tho I
cordially wish well to the progress of liberty in all nations, and

would forever give it the weight of our countenance, yet they are not to be touched without contamination from their other bad principles. Commerce with all nations, alliance with none, should be our motto."[13]

Jefferson would not have expressed himself thus in the early years of the decade; he would not have spoken in terms that implied the moral equivalence of "all nations," including France. Then, as Secretary of State, he had equated the fate of the French Revolution with that of liberty everywhere. Were France to go down before the might of the First Coalition, he had believed, its defeat could be expected to result in the permanent ascendancy of the enemies of liberty in America. The cause of republicanism would seriously decline and perhaps even perish. To ward off this dread prospect, Jefferson was willing to overlook the methods of the terrorists in France.[14]

It was largely out of this same conviction that Jefferson wished to intervene in the conflict, not by abandoning America's neutral status but by insisting on a view of neutral rights that would work to France's distinct advantage. That the course he favored would presumably also accomplish the goal of freeing this country from Britain's economic domination made it doubly desirable. But Washington did not respond in the way Jefferson desired. Unwilling to play a high-stakes game that might draw the country into war, he did not accept Jefferson's strategy of making American neutrality contingent upon a British acceptance of an expansive definition of neutral rights. Nor did Congress adopt the Republicans' proposals to discriminate against British commerce. Confronted by a possible war with Great Britain, the Washington administration ultimately settled on a less ambitious understanding of neutral rights than Jefferson desired and entered into the accommodation with England—the Jay Treaty—that he detested.

The year 1793 marks Jefferson's first attempt—tentative and qualified as it was—to intervene in the great conflict brought on by the French Revolution. The second occurred in 1807–9 and took the form of the embargo. As before, a particular interpretation of the rights of neutrals was championed that tilted heavily in the direction of France. In this second effort, however, the hope and expectation that informed the first was gone. With the coming to power of Napoleon, Jefferson no longer found moral importance in the European war. A conflict once endowed with profound moral significance, the outcome of which he had identified with the future of liberty, was now viewed as a mere struggle for power between the "tyrant on land" and the "tyrant of the ocean." How should the "common herd of cattle" look on the renewal of war in Europe, he asked Benjamin Rush in the fall of 1803: "With no partialities, certainly. If they [the belligerents] can so far worry one another as to destroy their power of tyrannizing, the one over the earth, the other the waters, the world may perhaps enjoy peace, till they recruit again." [15] These words were written scarcely six months after the "tyrant of the ocean" had been seen by the President as the last hope for preventing a French occupation of New Orleans.

Jefferson continued to insist on the moral equivalence of France and Great Britain to the end of their armed struggle. In his view, there was no room for the consideration that the one belligerent— the "tyrant of the ocean"—was defending the international order within which America could enjoy a basic security, whereas the other belligerent—the "tyrant of the land"—was intent on destroying this order and the security of the states that formed it. Nor was this refusal to differentiate between the belligerents essentially qualified by Jefferson's occasional bow to balance of power considerations. It was one thing to profess concern over the need to maintain a balance of the great European powers and another to reflect this concern in behavior. Whatever Jefferson's occasional words, his acts did not reveal a diplomacy based on the balance but one based on a particular interpretation of neutrality. In that interpretation, the neutral's rights were the analogue in war of a liberal system of trade in peace. Given that interpretation, there could be no other outcome than a policy which objectively favored the "tyrant on land."

Though separated by a decade and a half, the two critical episodes illuminate the tension in Jefferson between the desire to reform the international system and the desire to remain separate from it. Reformation could come only as a result of successfully imposing one's will on the system, or, at least, on a significant part of it. But this could prove difficult and dangerous. Men and nations being what they are, the world is resistant to reform. Whatever their initial intention, those bent on reform, and therefore on intervention, have generally had to resort to the sword. This Jefferson did not want and had no intention of doing. Force threatened the very interests for which he had sought reform.

The result of this reasoning could only lead to withdrawal. If the state system could not be reformed, it had to be abandoned. Then Jefferson could speak of the desirability of "Chinese isolation" and of the need to place "an ocean of fire between us and the old world," though he knew these metaphors expressed an impossible ideal. The need to trade would persist. Eventually, Jefferson believed, that need could be satisfied on America's terms. For the time would surely come, and in the not too distant future, when this country might impose its view of neutral rights on belligerents. Not long after he became President, Jefferson wrote these revealing words to one of his protégés:

> It would indeed be advantageous to us to have neutral rights established on a broad ground; but no dependence can be placed in any European coalition for that. They have so many other bye-interests of greater weight, that some one or other will always be bought off. To be entangled with them would be a much greater evil than a temporary acquiescence in the false principles which have prevailed. . . . If we can delay but a few years the necessity of vindicating the laws of nature on the ocean, we shall be the more sure of doing it with effect. The day is within my time as well as yours, when we may say by what laws other nations shall treat us on the sea. And we will say it. In the meantime, we wish to let every treaty we have drop off without renewal. We call in our diplomatic missions, barely keeping up those to the more important nations.[16]

Events were to show that it was not within Jefferson's time to say by what laws other nations should treat American commerce on the sea. The attempt to say just that a few years later in the form of the embargo failed. Nor did the subsequent attempt to say by war how American commerce should be treated succeed. The War of 1812 with Great Britain ended with the British reserving their position on every contested issue of neutral rights. Indeed, the "laws of nature on the ocean" were not vindicated a century later in World War I, when the United States clearly had achieved the position of dominance Jefferson had expected to occur in but a few years.

Throughout, Jefferson's isolationism reflected an outlook that was unwilling to come to terms with the political world of his time. In its essential features, this outlook expressed a true isolation, a real separateness, from the international system. In this vital respect, as in others, it was profoundly parochial. Unable to adjust to the existing world, it pointed either to the withdrawal from that world or to the attempt to reform it by imposing one's will on it. Jefferson's presidency illustrated the perils of both courses.

The perils of isolation were illustrated by the Louisiana crisis. A passion for peace reflecting a profound aversion to war led to the assumption that the external world did not impose a necessity that might require war. As matters turned out, Louisiana issued in a striking success, but the risk taken was great. At best, Louisiana taught that a strategy of playing for time sometimes works, despite the dangers it may incur. Contending forces may balance themselves out. A dangerous adversary may get overcommitted elsewhere. But this modest lesson was not the lesson Jefferson appeared to have learned from his one success in foreign policy. Instead, he learned that passivity and a reliance on the power of others was the course of wisdom. Not surprisingly, Louisiana was not followed by similar diplomatic triumphs.

The perils of intervention were illustrated by Jefferson's penchant, amounting at times to an obsession, to define American rights in such a manner as to persuade himself, and the nation, of their purity—but a definition that failed to take adequate account of the interests of others. Almost without realizing it, he defined neutral rights in an expansive way. Put forth usually in terms of abstract principle, they were difficult to compromise. Their very character, moreover, served to obscure the reality that, in the guise of neutrality, Jefferson in effect intervened in the war between France and England. Employed in part to force open the maritime system, the embargo was then persisted in as a means of withdrawal from the system. Thus the cycle of rejection, reform, and withdrawal was completed, with all of the dreadful consequences that followed.

Jefferson was the first president who sought to reform the international system. He was not the last. In the history of the nation's encounter with the world, the themes of withdrawal and reformation are deeply embedded and apparent throughout. They have formed the enduring characteristics of the American outlook on the world. For Jefferson, as for subsequent American statesmen, the desire to change the world was at war with the desire not to be corrupted by the world. The desire to change the world sprang, for the most part, from the conviction that only in a changed

world could republican institutions in America flourish and be secure. But the very attempt to change the world incurred the risk
of contamination by it, for the methods by which it had been
changed in the past were those that held out the greatest threat to
republican institutions.

26

Jefferson and Liberty:
Exemplar or Crusader?

O F THE NATION's enduring traditions that Jefferson so influenced, none is more striking in its significance than the deeply
ingrained inwardness of national feeling that marks to this day the
American outlook. Foreign observers have always been impressed
by this trait, which they have not hesitated to identify with the
parochialism of Americans. That it should be traced in part to so
cosmopolitan a figure as Jefferson cannot but appear paradoxical.
Yet in this as in so many other respects there were two quite different sides to Jefferson, just as there have been two quite different
sides to the nation. A vision of man's future that was as grand as
it was timeless was joined to a view that seemed not only unwilling but almost unable to transcend, however modestly, the particular interests of the state. This combination of the universal and
the particular is bound to create difficulties when applied to the
realm of diplomacy. For the outlook that informs a successful diplomacy must fall somewhere between these two perspectives. The
grand vision is too elevated to make meaningful and effective con-
tact with political reality, while the parochial view is too self-
centered to achieve the kind of compromise diplomatic solutions
normally require. Jefferson's diplomacy nevertheless encompassed
both of these perspectives and on more than one occasion sought
to combine them in a manner that would give the nation's particular interests a universal significance.

The maritime crisis illustrates these two sides of Jefferson's diplomatic outlook just as it illustrates his almost inveterate propensity to convert issues of interest into matters in which great moral principles were held to be at stake. The conflict with England over impressment and neutral rights was marked from first to last by a view that seemed incapable of giving serious consideration to any interests, let alone to the possible legitimacy of any interests, save those of the neutral. That England was engaged in a war for its very survival as a great power, that the ferocity of the hegemonic struggle with France made British adherence to neutral rights as defined by the Jefferson administration difficult to reconcile with survival, and that on the outcome of the conflict depended the preservation of the balance of power, which was the only safeguard, inadequate though it had always been, of any neutral rights—all of these considerations counted for very little against the interests America insistently advanced. Unwilling to consider these interests in relation to a larger diplomatic constellation, Jefferson remained equally unwilling to consider the kinds of compromise that might have substantially eased, if not entirely resolved, the conflict with Great Britain. Such compromise was viewed as a betrayal not merely of interest but of "the laws of nature on the ocean." Even had he not equated America's maritime interests with national honor and independence, there was no room for normal diplomatic give and take in the position Jefferson came to embrace. For that position identified the pursuit of self-interest with the vindication of sacred right.

It was not only the maritime crisis with England that revealed these traits. They were equally apparent in the diplomacy of territorial expansion. No chapter of American diplomacy would appear to lend itself less to the category of a morality tale than the diplomacy of the Floridas. From start to finish it was attended by threats made on behalf of what were, at best, dubious claims. Yet it is not so much the diplomatic tactics employed against Spain that are revealing but the moral judgment Jefferson made of those tactics (which had stopped short of the use of force). They reflected, he insisted, America's respect for the France of Napoleon and, of course, this nation's sense of forbearance even against a Spanish nation that had supposedly, in the course of the dispute over the Floridas, set a record in perfidy and injustice in resisting American demands to dispossess Spain of the disputed territory.

What can account for a diplomatic outlook that combined such disparate perspectives? In large measure, surely, the explanation must be found in the conviction that identified the nation's fate with the fate of freedom in the world. If the security and well-being of the United States were inseparable from the prospects of free government everywhere, as Jefferson was so deeply per-suaded, it followed that American interests were invested with a sanctity that exempted them from the kind of compromise en-demic to diplomacy. If America was the last best hope for the cause of freedom in the world, it was apparent that the justice and rectitude of its diplomatic behavior followed by virtue of this his-toric role. The combination of universalism and parochialism is the result of a self-consciousness over role that forms a constant in the nation's history.

A belief in the justice of American behavior does not settle the issue of what that behavior should be. A conviction that the role of the nation is to promote the cause of liberty does not settle the issue of how that cause is to be served. Nor was there need to resolve this matter in the early history of the republic. Whether the United States should serve as an example or a crusader on freedom's behalf could scarcely prove a meaningful question then, given the precarious position of the new nation and the modest power at its disposal. It is only in the twentieth century that this question has taken on a meaning and relevance it could not earlier possess.

If the question—exemplar or crusader—is nevertheless raised in considering the legacy of Jefferson to American foreign policy, it is because it is difficult to conclude that in this critical area his influence has been irrelevant. Most students of Jefferson's thought incline to the view that the crusading impulse was, in Gilbert Chi-nard's words, "utterly foreign to his temperament."[17] Julian Boyd has written in a similar, though less categorical vein in stating that the Empire of Liberty was intended to hold sway in the world not by arms or political power "but by the sheer majesty of ideas and ideals. The great republican experiment would be an example and a beacon, unsheltered and unafraid of the light of truth."[18] And a more recent work on Jefferson's attitude toward war concludes that his "strongly pragmatic approach to foreign policy meant that he eschewed international crusades."[19]

These judgments are broadly shared by Jefferson scholars. The man who championed the primacy of domestic over foreign policy, who harbored such profound isolationist tendencies, and who avowed that the attempt on our part to reform Europe by war would only "prove us maniacs of another character" would hardly seem disposed to support under any circumstances the role of crusader, even a crusader on behalf of freedom. And yet doubt must persist over the Jeffersonian view of the nation's role. He certainly appears to have eschewed the role of crusader at the end of his life. Earlier, however, during the first years of the French Revolution, he was caught up in the enthusiasm of the moment and explicitly made the connection that all American crusaders made: that the cause of liberty at home was indissolubly linked to its fortunes abroad.

The case for concluding that Jefferson intended the nation to serve in the role of exemplar of freedom rather than crusader for freedom may be summarized thus. Given what he saw as the conditions of other peoples and the circumstances marking their development, Jefferson became increasingly skeptical over the prospects for the spread of liberty. Freedom was a plant, he often observed, that could grow and flourish only in a favorable environment. That environment was present in America, but could it be found—or developed—elsewhere? As Jefferson grew older, he entertained growing doubt. That doubt was not consistently adhered to even in later years; from time to time his innate optimism would have out.[20] Still, his disillusionment over the course of the French Revolution ended the only real "enthusiasm" he ever entertained respecting the imminent prospects for liberty among the nations of Europe. Although he could still write in 1802 to Thomas Cooper, with respect to the course of events in France, that he was "willing to hope, as long as anybody will hope with me,"[21] in the years that followed he gave up much of his earlier hopes for, and certainly his expectations in, freedom's progress outside the boundaries of his own country. Of the nations of Europe, he wrote to Monroe in 1823, "All their energies are expended in the destruction of the labor, property and lives of their people."[22] Thus it was, thus it had always been, and, Jefferson finally came close to believing, thus it always would be. A continent whose nations were doomed—if only by virtue of their proximity—to never-ending rivalries and wars was one that also afforded small prospects for the development of free institutions.

Nor did he think these prospects any better among the peoples of this hemisphere. "What kind of government will they establish? How much liberty can they bear without intoxication?"[23] These questions Jefferson put to Alexander von Humboldt, at a time when the revolutionary movement in South America promised to result in several independent nations. The answer he eventually gave was pessimistic. While wars were the "natural state of man" in Europe, in South America the new nations were made up of "priest-ridden" peoples.[24] Writing in 1818 to John Adams, who shared his views on this matter, Jefferson observed that while the peoples to the south "will succeed against Spain . . . the dangerous enemy is within their own breasts. Ignorance and superstition will chain their minds and bodies under religious and military despotism."[25] At the same time, he went on to declare that "it is our duty to wish them independence and self-government, because they wish it themselves; and they have the right, and we none, to choose for themselves, and I wish, moreover, that our ideas may be erroneous, and theirs prove well-founded."

Even if Jefferson had seen the world as being more receptive to the institutions of freedom than he did in his later years, there still would have remained strong reservations on his part to taking up the role of crusader for freedom. Any role that went beyond example necessarily incurred the risk of war. But that risk in turn raised prospects—executive aggrandizement, debt, taxes, and so on—Jefferson believed were fatal to republican institutions. The role of the crusader was to change the world. Yet even for the powerful crusader, the effort to reform a world resistant to change raised the prospect of corruption. Jefferson's sensitivity to that prospect and the intensity of his desire to escape it were proverbial.

This is the crux of the case for concluding that Jefferson saw the nation in the role of exemplar rather than crusader. Indeed, there is some difficulty in equating these elements even with support for the role of example to the world. Jefferson was quite certain that the circumstances conditioning our existence as a people would be very difficult to approximate, let alone duplicate. This was, in large part, precisely why we were exceptional. It was also why we were to serve as an example to the world. Yet if being exceptional implied not only a state that was relatively better than others but one that, because of distinctive circumstances, was impossible to approximate elsewhere, it is difficult to see how the nation could even serve in the role of example. For then, there was no "elsewhere"; there was no place where the nation's example might prove relevant. This is why Jefferson could write, as he did in 1802 to Joseph Priestley, "It is impossible not to be sensible that we are acting for all mankind; that circumstances denied to others, but indulged to us, have imposed on us the duty of proving what is the degree of freedom and self-government in which a society may venture to leave its individual members."[26] This is also why some historians have insisted that our historic role as example has been misconstrued in the post–World War II period, that our example was initially confined within our borders, and that it was intended to be emulated only by those who chose to come here.[27]

These considerations weigh heavily against any attempt to make of Jefferson the crusader for freedom in the world. And yet a case for doing just that can be plausibly made. There is, to begin with, the consideration that Jefferson did not consistently find the circumstances of American society so exceptional as to preclude their emulation elsewhere. If he could express what he did to Joseph Priestley, he could also express only fifteen months before, in 1801, the conviction to John Dickinson that "a just and solid republican government maintained here, will be a standing monument and example for the aim and imitation of the people of other countries."[28] Although he did indeed become increasingly skeptical over the prospects for freedom elsewhere, his skepticism never quite overcame a deeper and congenital optimism. The experience of the French Revolution had been painful, in ways even shattering, but neither it nor subsequent events led Jefferson to abandon his deepest hopes and beliefs in the prospects for freedom. To John Adams he could write in 1816: "I steer my bark with Hope in the head, leaving Fear astern. My hopes indeed sometimes fail; but not oftener than the forebodings of the gloomy."[29] While declaring to one correspondent in 1820 that Europe was a region "where war seems to be the natural state of man,"[30] he could also write on the same day to La Fayette that he was still hopeful "the disease of liberty is catching."[31] And despite all his doubts about the peoples to the south being ready for freedom, he still thought that they might at least obtain freedom "by degrees . . . because that would . . . bring on light and information, and qualify them to take charge of themselves understandingly."[32]

Then, too, there were very important reasons for not abandoning hope in the prospects for freedom in the world. If one did concede that these prospects were negligible, could liberty be secure even in this country? In a world that was ruled by arbitrary power, the fate of free institutions in America would never be quite assured, however much we might try to isolate ourselves from the world. For monarchy meant war and, despite our best efforts, Europe's wars might always spread their deadly virus to this hemisphere. Only a world made up of republics would be a world where peace was truly possible. The great and indispensable step toward promoting a lasting peace, as Madison had written in his essay on "Universal Peace," was the replacement of monarchical governments by governments that rested on consent.[33] A world made safe for republican government was a world made safe for, as well as by, peace.

The logic of arbitrary power was not only that of war; it was also the logic of a closed system. By contrast, the logic of republics was not only that of peace; it was also the logic of an open system. The enduring issue that thrust an unwilling Jefferson onto the world was not political or ideological but commercial. Persuaded that the health and well-being of the American republic required an open—that is, free—trading system, Jefferson also believed that he could isolate the commercial interest from the political entanglement he was determined to avoid. But events were to show that the insistence on preserving an open trading system entailed the need to intervene against those whose efforts were directed to keeping the system closed. It did so then just as it did so again in the years that led to American intervention in the First World War.

Certainly Jefferson never consciously contemplated the role of crusader for the nation. He did not do so in his own day and for the most apparent and compelling of reasons. But what of the day that he was sure would come, a day when we might "shake a rod over the heads of all" and do so with relative impunity?[34] Why should we not do so then, if doing so might contribute to our security and well-being while also striking a blow for the cause of liberty? It was one thing to reject "a war to reform Europe" and quite another to eschew shaking a rod over the heads of all. The former implied a course of action that might always prove disastrous for us internally, a prospect that necessarily outweighed that of attempting to replace arbitrary power elsewhere with government based on consent. The latter held out a far different course, though, one that promised reformation of the international system at but modest cost. Such a course was in the tradition of "peaceable coercion," and it is difficult to understand why Jefferson should reject it.

To the degree that Jefferson never abandoned his earlier conviction of America serving as an example to the world, the temptation persisted to equate, at some point and in some circumstances, the role of example with more than just a passive stance. The conventional contrast of the roles of exemplar and crusader has often obscured the affinity that may also exist between the two. A marked self-consciousness about serving as an example to the world may well act, under the proper circumstances, as a standing temptation to go beyond that role. The same sentiments that find gratification or fulfillment by serving in the role of example also sustain at some point the role of crusader.

On the issue of the nation's proper role in the world, then, Jefferson's legacy remains ultimately ambiguous. It is this ambiguity, among others, that lends Jefferson's name to such conflicting uses in the never-ending debates over American foreign policy. Among the statesmen of the early republic, he is more responsible than any for warning of the hazards that must attend the role of crusader. Yet he is also the statesman who is more responsible than any for evoking the perennial attractions of this role.

Notes

Preface

1. Adam Smith, *An Inquiry into the Nature and Causes of the Wealth of Nations* (New York, 1937 [1776]), 587–88.

Part I. An American Statesman

1. Henry Adams, *History of the United States of America during the Administration of Jefferson and Madison* (New York, 1903 [1889–91]), I, 277.

2. The Examination No. 3, Dec. 24, 1801, Harold C. Syrett, ed., *The Papers of Alexander Hamilton* (New York, 1961–79), XXV, 467, in which Hamilton speaks of "indolent and temporising rulers, who love to loll in the lap of epicurean ease, and seem to imagine that to govern well, is to amuse the wondering multitude with sagacious aphorisms and oracular sayings."

3. Merrill D. Peterson, *Thomas Jefferson and the New Nation: A Biography* (Oxford, 1970), 822.

4. Dumas Malone, *Jefferson and His Time* (Boston, 1948–81), XVI, 79.

5. Gilbert Chinard, *Thomas Jefferson: The Apostle of Americanism,* 2nd ed. (Ann Arbor, Mich., 1957 [1st ed.]), 275, 382. See also Lawrence Kaplan, *Entangling Alliances with None: American Foreign Policy in the Age of Jefferson* (Kent, Ohio, 1987).

6. "His dream," Adams writes of the American, "was his whole existence. The men who denounced him admitted that they left him

in his forest swamp quaking with fever, but clinging in the delirium of death to the illusions of his dazzled brain." *History*, I, 174.

7. Jefferson to "Citizens of Washington," March 4, 1809, Andrew A. Lipscomb and Albert Bergh, eds., *The Writings of Thomas Jefferson* (Washington, D.C., 1904–5), XVI, 347–48.

8. Jefferson to William Branch Giles, Dec. 26, 1825, Lipscomb and Bergh, eds., *Writings*, XVI, 148.

9. On the relationship between ideals and institutions, see Samuel P. Huntington, *American Politics: The Promise of Disharmony* (Cambridge, Mass., 1980).

10. This is how Merrill Peterson characterizes the theme that suffuses the work of Julian Boyd's *Jefferson Papers*. Peterson, Book Review, *The William and Mary Quarterly*, 3rd ser., 23 (October 1975), 658.

11. Leonard Levy, *Jefferson and Civil Liberties: The Darker Side* (Cambridge, Mass., 1963).

12. Jefferson to Benjamin Rush, Sept. 23, 1800, Paul Leicester Ford, ed., *The Writings of Thomas Jefferson* (New York, 1892–99), VII, 460; Jefferson to Gov. Thomas McKean, Feb. 19, 1803, ibid., VIII, 218–19.

13. Jefferson to Anne Cary, Thomas Jefferson, and Ellen Wayles Randolph, March 2, 1802, Merrill D. Peterson, ed., *Thomas Jefferson: Writings* (New York, 1984), 1102.

14. Jefferson to Washington, Dec. 15, 1789, Ford, ed., *Writings of Jefferson*, V, 140.

15. Jefferson to Madison, June 9, 1793, Ford, ed., *Writings of Jefferson*, VI, 291–92.

16. Peterson, *Jefferson and the New Nation*, 544.

17. Jefferson to Madison, Aug. 28, 1789, Ford, ed., *Writings of Jefferson*, V, 111.

18. Jefferson to Joseph Priestley, June 19, 1802, Ford, ed., *Writings of Jefferson*, VIII, 158–59.

19. Second Inaugural Address, March 4, 1805, Ford, ed., *Writings of Jefferson*, VIII, 343.

20. Albert Sorel, *L'Europe et la Révolution Française* (Paris, 1908), Vol. I, Bk. I. The passages in the text from F. H. Herrick, trans., *Europe Under the Old Regime* (Los Angeles, Calif., 1947), 9, 24, 27.

21. Jefferson to Governor Rutledge, Aug. 6, 1787, Lipscomb and Bergh, eds., *Writings*, VI, 251.

22. Herrick, *Europe Under the Old Regime*, 47.

23. Felix Gilbert, *History: Choice and Commitment* (Cambridge, 1977), 169–70.

24. Reason of state provided no license for the taking of measures that escaped the constraints of positive law and morality in the nor-

mal conduct of statecraft. The argument of necessity was not directed to the normal conduct of statecraft but to the extraordinary or limiting situation. Though certainly abused in practice, as a principle necessity cannot be passed off as little more than the thinly disguised attempt to justify power and its works.

25. The claims to autonomy for the state and to the supremacy of its vital interests formed the core of the notion of reason of state. In acting consistently with these claims the statesman expressed by his actions the state's *ratio*. From this perspective the idea of the state's *ratio* was imply that there was at any particular moment in history a certain course or pattern of action best calculated to preserve the state's independence and continuity. This course *is* the state's reason and to act in accordance with it was considered a necessity imposed on the statesman. "Raison D'Etat," begins the modern classic on reason of state, "is the fundamental principle of national conduct, the State's first Law of Motion. It tells the statesman what he must do to preserve the health and strength of the State. . . . For each State at each particular moment there exists one ideal course of action, one ideal *raison d'état*. Only so long as the statesman is uncertain which is the true *raison d'état* is it possible for him to choose." Friedrich Meinecke, *Machiavellism: The Doctrine of Raison D'État and Its Place in Modern History*, trans. Douglas Scott (New Haven, Conn., 1957), 1–2. Elsewhere (p. 212), Meinecke speaks of "the natural basic task of *raison d'état*, namely the struggle for security and self-preservation at any price, and by any means."

26. Thomas Paine, "The Rights of Man" (1791), Moncure Daniel Conway, ed., *The Writings of Thomas Paine* (New York, 1967), II, 388.

27. *The Federalist*, ed. Jacob E. Cooke (Middletown, Conn., 1961), No. 6, 29.

28. Peterson, *Jefferson and the New Nation*, 754.

29. Jefferson to Madame La Duchesse D'Auville, April 2, 1790, Ford, ed., *Writings of Jefferson*, V, 153.

30. Jefferson to Thaddeus Kosciusko, April 13, 1811, Lipscomb and Bergh, eds., *Writings*, XIII, 40.

31. Draft of the Kentucky Resolutions of 1798. Fair Copy. Ford, ed., *Writings of Jefferson*, VII, 304–5.

32. "Political Observations" (1795), *Letters and Other Writings of James Madison*, 4 vols. (New York, 1884 [Philadelphia, 1865]), IV, 491–92.

33. Turreau to Talleyrand, July 9, 1805, cited in Adams, *History*, III, 85.

34. Jefferson to William Short, Nov. 28, 1814, Lipscomb and Bergh, eds., *Writings*, XIV, 214: "From a peaceable and agricul-

tural nation, [our enemy] makes us a military and manufacturing one."

35. Jefferson to Thomas Pinckney, May 29, 1797, Ford, ed., *Writings of Jefferson*, VII, 129.

36. Jonathan Elliot, ed., *The Debates in the Several State Conventions on the Adoption of the Federal Constitution* (Washington, D.C., 1836), III, 53.

37. Max Farrand, ed., *The Records of the Federal Convention of 1787*, 4 vols. (New Haven, Conn., 1937), I, 466–67. See the discussion in Gerald Stourzh, *Alexander Hamilton and the Idea of Republican Government* (Stanford, Calif., 1970), 126–30. See also Hans J. Morgenthau, *The Purpose of American Politics* (New York, 1960), 11–14.

38. First Inaugural Address, March 4, 1801, Lipscomb and Bergh, eds., *Writings*, III, 319–20.

Part II. The Development of Republican Statecraft (1783–1801)

1. On Shelburne's motivations, see Vincent T. Harlow, *The Founding of the Second British Empire: Discovery and Revolution*, 2 vols. (London, 1952–64).

2. On American expectations at the end of the war, see Jerald A. Combs, *The Jay Treaty: Political Battleground of the Founding Fathers* (Berkeley, Calif., 1970), 3–28; Frederick W. Marks III, *Independence on Trial: Foreign Affairs and the Making of the Constitution* (Baton Rouge, 1973), 52–95; and Samuel Flagg Bemis, *Jay's Treaty: A Study in Commerce and Diplomacy* (New York, 1923), 21–36.

3. "On the instructions given to the first delegation of Virginia to Congress," August 1774, Andrew A. Lipscomb and Albert Bergh, eds., *The Writings of Thomas Jefferson* (Washington, D.C., 1904–5), I, 190–91.

4. See, e.g., Jefferson to Thomas Pleasants, May 8, 1786, Julian P. Boyd, ed., *The Papers of Thomas Jefferson* (Princeton, N.J., 1950–   ), IX, 472, in which Jefferson speaks of "the torrent of wealth which we are pouring into the laps of our enemies: for such the British are more generally and more rootedly at this time than at any moment of the war."

5. See Plan of Government of the Western Territory, Boyd, ed., *Jefferson Papers*, VI, 581–617.

6. See Jefferson to Madison, Jan. 30, 1787, Boyd, ed., *Jefferson Papers*, XI, 93–94.

7. Jefferson to Archibald Stuart, Jan. 25, 1786, Boyd, ed., *Jefferson Papers*, IX, 218.

8. Cabinet Minutes, Oct. 22, 1808, Paul Leicester Ford, ed., *The Writings of Thomas Jefferson* (New York, 1892–99), I, 424. A week later, Jefferson wrote to Governor Claiborne of Louisiana that the United States and the advocates of Spanish American independence had the same interests, "and that the object of both must be to exclude all European influence from this hemisphere." Ibid., XI, 55. On the "large policy of 1808," see the excellent discussion in Arthur P. Whitaker, *The United States and the Independence of Latin America, 1800–1830* (New York, 1964 [1941]), 39–60.

9. See the valuable study of Drew R. McCoy, *The Elusive Republic: Political Economy in Jeffersonian America* (New York, 1980), 201.

10. Gilbert Chinard, *Thomas Jefferson: The Apostle of Americanism*, 2nd ed. (Ann Arbor, Mich., 1957 [1st ed., 1929]), 327–29.

11. Queries 19 and 22, *Notes on the State of Virginia*, Merrill D. Peterson, ed., *Thomas Jefferson: Writings* (New York, 1984), 290–91, 301.

12. Jefferson to G. K. van Hogendorp, Oct. 13, 1785, Boyd, ed., *Jefferson Papers*, VIII, 633. See also Jefferson to Brissot de Warville, Aug. 16, 1786, ibid., X, 262. In selecting the "particular passages" from Brissot's work that gave him "particular satisfaction," Jefferson chose those in which the author "prove to the United States that they will be more virtuous, more free, and more happy, emploied in agriculture, than as carriers or manufacturers. It is a truth, and a precious one for them, if they could be persuaded of it."

13. Jefferson to Edward Rutledge, July 4, 1790, Boyd, ed., *Jefferson Papers*, XVI, 601.

14. Jefferson to Washington, Dec. 4, 1788, Boyd, ed., *Jefferson Papers*, XIV, 328.

15. Jefferson to van Hogendorp, Oct. 13, 1785, Boyd, ed., *Jefferson Papers*, VIII, 633.

16. Jefferson to Washington, Sept. 9, 1792, Lipscomb and Bergh, eds., *Writings*, VIII, 407.

17. For the general contours of Hamilton's financial program, see Forrest McDonald, *Alexander Hamilton* (New York, 1979), and McCoy, *The Elusive Republic.*

18. Report on Manufactures, Hamilton's Final Version, Harold C. Syrett, ed., *The Papers of Alexander Hamilton* (New York, 1961–79), X, 262–64.

19. See "Political Observations" (1795), *Letters and Other Writings of James Madison*, 4 vols. (New York, 1884 [Philadelphia, 1865]), IV, 488–89. The most complete exposition of these "inequities" is to be found in Jefferson's Report on the Privileges and Restrictions on

the Commerce of the United States in Foreign Countries," Dec. 13, 1793, Peterson, ed., *Jefferson: Writings*, 435–48. See also Madison's speeches in Congress, *Annals of Congress*, 3rd Cong., 1st sess., Jan. 3, 1794, Jan. 14, 1794, 155–58, 209–25. For the Federalist reply, see ibid., Jan. 13, 1794, 174–209 (Smith).

20. On "asymmetrical interdependence," see Klaus Knorr, *The Power of Nations: The Political Economy of International Relations* (New York, 1975), 207–38.

21. *Annals of Congress*, 3rd Cong., 1st sess., Jan. 15, 1794, 227–28 (Fisher Ames). "Let those," Ames noted, "who rely upon the effect it will have on the English manufacturers and artisans, look back to 1773 and 1774, and recollect the effect it then had."

22. *Annals of Congress*, 3rd Cong., 1st sess., Jan. 23, 1794, 274.

23. The intellectual background of this argument is traced in Albert O. Hirschmann, *The Passions and the Interests: Political Arguments for Capitalism Before Its Triumph* (Princeton, N.J., 1977). See also the discussion in Daniel G. Lang, *Foreign Policy in the Early Republic: The Law of Nations and the Balance of Power* (Baton Rouge, 1985), 146–51. Madison's appeal to virtue at home and interest abroad is perhaps best illustrated by his response to the suggestion that Congress might be obliged to recede from popular outcry before the British government felt "the necessity of doing so." Madison felt no such apprehension. "He thought more favorably of the good sense as well as virtue of his fellow citizens. On the side of Great Britain it had been shown there would be the greatest distress, and the least ability to bear it. The people there were not accustomed, like the people of the United States, to self-denying regulations." *Annals of Congress*, 3rd Cong., 1st sess., Jan. 29, 1794, 382. Hamilton, by contrast, was skeptical of the belief in the superior virtue of the American people, and he ridiculed the Republicans' belief that British conduct would be regulated wholly by interest. "National pride," as he would later write, "is generally a very intractable thing. In the councils of no country does it act with greater force, than in those of Great Britain." In defense of the policy that sent John Jay to London and avoided the commercial retaliations favored by the Republicans, Hamilton argued that though Britain might "yield to negotiation, she could have yielded nothing to compulsion, without self-degradation, and without the sacrifice of that political consequence which, at all times very important to a nation, was peculiarly so to her at the juncture in question. It must be remembered too that from the relations in which the two Countries have stood to each other it must have cost more to the pride of Great Britain to have received the law from us than from any other power." The Defense No. 2, July 25, 1795, Syrett, ed., *Hamilton Papers*, XVIII, 494.

24. See Lance Banning, *The Jeffersonian Persuasion: Evolution of a Party Ideology* (Ithaca, N.Y., 1978) and Joyce Appleby, *Capitalism and a New Social Order: The Republican Vision of the 1790s* (New York, 1984). See also the excellent survey of the debate by John Ashworth, "The Jeffersonians: Classical Republicans or Liberal Capitalists?," *Journal of American Studies,* 18 (1984), 425–35.

25. Jefferson to Madison, Sept. 21, 1795, Lipscomb and Bergh, eds., *Writings,* IX, 309–10.

26. Jefferson to Thomas Gates, May 30, 1797, Ford, ed., *Writings of Jefferson,* VII, 130.

27. This theme is further developed in Banning, *The Jeffersonian Persuasion;* McCoy, *The Elusive Republic;* and Gerald Stourzh, *Alexander Hamilton and the Idea of Republican Government* (Stanford, Calif., 1970).

28. The Continentalist No. 6, July 4, 1782, Syrett, ed., *Hamilton Papers,* III, 103.

29. Hamilton to Edward Carrington, May 26, 1792, Syrett ed., *Hamilton Papers,* XI, 429.

30. See Jefferson to Dr. George Gilmer, June 28, 1793, Lipscomb and Bergh, eds., *Writings,* IX, 143–44; Jefferson to Monroe, June 28, 1793, ibid., 145; Jefferson to Madison, June 29, 1793, ibid., 147–48; Jefferson to Colonel Bell, May 18, 1797, ibid., 386.

31. The Defense No. 7, Aug. 12, 1795, Syrett, ed., *Hamilton Papers,* XIX, 120–21.

32. Jefferson, writes Chinard, "was quite sincere in his dislike of Hamilton's budgets, for the simple reason that he did not understand them himself. The master financier and expert was beyond Jefferson's comprehension; in many respects he was even far ahead of his own time, while Jefferson, in matters of finance at least, remained all his life an eighteenth century man." *Jefferson,* 310–11.

33. Opinion on the Constitutionality of an Act to Establish a Bank, Feb. 23, 1791, Syrett, ed., *Hamilton Papers,* VIII, 98.

34. Opinion on the Constitutionality of National Bank, Peterson, ed., *Jefferson: Writings,* 416.

35. Letters of Helvidius (Aug., Sept. 1793), No. IV, Gaillard Hunt, ed., *The Writings of James Madison* (New York, 1900–1910), VI, 171–72. That Madison and Jefferson did not adhere strictly to their narrow theories of constitutional interpretation is noted by McDonald, *Hamilton,* 201–22. Strict adherence to this doctrine would have paralyzed the government during the 1793 crisis over neutrality. Jefferson, though initially doubting the constitutionality of the proclamation, later told Madison that "the declaration of the disposition of the United States can hardly be called illegal, though it was certainly

officious and improper." June 29, 1793, Ford, ed., *Writings of Jefferson*, VI, 328.

36. The Defense No. 18, Oct. 6, 1795, Syrett, ed., *Hamilton Papers*, XIX, 300. Hamilton's dread of the domestic consequences of war applied not only to a British war; indeed, he thought a war with France would be "a more unmanageable business, than war with Great Britain," because of the "sentiments and prejudices of our country." Letter from Alexander Hamilton, Concerning the Public Conduct and Character of John Adams, Esq., President of the United States, Oct. 24, 1800, ibid., XXV, 232.

37. Letters of Helvidius, No. IV, Hunt, ed., *Writings of Madison*, VI, 174.

38. Query 22, *Notes on the State of Virginia*, in Peterson, ed., *Jefferson: Writings*, 301. Jefferson consistently recommended the use of force against the Algerines while Minister to France in the late 1780s: "The motives pleading for war rather than tribute are numerous and honourable, those opposing them are mean and shortsighted." Since a naval force was necessary "if we mean to be commercial," he thought there was no better occasion of beginning one. Jefferson to Monroe, Feb. 6, 1785, Boyd, ed., *Jefferson Papers*, VII, 639. When the Federalists did seize the occasion of an Algerine war for beginning a navy in 1794, however, they were opposed by Madison, who thought "this expedient unlikely to answer the purpose, and liable to many objections." See *Annals of Congress*, 3rd Cong., 1st sess., Feb. 7, 1794, 438.

39. The Republicans' attitude toward a navy is sympathetically examined in Craig L. Symonds, *Navalists and Antinavalists: The Naval Policy Debate in the United States, 1785–1827* (Newark, N.J., 1980). Symonds argues that Jefferson, even in the 1780s, "desired a national navy that could be used to pacify pirates if necessary, but that would be no greater financial or constitutional burden to the country than the disease it was designed to cure. He did not envision any diplomatic function for it, either as a tool for political involvement or as a means of protecting America's already far-flung commerce" (p. 25). This is an accurate description of Jefferson's mature views; such a conception of naval power, however, did represent a substantial retreat from Jefferson's outlook in the 1780s, when he explicitly conceived of the navy as a "tool for political involvement" and as a "means of protecting America's already far-flung commerce." See Query 22, *Notes on the State of Virginia*, in Peterson, ed., *Jefferson: Writings*, 301–2.

40. Peterson, ed., *Jefferson: Writings*, 214–15; Jefferson to George Rogers Clark, Feb. 19, 1781, Boyd, ed., *Jefferson Papers*, IV, 653; Lawrence Delbert Cress, *Citizens in Arms: The Army and the Militia*

in American Society to the War of 1812 (Chapel Hill, N.C., 1982), 152, 212.

41. Jefferson to John Brown, May 28, 1788, Boyd, ed., *Jefferson Papers,* XIII, 212.

42. Jefferson to Elbridge Gerry, Jan. 26, 1799, Lipscomb and Bergh, eds., *Writings,* X, 77. On Federalist and Republican military thought, see Cress, *Citizens in Arms,* and Reginald C. Stuart, *War and American Thought from the Revolution to the Monroe Doctrine* (Kent, Ohio, 1982). See also Theodore J. Crackel, *Mr. Jefferson's Army: Political and Social Reform of the Military Establishment, 1801–1809* (New York, 1987), who deprecates—unconvincingly—the existence of Jefferson's "supposed antiarmy bias." Jefferson's "profession of political faith" to Gerry cannot be passed off, as Crackel does, as merely the "appeasing" acknowledgment of the views of a "prominent New England Republican." Jefferson's sentiments were genuine and, as Crackel himself acknowledges, appeared in Jefferson's First Inaugural as well as his First Annual Message. Nor can Jefferson's support of an enlarged regular army in 1808, when his whole system lay in ruins, be adduced as evidence for his views in the 1790s.

43. Jefferson to Gouverneur Morris, Aug. 12, 1790, Boyd, ed., *Jefferson Papers,* XVII, 127, and Outline of Policy on the Mississippi Question, Aug. 2, 1790, ibid., 113–16.

44. First Opinion of the Secretary of State, Aug. 27, 1790, Boyd, ed., *Jefferson Papers,* XVII, 129

45. Alexander Hamilton et al., *The Federalist Papers,* ed. Jacob E. Cooke (Middletown, Conn., 1961 [1787–88]), No. 9, 50. Edward Gibbon, *The Decline and Fall of the Roman Empire,* ed. J. B. Bury (London, 1930 [1776–88]), I, 8–9. On the candid nature of Hamilton's diplomatic style, one very far removed from Jefferson's own, see McDonald, *Hamilton,* 134–35, 267–70. Hamilton's conversations with Britain's representatives to the United States, George Beckwith and, later, George Hammond, have often been regarded as treasonable by later historians. The indictment is accepted even by historians relatively sympathetic to Hamilton (see, e.g., Stourzh, *Hamilton and the Idea of Republican Government,* 203, and A. B. Darling, *Our Rising Empire: 1763–1803* [New Haven, Conn., 1940], 184). The charge was most fully set forth by Julian P. Boyd, who hated Hamilton more than he loved Jefferson. See *Number 7, Alexander Hamilton's Secret Attempts to Control American Foreign Policy* (Princeton, N.J., 1964). For a devastating critique of Boyd's thesis, see Charles R. Ritcheson's review of *Number 7, Journal of Southern History,* 31 (1965), 202–3. See also Gilbert L. Lycan, *Alexander Hamilton and American Foreign Policy: A Design for Greatness* (Norman, Okla., 1970), 121–23, 394.

46. Hamilton et al., *The Federalist*, No. 6, Cooke, ed., 32.

47. Thomas Paine, "The Rights of Man" (1791), Moncure Daniel Conway, ed., *The Writings of Thomas Paine* (New York, 1967), II, 316.

48. Madison parted company with Paine in one respect, and the manner in which he did so illuminates the special character of the Republican opposition to Hamiltonianism. Paine had argued that once monarchical sovereignty was abolished throughout Europe, "the cause of wars would be taken away." Madison, however, acknowledged that popular wars might still exist. Debt was the instrument of popular temptation, for it offered a means of carrying on wars "at the expense of other generations." The remedy was to make each generation bear the burden of its own wars by restricting the nation's capacity to borrow. The more onerous and direct the taxes, the better; only in this way might the people be kept alive to "misapplications of their money." Here, he thought, was the real key to perpetual peace, a resolution that had special relevance for America and which, if extended to Europe, would in all likelihood eliminate the scourge of war. "Were a nation to impose such restraints on itself," he argued, "avarice would be sure to calculate the expenses of ambition; in the equipoise of these passions, reason would be free to decide for the public good." The proposal did not depend on its adoption by all states—that was its peculiar merit. Individual states would gain because they would be able to avoid "wars of folly" and because their resources would remain unimpaired "for wars of necessity and defense." And if all nations followed this course of action, Madison thought, "the temple of Janus might be shut, never to be opened more." Hunt, ed., *Writings of Madison*, VI, 88–91. Hamilton took note of this idea in his "Defense of the Funding System," July 1795, Syrett, ed., *Hamilton Papers*, XIX, 56. Drawing on the experience of the European powers, he held the truth to be "not that funding systems produce wars expenses and debts, but that the ambition, avarice revenge and injustice of man produce them. The seeds of war are sown thickly in the human breast. It is astonishing, after the experience of its having deluged the world with calamities for so many ages with how much precipitance and levity nations still rush to arms against each other."

Madison's use of Rousseau also deserves comment. What Madison called "Rousseau's plan" was in reality that of Saint-Pierre, whose views on European federation Rousseau popularized in the *Extrait du Project de Paix Perpétuelle de Monsieur l'Abbé de Saint-Pierre*, published in 1761. In a work written contemporaneously with the *Extrait* but not published until 1782, *Jugement sur la Paix Perpétuelle*, Rousseau himself found much that was utopian as well as dangerous in the whole enterprise. The *Extrait* and the *Jugement* are available in English translation in M. G. Forsyth, H. M. A. Keens-Soper, and P. Savigear, eds., *The Theory of International Relations: Selected Texts from Gentili to Treitschke* (New York, 1970 [1761]). Two valuable critical studies of Rousseau's thoughts on war and peace, though not agreeing in all particulars, are F. H. Hinsley, *Power and the Pursuit of Peace: Theory and Practice in the History of Relations between States* (Cambridge, 1967), 46–61; and Stanley Hoffmann, *The State of War: Essays in the Theory and Practice of International Politics* (New York, 1965), 54–87.

49. Jefferson to William Short, Jan. 3, 1793, Ford, ed., *Writings of Jefferson,* VI, 154.

50. Jefferson to George Mason, Feb. 4, 1791, Ford, ed., *Writings of Jefferson,* V, 274–75.

51. Jefferson to William Carmichael, Dec. 15, 1787, Boyd, ed., *Jefferson Papers,* XIII, 424–25.

52. "Political Observations," April 20, 1795, *Letters of Madison,* IV, 491.

53. In this respect, Hamilton followed the argument of David Hume, "Of the Independency of Parliament," *Essays: Moral, Political, and Literary,* ed. Eugene F. Miller (Indianapolis, Ind., 1985), 42–46. See the discussion in Stourzh, *Hamilton and the Idea of Republican Government,* 76–90, and Merrill D. Peterson, *Adams and Jefferson: A Revolutionary Dialogue* (Athens, Ga., 1976), 60–61.

54. Political parties in the modern sense took shape well after the initial clashes in the cabinet. See, among many studies of this development, Joseph Charles, *The Origins of the American Party System* (New York, 1956). See also the fine work of Ralph Ketcham, *Presidents Above Party: The First American Presidency, 1789–1829* (Chapel Hill, N.C., 1984). The term party is used in our work in the Burkean sense of "a body of men united, for promoting by their joint endeavors the national interest, upon some particular principle in which they are all agreed."

55. Jefferson to Madison, March 1793, Lipscomb and Bergh, eds., *Writings,* IX, 33–34.

56. See The Defense No. 20, Oct. 23 and 24, 1795, Syrett, ed., *Hamilton Papers,* XIX, 341, with accompanying citations. A careful examination of the influence of the "modern law of nations" on the Founding Fathers is contained in Daniel G. Lang, *Foreign Policy in the Early Republic: The Law of Nations and the Balance of Power* (Baton Rouge, 1985). The "elements, or ingredients" composing "what is called the laws of Nations," as Hamilton said in a classic summary of eighteenth-century jurisprudence, are

1. The *necessary* or *internal,* law, which is the *law of Nature* applied to nations; or that system of rules for regulating the conduct of Nation to Nation which reason deduces from the principles of natural right as relative to political societies or States. 2. The *voluntary* law, which is a system of rules resulting from the equality and independence of nations, and which, in the administration of their affairs and the pursuit of their pretensions, proceeding on the principle of their having no common judge upon earth, attributes equal validity, as to external effects, to the measures or conduct of one as of another without regard to the intrinsic justice of those measures or that conduct. . . . 3. The *Pactitious* or *Conventional* law, or that law which results from a Treaty between two or more nations. This is evidently a particular not a general law; since a treaty or pact can only bind the contracting parties: Yet when we find a provision universally pervading the Treaties between nations for a length of time, as a kind of formula, it is high evidence of the general law of Nations. 4. The *customary law,* which consists of those rules of conduct that in practice are respected and observed among Nations. Its authority depends on usage implying a tacit consent and agreement. This also is a particular not a general law, obligatory only on those nations whose acquiescence has appeared or from circumstances may fairly be presumed. Thus the customary law of Europe may not be that of a different quarter of the Globe. The three last branches are sometimes aggregately denominated the *positive law* of Nations. The two first are discoverable by reason; the two last depend on proof as matters of fact.

57. See, e.g., Jefferson to Madison, Aug. 3, 1793, Ford, ed., *Writings of Jefferson,* VI, 362. Of Genet, Jefferson wrote that "his ignorance of everything written on the subject [of the law of nations] is astonishing. I think he has never read a book of any sort in that branch of science." See also Harry Ammon, *The Genet Mission* (New York, 1973).

58. Jefferson to Morris, Nov. 7, 1792, Ford, ed., *Writings of Jefferson,* VI, 120; Jefferson to Morris, March 12, 1793, ibid., VI, 199.

59. The traditional American policy of recognition may be traced in John Bassett Moore, *A Digest of International Law* (Washington, D.C., 1906), I, 72–164. The clearest statement in the development of this policy was made by Daniel Webster in 1852: "From President Washington's time down to the present day it has been a principle, always acknowledged by the United States, that every nation possesses a right to govern itself according to its own will, to change institutions at discretion, and to transact its business through whatever agents it may think proper to employ. This cardinal point in our policy has been strongly illustrated by recognizing the many forms of

political power which have been successively adopted by France in the series of revolutions with which that country has been visited" (p. 126).

60. See Julius Goebel, *The Recognition Policy of the United States* (New York, 1915), 98. "No doctrine," Goebel held, "bears more deeply the imprint of [Jefferson's] political thinking than does our recognition policy. Indeed, so far removed were his doctrines from the accepted canons of international law . . . that it is impossible to trace any relationship between the two." See also Francis L. Loewenheim, "A Legacy of Hope and a Legacy of Doubt: Reflections on the Role of History and Historians in American Foreign Policy Since the Eighteenth Century," Loewenheim, ed., *The Historian and the Diplomat: The Role of History and Historians in American Foreign Policy* (New York, 1967), 7. Loewenheim says of the Founders that if they were "truly revolutionary, they were never more so than in their attitude toward foreign policy." In Loewenheim's view, that revolutionary attitude was reflected in Jefferson's recognition policy and in his "strong and unmistakable insistence that one state not interfere in the internal affairs of another."

61. Dumas Malone, *Jefferson and His Times* (Boston, 1948–81), III, 43. See also Albert H. Bowman, *The Struggle for Neutrality: Franco-American Diplomacy During the Federalist Era* (Knoxville, Tenn., 1974), 63n.

62. Emmerich de Vattel, *The Law of Nations or the Principles of Natural Law applied to the Conduct and to the Affairs of Nations and of Sovereigns*, ed. Charles G. Fenwick, 3 vols. (Washington, D.C., 1916 [1758]), III, 19 (Book I, no. 36–37). See also Book IV, no. 68 (365–66). On British doctrine, see R. J. Vincent, *Nonintervention and International Order* (Princeton, N.J., 1974), 70–73. See also W. E. H. Lecky, *The French Revolution: Chapters from the Author's History of England During the Eighteenth Century* (New York, 1904), 421–23.

63. Hamilton and Henry Knox to Washington, May 2, 1793 (Enclosure), Syrett, ed., *Hamilton Papers*, XIV, 374–75, 368, 394–95. See also Pacificus No. 1, June 29, 1793, ibid., XV, 41, where Hamilton speaks of judging, "in the case of a Revolution of Government in a foreign Country, whether the new rulers are *competent organs of the National Will* and ought to [be] recognized or not" (emphasis added).

64. Pacificus No. 2, July 3, 1793, Syrett, ed., *Hamilton Papers*, XV, 59–62. French decrees in F. M. Anderson, ed., *The Constitutions and Other Select Documents Illustrative of the History of France, 1789–1907* (New York, 1908 [1904]), 130–34. Cambon argued before the Convention that the November 19 decree was too generous

and was thus "une oeuvre de dupes"; he proposed that the privileged classes of occupied countries pay for the war, and thus altered, with the decree of December 15, the thrust of French foreign policy from proselytism to revolution and conquest. See Albert Sorel, *L'Europe et la Révolution Française*, 7th ed., 8 vols. (Paris, 1908), III, 232–36. The December 15 decree was subsequently modified by the Decree of Nonintervention of April 13, 1793, engineered by Danton. In the Constitution rejected by the Convention, the decrees of December 15 and April 13 appear side by side, "with no explanation given for their apparent inconsistency." Vincent, *Nonintervention and International Order*, 69.

In 1798, Hamilton recalled the initial decrees "vomited forth by the Convention" and the subsequent mellowing of French foreign policy: "It has since clearly appeared, that at the very moment she was making these professions, and while her diplomatic agents were hypocritically amusing foreign courts with conciliatory explanations and promises of moderation, she was exerting every faculty, by force and fraud, to accomplish the very conquest and aggrandizement she insidiously disavowed." The Stand No. 3, April 7, 1798, Syrett, ed., *Hamilton Papers*, XXI, 405. These French decrees (and the rejection of them by the British government) were of central importance in the years to come. Madison would later recall that "the British Government thought a war of more than 20 years called for against France by an edict, afterwards disavowed, which assumed the policy of propagating changes of Government in other Countries." Madison to Richard Rush, July 22, 1823, *Letters of Madison*, III, 330–31.

65. Jefferson to C. W. F. Dumas, March 24, 1793, Lipscomb and Bergh, eds., *Writings*, IX, 56.

66. Jefferson to Thomas Mann Randolph, June 24, 1793, Ford, ed., *Writings of Jefferson*, VI, 318. "They seem," he added, "to be correcting themselves in the latter point."

67. Jefferson to Madison, May 19, 1793, Lipscomb and Bergh, eds., *Writings*, IX, 97.

68. Jefferson to Tench Coxe, May 1, 1794, Ford, ed., *Writings of Jefferson*, VI, 507–8.

69. Jefferson to William B. Giles, April 27, 1795, Lipscomb and Bergh, eds., *Writings*, IX, 305. In early 1795, General Jean Charles Pichegru was commander of the French Army of the North. To Adet, Genet's replacement as French minister, Jefferson wrote that "Your struggle for liberty keep alive the only sparks of sensation which public affairs now excite in me." Oct. 14, 1795, ibid., 312. See also Jefferson to Peregrine Fitzhugh, February 23, 1798, Ford, *Writings of Jefferson*, VII, 211. "I do not indeed wish to see any nation have a

form of government forced on them; but if it is to be done, I should rejoice at it's being a freer one."

70. Jefferson to Madison, June 23, 1793, Lipscomb and Bergh, eds., *Writings*, IX, 138.

71. James D. Richardson, ed., *A Compilation of the Messages and Papers of the Presidents, 1789–1897* (Washington, D.C., 1896), I, 156–57.

72. Madison to Jefferson, June 10, 1793, Hunt, ed., *Writings of Madison*, VI, 127.

73. Article XI of Treaty of Alliance (1778), in William M. Malloy, ed., *Treaties, Conventions, International Acts, Protocols and Agreements Between the United States of America and Other Powers, 1776–1909*, 2 vols. (Washington, D.C., 1910), I, 481–82.

74. Opinion on French Treaties, April 28, 1793, Ford, ed., *Writings of Jefferson*, VII, 284; Jefferson to Morris, April 20, 1793, ibid., VII, 282.

75. Jefferson to Madison, June 2, 1793, Lipscomb and Bergh, eds., *Writings*, IX, 105–6.

76. Jefferson to Madison, May 19, 1793, Lipscomb and Bergh, eds., *Writings*, IX, 97. "In short," Jefferson said, "[Genet] offers everything, and asks nothing." Genet's instructions were that the guarantee of the West India islands "was not to be demanded *until after it should have been incorporated in the new treaty to be proposed*"—a feature of the instructions, as Bemis comments, "which frequently escapes the emphasis of writers on the period." Bemis, *Jay's Treaty*, 143–44. These instructions also made the guarantee "une condition *sine qua non*" of free commerce between the United States and the French West Indies. See "Supplément Aux Instructions Données Au Citoyen Genet," in Frederick J. Turner, ed., "Correspondence of the French Ministers to the United States, 1791–1797," *Annual Report of the American Historical Association for the Year 1903*, 2 vols. (Washington, D.C., 1904), II, 209–10.

77. Explanation of the Origin of the Principle that "Free bottoms make free goods," Dec. 20, 1793, Lipscomb and Bergh, eds., *Writings*, XVII, 350–51.

78. Jefferson to Madison, April 3, 1794, Lipscomb and Bergh, eds., *Writings*, IX, 282.

79. "Explanation," Lipscomb and Bergh, eds., *Writings*, XVII, 350; Jefferson to Madison, April 3, 1794, ibid., IX, 282. Washington would doubtless have opposed this construction, which accounts for Jefferson's silence. Hamilton took note of Jefferson's idea in his first Pacificus paper, where he noted that "even to furnish *determinate* succours, of a certain number of Ships or troops, to a Power at War, in

consequence of *antecedent treaties having no particular reference to the existing war,* is not inconsistent with neutrality; a position well established by the doctrines of Writers and the practice of Nations." (Here Hamilton made reference to Vattel, Book III, Chap. VI, sec. 101, as Jefferson might well have done.) Hamilton went on to say, however, that execution of the clause of guarantee "would be contrary to the sense and spirit of the Proclamation; because it would engage us with our whole force as an *associate* or *auxiliary* in the War; it would be much more than the case of a definite limited succour, previously ascertained." Pacificus No. 1, June 29, 1793, Syrett, ed., *Hamilton Papers,* XV, 35–36.

80. Speaking to him as Mr. Jefferson and not as the Secretary of State, Genet detailed his plans for the liberation of New Orleans, with soldiers to be raised in Kentucky and Louisiana but to rendezvous out of the territory of the United States. "I told him," Jefferson recorded in the *ANAS,* "that his enticing officers and soldiers from Kentucky to go against Spain, was really putting a halter about their necks; for that they would assuredly be hung if they commenced hostilities against a nation at peace with the United States." Thus far Jefferson had adhered to the line of neutrality; he added, however, that "leaving out that article I did not care what insurrections should be excited in Louisiana," and he honored Genet's request for a letter of introduction to Governor Shelby of Kentucky for Andre Michaux, who was to arrange the filibustering expedition for France. Genet clearly believed that he had been given a green light, and said so in later charges against Jefferson. Even on Jefferson's own evidence, moreover, it appears that the message he delivered was, in effect, to go ahead but don't get caught. July 5, 1793, Franklin B. Sawvel, ed., *The Complete ANAS of Thomas Jefferson* (New York, 1903), 130. See also the discussion in Ammon, *Genet Mission,* 84–85, and McDonald, *Hamilton,* 277–78, 426.

81. The complexity of this task, and the need for executive initiative to contend with it, is indicated by the set of twenty-nine questions—twenty-one drawn up by Hamilton, seven by Jefferson, and one by Washington—each of which required an answer. See "Queries as to the Rights and Duties of the United States under her Treaties with France, and the Laws of Neutrality," Lipscomb and Bergh, eds., *Writings,* XVII, 293–99. See also Washington to Hamilton et al., April 18, 1793, Syrett, ed., *Hamilton Papers,* XIV, 326–27.

82. See Jefferson to George Hammond, Sept. 9, 1793, Walter Lowrie and Matthew St. Clair Clarke, eds., *American State Papers: Foreign Relations* (Washington, D.C., 1832–61), I, 176. Hereafter cited as *A.S.P.F.R.*

83. Jefferson to Morris, Aug. 16, 1793, *A.S.P.F.R.*, I, 168.

84. Jefferson to Genet, Aug. 7, 1793, *A.S.P.F.R.*, I, 167. Jefferson dissented from Hamilton's recommendation that compensation be offered to all British vessels illegally seized. The question was subsequently adjusted in the Jay Treaty. See Syrett, ed., *Hamilton Papers*, XV, 183–84n.

85. Hammond to Grenville (date unclear), cited in Charles R. Ritcheson, *Aftermath of Revolution: British Policy toward the United States, 1783–1795* (New York, 1971), 286.

86. Plan of Treaties, July 18, 1776, Worthington C. Ford et al., eds., *Journals of the Continental Congress* (Washington, D.C., 1904–37), V, 576–89. See also Gregg L. Lint, "The American Revolution and the Law of Nations, 1776–1789," *Diplomatic History*, 1 (Winter 1977), 20–34. The same principles were incorporated in the treaty of the United States with Holland. The 1785 treaty with Prussia went even further by declaring that nothing shall be contraband. Other disputes, concerning particularly the rule of 1756 and the requirements of an effective blockade, are discussed in Part IV.

87. Report on American participation in a European neutral confederation, June 12, 1783, Worthington C. Ford et al., eds., *Journals of the Continental Congress*, XXIV, 392–94, discussed in Lint, "American Revolution and Law of Nations," 28. The authors of the committee report were James Madison, Oliver Ellsworth, and Alexander Hamilton.

88. See Ritcheson, *Aftermath of Revolution*, 376–77.

89. Adams to Marshall (then Secretary of State), Oct. 3, 1800, John Bassett Moore, *A Digest of International Law* (Washington, D.C., 1906), VII, 439.

90. Hammond to Jefferson, Sept. 12, 1793, *A.S.P.F.R.*, I, 240. See also W. Alison Phillips and Arthur H Reede, *Neutrality: Its History, Economics and Law, II: The Napoleonic Period* (New York, 1936), 38.

91. Grenville to Hammond, Sept. 17, 1793, cited in Ritcheson, *Aftermath of Revolution*, 280. See also Bemis, *Jay's Treaty*, 156. In an earlier dispatch to Hammond, Grenville had sketched the outlines of the policy that Britain was likely to follow, defining contraband as including all things "of such a nature as to enable the enemies of this country to carry on the war against us." Bemis comments that "in the letter and the latitude of this definition one reads an intention to include foodstuffs," and he cites Hammond to the effect that Hamilton "personally agreed as to the justice of such measures but would not be responsible for the opinion of his colleagues." This view is not persuasive. Hamilton's reaction to the provision order of June 8, noted in the text, supports an opposite construction: that Hammond, in his

initial comments to Hamilton, did not go beyond a very general reference to contraband. This view is also supported by Hamilton's subsequent record, in which he clearly stipulated the right to send provisions to France and made it a cause of war if the British refused. See Hamilton to Washington, Sept. 4, 1795, Syrett, ed., *Hamilton Papers,* XIX, 235; The Defense No. 31, ibid., XIX, 473–81.

92. Jefferson to Pinckney, Sept. 7, 1793, *A.S.P.F.R.,* I, 239.

93. See The Defense No. 31, Syrett, ed., *Hamilton Papers,* XIX, 473–81; and Pinckney, Marshall, and Gerry to Talleyrand, Feb. 7, 1798, *A.S.P.F.R.,* II, 172. (This latter statement was drafted by Marshall.) See also Pickering to J. Q. Adams, July 17, 1797, ibid., 250; and Moore, *Digest of International Law,* VII, 434–39.

94. Jefferson to Pinckney, May 7, 1793, Lipscomb and Bergh, eds., *Writings,* IX, 80. Jefferson noted here that similar principles were to be found in the American treaty with Holland and in the Anglo-French treaty of 1786. (On the significance of the latter, see Ritcheson, *Aftermath of Revolution,* 376–77.)

95. Explanation of the Origin of the Principle that "Free bottoms make free goods," Dec. 20, 1793, Lipscomb and Bergh, eds., *Writings,* XVII, 350.

96. The abandonment, Hamilton later said, was made "on full & mature examination and reflection by an unanimous opinion of those consulted." Remarks on the Treaty of Amity Commerce and Navigation lately made between the United States and Great Britain, Syrett, ed., *Hamilton Papers,* XVIII, 438.

97. See Jefferson to Genet, July 24, 1793, *A.S.P.F.R.,* I, 166; Jefferson to Morris, Aug. 16, 1793, ibid., I, 170.

98. See Ritcheson, *Aftermath of Revolution,* 285, who notes that "Jefferson's abandonment of the 'free ships, free goods' doctrine might have been written by Lord Hawkesbury himself." Federalist spokesmen frequently cited with approval Jefferson's remarks on the doctrine. See above, note 93.

99. Explanation, Dec. 20, 1793, Lipscomb and Bergh, eds., *Writings,* XVII, 351–52. See also Jefferson to Edward Everett, Feb. 24, 1823, ibid., XV, 410–12.

100. Jefferson to Madison, Jan. 1, 1797, Ford, ed., *Writings of Jefferson,* VII, 100.

101. The Defense No. 20, Oct. 23 and 24, 1795, Syrett, ed., *Hamilton Papers,* XIX, 332.

102. See *Annals of Congress,* 3rd Cong., 3rd sess., March 1794, 535–37 (Dayton), 542–52 (Giles). See also "The Jay Treaty. Speech in the 4th Congress, April 6, 1796," Hunt, ed., *Writings of Madison,* VI, 288. In his speech opposing the Jay Treaty, Madison held that "he should probably be among the last who would be disposed to

resort to such an expedient [sequestration of debts] for redress." But Madison would not approve "a perpetual and irrecoverable abandonment of a defensive weapon." Without "fleets and armies to command a respect" for American rights, "we ought to keep in our hands all such means as our situation gave us."

103. See The Defense No. 20, Syrett, ed., *Hamilton Papers,* XIX, 342, 336.

104. Jefferson to Mitchell, June 13, 1800, Jefferson Papers, Library of Congress, Reel 22. In this respect, as in others, the conflict between Jefferson and Hamilton was an instance, in Sorel's words, of "the eternal dispute between those who imagine the world to suit their policy, and those who arrange their policy to suit the realities of the world." Cited in E. H. Carr, *The Twenty Years' Crisis, 1919–1939,* 2nd ed. (New York, 1946), 11.

105. Jefferson to Madison, Aug. 28, 1789, Ford, ed., *Writings of Jefferson,* V, 111.

106. Jefferson to Madison, Aug. 28, 1789, Ford, ed., *Writings of Jefferson,* V, 111.

107. Pacificus No. 4, July 10, 1793, Syrett, ed., *Hamilton Papers,* XV, 86.

108. Report on Negotiation With Spain, March 18, 1792, Ford, ed., *Writings of Jefferson,* V, 468, 470.

109. Jefferson to Short, July 28, 1791, Ford, ed., *Writings of Jefferson,* V, 364.

110. Jefferson to Pinckney, Sept. 7, 1793, Ford, ed., *Writings of Jefferson,* VI, 413.

111. Jefferson to Short, July 28, 1791, Ford, ed., *Writings of Jefferson,* V, 364.

112. Cf. Hans J. Morgenthau, *In Defense of the National Interest: A Critical Examination of American Foreign Policy* (New York, 1951), 22.

113. On this point, see Burton Spivak, "Thomas Jefferson, Republican Values, and Foreign Commerce," in *Traditions and Values: American Diplomacy, 1790–1865,* ed. Norman A. Graebner (Lanham, Md., 1985); and Paul Varg, *Foreign Policies of the Founding Fathers* (East Lansing, Mich., 1963).

114. The Defense No. 7, Aug. 12, 1795, Syrett, ed., *Hamilton Papers,* XIX, 116.

115. The Defense No. 7, Aug. 12, 1795, Syrett, ed., *Hamilton Papers,* XIX, 117–18.

116. The Defense No. 5, Aug. 5, 1795, Syrett, ed., *Hamilton Papers,* XIX, 90.

117. Jefferson to Thomas Mann Randolph, Aug. 11, 1795, cited in Dumas Malone, *Jefferson and His Time* (Boston, 1948–81), III,

247; and Jefferson to Rutledge, Nov. 30, 1795, Lipscomb and Bergh, eds., *Writings*, IX, 314.

118. The Defense No. 31, Dec. 12, 1795, Syrett, ed., *Hamilton Papers*, XIX, 478–80.

119. *Annals of Congress*, 4th Cong., 1st sess., April 26, 1796, 1183–1202.

120. "The Jay Treaty. Speech in the 4th Congress, April 6, 1796," and Madison to _____, Aug. 23, 1795, Hunt, ed., *Writings of Madison*, VI, 280, 243, 249–50, 295, 256.

121. The Defense No. 5, Aug. 5, 1795, Syrett, ed., *Hamilton Papers*, XIX, 95.

122. That a war with Great Britain was the likely result of a rejection of Jay's Treaty is the considered judgment of most recent historians of the controversy. See, e.g., Ritcheson, *Aftermath of Revolution*, 326; Combs, *Jay Treaty*, 182–83; Bemis, *Jay's Treaty*, 270–71; A. L. Burt, *The United States, Great Britain and British North America: From the Revolution to the Establishment of Peace After the War of 1812* (New Haven, Conn., 1940), 152; and A. B. Darling, *Our Rising Empire: 1763–1803* (New Haven, Conn., 1940), 204–5. See also Henry Adams, *The Life of Albert Gallatin* (New York, 1943 [1879]), 158: "That Mr. Jay's treaty was a bad one," Adams held, "few persons even then ventured to dispute; no one would venture on its merits to defend it now." Yet Adams, too, believed that war alone might have secured American rights against England: "No considerations of a possible war with England averted or postponed can blind history to the fact that this blessing of peace was obtained by the sacrifice of national consistency and by the violation of neutrality towards France."

123. On Republican expectations, see Arthur P. Whitaker, *The Spanish–American Frontier* (Boston, 1927), 203–4. James Monroe, then Minister to France, believed that a "good understanding" existed between France and the United States on the navigation of the Mississippi. Monroe to Secretary of State, March 17, 1795, *A.S.P.F.R.*, I, 701. His confidence on this point is to be contrasted with France's determination to secure the retrocession of Louisiana from Spain in the negotiations leading up to the Treaty of Basel. Whitaker, *Spanish–American Frontier*, not only insists that it was "fatuous" to expect that France would tender her good offices to assist in a satisfactory Spanish settlement but that American diplomats—Randolph, Pinckney, and Jay, in particular—were "maladroit" in not speeding Pinckney's mission to Spain while it was still possible to play upon Spanish fears of an Anglo-American alliance. By the time the Treaty of San Lorenzo (Pinckney's Treaty) was signed, Whitaker insists, the text of Jay's Treaty had already reached Spain and served to prejudice the

negotiation (pp. 204–5). "It was not Jay's treaty with England," he concludes, "but Godoy's treaty with France that made possible Pinckney's triumph at San Lorenzo" (p. 207).

In contrast, Samuel Flagg Bemis (*Pinckney's Treaty: A Study of America's Advantage from Europe's Distress, 1783–1800* [New York, 1926], 287) argued that the full terms of Jay's Treaty were not known by Godoy; had they been, "it is doubtful whether he could have achieved such a successful ending of his Spanish mission." (This opinion was adhered to in the revised edition of Bemis's work, published in 1960.) Bemis and Whitaker agree that the prospect of an Anglo-American alliance struck terror into Godoy's heart, and that knowledge of the text of the Jay Treaty would have eliminated that danger for Godoy. But whereas Whitaker insists that Godoy's anxiety over an Anglo-American alliance had disappeared by the time San Lorenzo was signed and that the triumph was due to Godoy's fear that Britain would attack Spain's American dominions on learning of the Franco-Spanish reconciliation at Basel, Bemis concludes that the operative factor was Spain's ungrounded and as yet undispelled fear of an Anglo-American alliance. Both writers leave the impression that the successful conclusion of Pinckney's mission was far more a matter of luck— an "accident" of European diplomacy—than of skill.

That impression is misleading. Whatever Godoy's knowledge of the text of Jay's Treaty, he could not know if a private understanding had been reached between Jay and Grenville. The important factor, as A. B. Darling notes, "was not what the treaty said in whole or in part—or what people said that it said—but that Britain and the United States had come to terms again. As Spain was leaving her alliance with Britain and returning to Britain's enemy, Britain was winning the United States from that enemy" (*Our Rising Empire*, p. 215). If Britain was unwilling to contest America's claim to the upper eastern Mississippi valley, Spain was in no position to deprive the United States of the lower part. At the very least, the United States, having eliminated one enemy in the interior, was in a far stronger position to contend with Spain—as Hamilton emphasized in his defense of the Jay Treaty. See The Defense No. 2, July 25, 1795, Syrett ed., *Hamilton Papers*, XVIII, 499. See also Hamilton to Jay, May 6, 1794, ibid., XVI, 384–85, and "The Warning No. 4," Feb. 27, 1797, ibid., XX, 526.

Even more misleading is the argument of Julian P. Boyd, ed., *Jefferson Papers*, XIX, 518 (The Threat of Disunion in the West), that "just as Alexander Hamilton was the architect of the treaty that bears the name of John Jay, so Thomas Jefferson was the one who planted the seed of that negotiated by Pinckney." He did so, Boyd believes, by insisting to Spain, early in 1791, that "should any spark kindle

. . . our borderers into a flame, we are involved beyond recall by the eternal principles of justice to our citizens, which we will never abandon." Jefferson to Carmichael, March 7, 1791, ibid., XIX, 523–24. Jefferson's determination to secure the navigation of the Mississippi and the 31st parallel while Secretary of State is clear; that he would have been able to do so if caught up in a bitter conflict with England is much less so.

124. The Defense No. 5, Aug. 5, 1795, Syrett, ed., *Hamilton Papers,* XIX, 96.

125. Decree of Executive Directory, July 2, 1796, *A.S.P.F.R.,* I, 577.

126. See Hamilton to Oliver Wolcott, Aug. 10, 1795, Syrett, ed., *Hamilton Papers,* XIX, 111–13. "The possibility of abuse from a doubtful construction of a Treaty between two powers," Hamilton argued, "is no subject of offense to a third. It is the fact which must govern. According to this indisputable criterion, France has had no cause to complain on this account; for since the ratification of the Treaty no instance of the seizure of provisions has occurred & it is known that our government protested against such a construction." The Stand No. 5, April 16, 1798, ibid., XXI, 426. See also Pickering to Adet, Nov. 1, 1796, *A.S.P.F.R.,* I, 576–78; and Marshall to Talleyrand, ibid., I, 176.

127. The conjunction of this ambitious stance with the insistence on pure principle also recalls Meinecke's warning that "every influx of unpolitical motives into the province of pure conflicts of power and interest brings with it the danger that these motives will be misused and debased by the naturally stronger motives of mere profit, of *raison d'état.* The latter resembles some mud-coloured stream that swiftly changes all the pure waters flowing into it into its own murky colour." *Machiavellism,* 210.

128. Samuel Flagg Bemis, ed., *The American Secretaries of State and Their Diplomacy* (New York, 1958 [1928]), II, 12.

129. The indictment of Adams is set out at length in Hamilton's own "Letter from Alexander Hamilton, Concerning the Public Conduct and Character of John Adams, Esq. President of the United States," Oct. 24, 1800, Syrett, ed., *Hamilton Papers,* XXV, 186–234. See also Jacob E. Cooke, "Country Above Party: John Adams and the 1799 Mission to France," *Fame and the Founding Fathers,* ed. Edmund P. Willis (Bethlehem, Pa., 1967); McDonald, *Hamilton,* 329–53; and Lycan, *Alexander Hamilton and American Foreign Policy.* Adams's defense of his own conduct—along with a biting (and sometimes wickedly funny) condemnation of Hamilton—is contained in "Correspondence Originally Published in the Boston Patriot" (1809), in Charles Francis Adams, ed., *The Works of John Adams, Second Pres-*

ident of the United States. . . , 10 vols. (Boston, 1854), IX, 239–311. Secondary accounts favorable to Adams include Stephen G. Kurtz, *The Presidency of John Adams: The Collapse of Federalism, 1795–1800* (Philadelphia, 1957), and Alexander DeConde, *The Quasi-War: The Politics and Diplomacy of the Undeclared War with France, 1797–1801* (New York, 1966).

130. John Adams to James Lloyd, March 29, 1815, *Works of Adams,* X, 149. The project of Francesco de Miranda to revolutionize South America attracted the interest of both Hamilton and the British government. Its ardent champion was the American minister in London, Rufus King. Adams scorned the project, however, in part because he thought the Latin Americans were incapable of free government. Thus he considered Miranda's project "visionary, though far less innocent than that of his countryman Gonzalez, of an excursion to the moon, in a car drawn by geese trained and disciplined for the purpose." Adams to James Lloyd, March 27, 1815, ibid., 146.

131. *Annals of Congress,* 5th Cong., 2nd sess., July 5, 1798, 2110. The Federalists, in any case, made their own full contribution to the vicious and reckless charges that then passed for political discourse. In ridiculing Jefferson's "several ingenious distinctions"—admittedly foolish—to rescue the Directory "from the odium and corruption unfolded by" the X, Y, Z dispatches, Hamilton detected a more sinister design: "To be the pro-consul of a despotic Directory over the United States, degraded to the condition of a province, can alone be the criminal, the ignoble aim of so seditious, so prostitute a character." The Stand No. 7, April 21, 1798, Syrett, ed., *Hamilton Papers,* XXI, 442. It is difficult to see why this was any different, in point of sheer calumny, from the equally baseless charges brought against Hamilton both at the time and subsequently—that he was a British agent, that he desperately coveted a war with France and a permanent connection with Great Britain, that he aimed to establish a military dictatorship in the United States, etc., etc.

132. Jefferson to John Taylor, June 1, 1798, Ford, ed., *Writings of Jefferson,* VII, 263–65.

133. Drafts of the Kentucky Resolutions of 1798. Fair Copy. Nov. 1798, Ford, ed., *Writings of Jefferson,* 300–308.

134. Malone, *Jefferson and His Time,* III, 404.

135. Hamilton to Theodore Sedgwick, Feb. 2, 1799, Syrett, ed., *Hamilton Papers,* XXII, 453.

136. The Stand No. 1, March 30, 1798, Syrett, ed., *Hamilton Papers,* XXI, 382; The Stand No. 3, April 7, 1798, ibid., 408.

137. The Stand No. 4, April 12, 1798, Syrett, ed., *Hamilton Papers,* XXI, 412–13.

138. Washington to James McHenry (drafted by Hamilton), Dec.

13, 1798, Syrett, ed., *Hamilton Papers,* XXII, 345. Rufus King had sounded the same note in a dispatch written in 1797:

> Nothing can be more erroneous in the Politicks of a Nation than to omit on any consideration, much less on the score of a mistaken and dangerous economy, those Preparations for self defence and national safety, which, even in days of profound Peace, ought not to be neglected, and of which, in this mighty crisis of human affairs, he must be blind who does not see the absolute necessity. It is a false, and will prove a fatal, security, if we allow ourselves to be persuaded that we are either too just, too remote, too wise, or too powerful to be drawn into the present war, which continues to exhaust and threatens to change the face of Europe, and the results of which, in reference to the whole world, were never more uncertain or beyond the reach of human foresight, than at the present moment. (King to Secretary of State, Sept. 1, 1797, Charles R. King, ed., *The Life and Correspondence of Rufus King* [New York, 1894–1900], II, 221)

139. *Annals of Congress,* 5th Cong., 2nd sess., May 8, 1798, 1632–33.

140. *Annals of Congress,* 5th Cong., 2nd sess., March 1, 1798, 1118–39.

141. Jefferson to Samuel Smith, Aug. 22, 1798, Ford, ed., *Writings of Jefferson,* VII, 277–78.

142. Jefferson to Joseph Priestley, Jan. 18, 1800, Ford, ed., *Writings of Jefferson,* VII, 406.

143. See McCoy, *The Elusive Republic,* 176–77.

144. Jefferson to Colonel Arthur Campbell, Sept. 1, 1797, Ford, ed., *Writings of Jefferson,* VII, 170.

145. Jefferson to Gerry, June 21, 1797, Ford, ed., *Writings of Jefferson,* VII, 149–50.

146. Jefferson to Thomas Lomax, March 12, 1799, Ford, ed., *Writings of Jefferson,* VII, 374.

147. Speech to Both Houses of Congress, May 16, 1797, Charles F. Adams, ed., *Works of Adams,* IX, 117.

148. Jefferson to Carrington, Dec. 21, 1787, Boyd, ed., *Jefferson Papers,* XII, 447.

149. Jefferson to Short, Oct. 3, 1801, Ford, ed., *Writings of Jefferson,* VIII, 98; Jefferson to Gerry, Jan. 26, 1799, Ford, ed., *Writings of Jefferson,* VII, 328.

150. Jefferson to Lomax, March 12, 1799, Ford, ed., *Writings of Jefferson,* VII, 374.

151. Jefferson to Samuel Adams, Feb. 26, 1800, Ford, ed., *Writings of Jefferson,* VII, 425–26.

152. Jefferson to John Breckenridge, Jan. 29, 1800, Ford, ed.,

Writings of Jefferson, VII, 418. For a lyrical evocation of this theme, see Peterson, *Jefferson and the New Nation*, 628. "Darkness had fallen in Europe," Peterson writes, "and Jefferson was resigned, at last, to an American destiny in a world of its own."

153. First Inaugural, March 4, 1801, Lipscomb and Bergh, eds., *Writings*, III, 321.

154. Jefferson to John Dickinson, March 6, 1801, Ford, ed., *Writings of Jefferson*, VIII, 7–8.

Part III. The Diplomacy of Expansion (1801–5)

1. Jefferson to Spencer Roane, Sept. 6, 1819, Paul Leicester Ford, ed., *The Writings of Thomas Jefferson* (New York, 1892–99), X, 140.

2. Inaugural Address, March 4, 1801, Ford, ed., *Writings of Jefferson*, VIII, 4.

3. Allan Bowie Magruder, *Political, Commercial and Moral Reflections on the Late Cession of Louisiana to the United States* (Lexington, Ky., 1803), 7.

4. David Ramsay, *An Oration, on the Cession of Louisiana, to the United States* (Charleston, S.C., 1804), 6.

5. *National Intelligencer*, July 8, 1803, cited in Dumas Malone, *Jefferson and His Time* (Boston, 1948–81), IV, 297.

6. Arthur P. Whitaker, *The Mississippi Question 1795–1803: A Study in Trade, Politics, and Diplomacy* (Gloucester, Mass., 1962 [New York, 1934]), 202–7.

7. Malone, *Jefferson and His Time*, IV, 271–72, 285–86.

8. Merrill D. Peterson, *Thomas Jefferson and the New Nation: A Biography* (New York, 1970), 761–62.

9. "Pericles," *New-York Evening Post*, Feb. 8, 1803, Harold C. Syrett, ed., *The Papers of Alexander Hamilton* (New York, 1961–79), XXVI, 82–85.

10. "Purchase of Louisiana," *New-York Evening Post*, July 5, 1803, Syrett, ed., *Hamilton Papers*, XXVI, 129–36. This editorial is reviewed with comment by Douglas Adair, "Hamilton on the Louisiana Purchase: A Newly Identified Editorial from the *New-York Evening Post*," *William and Mary Quarterly*, 3d ser., 12 (1955), 268–81.

11. The central importance of Santo Domingo for the outcome of Louisiana was also recognized at the time by the editors of the *Annual Register*, who wrote: "Had Bonaparte completely succeeded in St. Domingo, then the armaments he would have constantly kept up in that island and Louisiana, would have protected New Orleans for ever against any American force which could be brought against it, while the coasts of the republic would be constantly exposed to the

depredations of French and Spanish invaders," *Cobbett's Annual Register* (London, 1804), 339.

12. Henry Adams, *History of the United States of America during the Administrations of Jefferson and Madison* (New York, 1903 [1889–91]), II, 21.

13. Adams, *History,* I, 445.

14. "Heads of consideration on the Navigation of the Mississippi for Mr. Carmichael," Aug. 2, 1790, Julian P. Boyd, ed., *The Papers of Thomas Jefferson* (Princeton, N.J., 1950–   ), XVII, 113.

15. Hamilton to Washington, Sept. 15, 1790, Syrett, ed., *Hamilton Papers,* VII, 51–53.

16. In this case the Spanish policy pertained to Florida, leading Jefferson to offer this advice to the President: "I wish a hundred thousand of our inhabitants would accept the invitation. It will be the means of delivering to us peaceably, what may otherwise cost us a war." Jefferson to Washington, April 2, 1791, Boyd, ed., *Jefferson Papers,* XX, 97. See also Jefferson to Archibald Stuart, Jan. 25, 1786, ibid., IX, 218. Isaac Joslin Cox, *The West Florida Controversy, 1798–1813* (Baltimore, 1918), particularly 22–24, 58–59, presents an interesting discussion of Spanish policy. The dilemmas of confronting an encroaching American population caused the Spanish, and later the Mexicans with regard to Texas, to shift between policies of exclusion and assimilation in their efforts to secure their territories.

17. Hamilton and much of the Federalist party dissented from this connection between expansion and the security of the union. Whereas New Orleans was indispensable as a commercial outlet for the western states, Hamilton concluded that the newly purchased territory west of the Mississippi was "not valuable to the United States for settlement." Indeed, the immense region raised the prospect of "a too widely dispersed population," and "by adding to the great weight of the western part of our territory, must hasten the dismemberment of a large portion of our country, or a dissolution of the Government." Against these dangers, Hamilton suggested that the western territory might be bartered with Spain for the Floridas. "Purchase of Louisiana," Syrett, ed., *Hamilton Papers* XXVI, 129–36.

18. *Annals of Congress,* 7th Cong., 2nd sess., 83–255. The Ross resolutions were introduced on Feb. 16, 1803, and debated on Feb. 23–25. They were defeated, 15 to 11, on a straight party vote.

19. On the American suspicion of a French design and for evidence of the actual Spanish motivations, see E. Wilson Lyon, *Louisiana in French Diplomacy, 1759–1804* (Norman, Okla., 1934), 167–73. Several senators gave voice to the suspicion that France was behind the suspension of the deposit during the debate over the Ross Resolutions. *Annals of Congress,* 7th Cong., 2nd sess., Feb. 16, 1803, 92; Feb. 23, 1803), 109, 144.

20. By the terms of the secret treaty, the retrocession of Louisiana was conditional on France creating an Italian kingdom in Tuscany (Etruria) for the Duke of Parma, brother of the Queen of Spain, on France evacuating her forces from that kingdom, and on recognition of the new sovereign by a major European power other than France. As evacuation and recognition were slow in coming, the Spanish would delay the official transfer of Louisiana until mid-1802, a matter of no small irritation to Napoleon.

21. Jefferson to Robert Livingston, April 18, 1802, Ford, ed., *Writings of Jefferson,* VIII, 144–45.

22. Jefferson to Livingston, April 18, 1802, Ford, ed., *Writings of Jefferson,* VIII, 144–45.

23. Among the many appreciations of the lasting significance of the Louisiana Purchase, perhaps the best known—certainly the most fulsome—is that of Frederick Jackson Turner:

> The acquisition of Florida, Texas, California, and the posses-
> sions won by the United States in the recent Spanish–American
> war, are in a sense the corollaries of this great event. France,
> England, and Spain, removed from the strategic points on our
> border, were prevented from occupying the controlling position
> in determining the destiny of the American provinces which so
> soon revolted from the empire of Spain. The Monroe Doctrine
> would not have been possible except for the Louisiana Purchase.
> It was the logical outcome of that acquisition. Having taken her
> decisive stride across the Mississippi, the United States enlarged
> the horizon of her views and marched steadily forward to the
> possession of the Pacific Ocean. From this event dates the rise of
> the United States into the position of a world power. (Turner,
> "The Significance of the Mississippi Valley in American His-
> tory," in Turner, *The Frontier in American History* [New York,
> 1920], 100)

24. Livingston to Madison, Sept. 1, 1802, Walter Lowrie and Matthew St. Clair Clarke, eds., *American State Papers: Foreign Relations* (Washington, D.C., 1832–61), II, 525. Hereafter cited as *A.S.P.F.R.*

25. The classic work on French policy in Louisiana is Lyon, *Louisiana in French Diplomacy, 1759–1804.* Volumes I to III of Henry Adams's *History of the United States of America during the Administrations of Jefferson and Madison* also remain indispensable in the consideration of French policy. Arthur P. Whitaker, *The Mississippi Question, 1795–1803* (Gloucester, Mass., 1962 [New York, 1934]), emphasizes the Spanish side of the Louisiana story and draws heavily from Spanish sources; it is also excellent on the general background of the purchase. Still useful is Gustav Roloff, *Die Kolonialpolitik Napoleons I* (Munich, 1899) and Carl L. Lokke, *France and the Colonial*

Question: A Study of Contemporary French Opinion, 1763–1801 (New York, 1932).

26. Santo Domingo was both the Spanish name for the island and the common name. The French called it Saint Domingue. A contemporary English appreciation of the island's significance as a colony reads:

> As to the intrinsic value and importance of St. Domingo as a colony, it is almost beyond the power of calculation. That part of it which belonged to France before the war, which was barely one third of the island, and by far the least fertile, was more productive and profitable, in every point of view, than all the British West India islands taken together: the value of its annual exports were above 7,000,000 £. sterling, which employed 1640 ships, and 26,770 seamen . . . if France could only hold St. Domingo as a colony, she need hardly wish for more foreign possessions, as that island alone would be worth all the colonies which the other European states possess (taken collectively), both for intrinsic value and from the number of ships and seamen it would employ in time of peace, which would at once lay the firm foundation of a commerce and a navy, that at no very distant day must be superior to that of any other nation. (*Cobbett's Annual Register* [London, 1803], 210)

27. Mary W. Treudley, "The United States and Santo Domingo, 1789–1866," *Journal of Race Development,* 7 (July 1916), 84–97.

28. The secret instructions for General Victor, the intended Captain-General of Louisiana, stipulated: "The system of that colony, as in all colonies which we own, must be to aim to concentrate its commerce into the national commerce. It must be the special aim to establish relations between it and our Antilles, so that they may replace the American commerce in the later colonies in all articles whose importation and exportation are allowed to the Americans." Instructions for the Captain-General, Nov. 26, 1802, James Alexander Robertson, ed., *Louisiana Under the Rule of Spain, France, and the United States, 1785–1807* (Cleveland, 1911), I, 366. See also François Barbé-Marbois, *The History of Louisiana,* ed. E. Wilson Lyon (Baton Rouge and London, 1977 [1830]), 200, and Lyon, *Louisiana in French Diplomacy,* 113. Senator Ross, in urging that the eastern states forcefully assist the westerners in liberating New Orleans, identified the probable costs to the United States under the French colonial system: "Bonaparte will then say to you, my French West India colonies, and those of my allies, can be supplied from my colony of Louisiana, with flour, pork, beef, lumber, and any other necessary. These articles can be carried by my own ships, navigated by my own sailors. If you, on the Atlantic coast, wish to trade with my colonies in those articles, you must pay

fifteen or twenty per cent of an impost. We want no further supplies from you, and revenue to France must be the condition of all future intercourse." *Annals of Congress,* 7th Cong., 2nd sess., Feb. 14, 1803, 88.

29. Laussat to Decrès, April 18, 1803, Robertson, ed., *Louisiana,* II, 37. In the various proposals put forward in the 1790s for a French return to Louisiana, the dismemberment of the Union and attachment of the western states to Louisiana were often portrayed as a feasible and desirable goal. In the formulation of General Victor Collot, this detachment was indispensable: "I conclude that the western states of the North American republic must unite themselves with Louisiana and form in the future one single compact nation; else that colony, to whatever power it shall belong, will be conquered or devoured." Such a Louisiana, stretching east to the Alleghenies, would also serve as a tool for manipulating the politics of the United States. Frederick J. Turner, "The Diplomatic Contest for the Mississippi Valley," *Atlantic Monthly,* 93 (June 1904), 812.

30. The French Minister of Marine, Decrès, emphasized this protective role in his instructions for Victor: "Since Louisiana is the bulwark of Mexico, that consideration alone guarantees to the captain-general a reciprocity of interests from the governor of Florida." Instructions to the Captain-General, Nov. 26, 1802, Robertson, ed., *Louisiana,* I, 364.

31. The episode is recounted by Carl L. Lokke, "The Leclerc Instructions," *Journal of Negro History,* 10 (January 1925), 80–81.

32. Carl L. Lokke, "Jefferson and the Leclerc Expedition," *American Historical Review,* 33 (October 1927), 326. The understanding between Great Britain and France was confirmed by Lord Hawkesbury, the British Foreign Secretary, in a conversation with the American minister, Rufus King, on November 25, 1801. Charles R. King, ed., *The Life and Correspondence of Rufus King . . .* (New York, 1894–1900), IV, 18.

33. During the period of the quasi war between the United States and France (1798–1800), the Adams administration had formed a close relationship with Toussaint. Although the French had been expelled from Santo Domingo, and the island was for all practical purposes independent of French authority. Toussaint's power was by no means uncontested. The mulattoes, led by Andre Rigaud, controlled a major portion of the island and were backed by the French government, whose policy was to keep Santo Domingo in a condition of chaos until such time as it might be reconquered by French forces. In June 1798, the American government passed an embargo act that put an end to trade with French possessions. To escape its operation, which might well have proved fatal to his cause, Toussaint promised the

American government that if trade relations were restored with the island, he would undertake to set aside French regulations respecting American commerce and to guarantee the safety of American vessels trading with Santo Domingo. In February 1799, an act of Congress exempted the island from the embargo and trade relations with the United States were normalized. In turn, Toussaint acted to suppress privateering and to facilitate the entry of American vessels to the island. The English government was subsequently drawn into this arrangement, London's cooperation being indispensable to its operation. But the English role in the tripartite relationship was never the same as the American, since London's fear of a black rebellion in its West Indian possessions set severe limits on the aid it was willing to extend to Toussaint. By contrast, the American government fully supported Toussaint in his struggle with Rigaud. Not only did the Adams administration facilitate the supply of food and munitions of war to Toussaint's forces; it extended naval aid to him in the siege of Jacmel. As a result of this aid, Jacmel fell to the blacks in March 1800 and thereby ensured Toussaint's victory in his campaign to control the island. Great was Toussaint's gratitude to his American benefactors. The entire island was opened up to American trade. But this now intimate relationship was suddenly cut short in the fall of 1800 by the convention of Mortfontaine. In effect, the agreement signaled the abandonment of Toussaint to the designs of Napoleon. Once the British too concluded peace with the French, as they did a year later, the way was open for Napoleon to move to subjugate the island. Accounts of American relations with Santo Domingo may be found in Rayford W. Logan, *The Diplomatic Relations of the United States with Haiti, 1776–1891* (Chapel Hill, N.C., 1941) and Treudley, "The United States and Santo Domingo," 83–145.

34. Lokke, "The Leclerc Instructions," 93.

35. The conversation was reported by Pichon to Talleyrand on July 22, 1801. *Affaires Etrangères, Correspondance Politique, Etats-Unis,* LIII, 178. Transcripts from French Archives, Library of Congress.

36. Lokke, "The Leclerc Instructions," 93.

37. Madison to Livingston, Sept. 28, 1801, *A.S.P.F.R.,* II, 510–11.

38. Jefferson to Livingston, April 18, 1802, Ford, ed., *Writings of Jefferson,* IX, 363–69.

39. It is commonly assumed that the strongly worded contents of Jefferson's April 18 letter were made known to the French. Frederick Jackson Turner went so far as to assert that it was "intended also for the perusal of Napoleon." Turner, "The Diplomatic Contest for the Mississippi Valley," 816. Proof of this conclusion has never been sat-

isfactorily provided, and the weight of the evidence shows otherwise. Intended for the edification of Livingston, this unofficial letter advised: "I should suppose that all these considerations might in some proper form be brought into view of the government of France." This injunction was further weakened by Jefferson himself in a letter written to Livingston on May 5, 1802, a letter that would have arrived through Du Pont simultaneously with that of April 18. Addressing Du Pont's objections to the tone of his first letter, the President wrote: "I found he received false impressions of the scope of the letter. I have written him therefore an explaining one this moment, and being much hurried I have not time to copy it to you, but have desired him to communicate it to you." "I have got further into this matter than I meant when I began my letter of April 18," the President concluded, "not having deliberately intended to volunteer so far into the field of the Secretary of State," Gilbert Chinard, ed., *The Correspondence of Jefferson and Du Pont de Nemours* (Washington, D.C., 1939), 54. In this frequently neglected letter, Jefferson made reference to the less belligerent tone of his letter to Du Pont and emphasized the primacy of Livingston's official instructions from Madison. Explaining himself to Du Pont, Jefferson had declared that the idea of conquering Louisiana was "not that of a single reasonable and reflecting man in the U.S." Still, he persisted in his warnings, however remote, "that the day may come, when [conquest] would be thought of, is possible. But it is a very distant one." Jefferson to Du Point, May 5, 1802. Courtesy of Jefferson Papers, Princeton University. The May 1 instructions from Madison to Livingston staked out even more pacific ground, lacking any of the strong language characterizing Jefferson's April 18 letter and making no mention of the prospect of alliance with Britain. Madison to Livingston, May 1, 1802, Gaillard Hunt, ed., *The Writings of James Madison* (New York, 1900–1910), VI, 450–54.

Given such an inconsistent combination of instructions and hesitant advice, it would have been remarkable indeed had Livingston delivered the full force of Jefferson's letter to Talleyrand or Napoleon. A review of Livingston's dispatches from Paris indicates that he did not follow such a bold course. The minister to France sought primarily to confirm American rights to navigation and deposit on the Mississippi and to persuade the French that their colonial scheme was misguided. Although Livingston does make reference to his British card on December 23, 1802, one may question the forcefulness of what was said to the French. In a January 1803 memoir submitted by Livingston to the Emperor, the prospect of a "close and intimate connexion with Britain" is raised, but only as a potential result of the Spanish revocation of the deposit. *A.S.P.F.R.*, II, 520–40 (minor errors in *State Papers* checked against originals, Records of the Depart-

ment of State, National Archives). Irving Brant, making little allow-
ance for the nature of Livingston's instructions, concludes that the
minister's diplomacy actually undermined Jefferson's efforts. *James
Madison: Secretary of State, 1800–1809* (Indianapolis, Ind., and New
York, 1953), 59–140.

40. Du Pont to Jefferson, April 30, 1802, Dumas Malone, ed.,
*Correspondence Between Thomas Jefferson and Pierre Samuel du Pont
de Nemours, 1798–1817* (Boston and New York, 1930), 52–61.

41. Jefferson to Du Pont, May 5, 1802, courtesy of Jefferson Papers, Princeton University.

42. Rufus King to Madison, June 1, 1801, King, ed., *Correspondence of King*, III, 469.

43. Madison to King, July 24, 1801, Hunt, ed., *Writings of Madison*, VI, 435.

44. King to Madison, April 2, 1803, King, ed., *Correspondence
of King*, IV, 241.

45. Madison to Livingston and Monroe, May 28, 1803, Hunt,
ed., *Writings of Madison*, VII, 49.

46. Conference with Lord Hawkesbury, Nov. 25, 1801, from King's
Memorandum Book, in King, ed., *Correspondence of King*, IV, 18–
19.

47. King to Madison, Jan. 15, 1802, King, ed., *Correspondence
of King*, IV, 56.

48. *Cobbett's Annual Register*, 264.

49. Jefferson to Livingston, April 18, 1802, Ford, ed., *Writings of
Jefferson*, IX, 365.

50. J. Leitch Wright observes a very different British policy during
this interim period of peace. While concluding that London would
have been unlikely to enter into an alliance with America for the purpose of excluding France from the Mississippi, he maintains that the
Addison ministry held a unilateral policy of preventing a French occupation, and that the British would have employed force to secure
this end. J. Leitch Wright, Jr., *Britain and the American Frontier, 1783–
1815* (Athens, Ga., 1975), 131–33. The evidence Wright presents,
however, does not bear this out; indeed, it indicates just the opposite.
Wright points to a May 7, 1802, letter from Lord Hawkesbury to
Rufus King in which Hawkesbury states that Britain has not given
positive sanction to the retrocession. This is much less than a determination to forcefully oppose France. In fact, Hawkesbury betrayed
a certain willingness to countenance the French occupation when he
affirmed that a France possessed of the Mississippi must be bound to
uphold free navigation of the river to American and British subjects.
William R. Manning, ed., *Diplomatic Correspondence of the United
States: Canadian Relations, 1784–1860* (Washington, D.C., 1940), I,

532. In light of the great sacrifices made by Britain in concluding the Peace of Amiens, it strains credulity to believe that Britain would have destroyed this peace over Louisiana. East Florida, of course, was of more vital interest to the British, for in possession of this France might have mounted a serious threat to the sea lanes between the Caribbean and the British Isles. Britain took an active role in pressuring Spain against including the Floridas in the retrocession. On this role see Adams, *History,* I, 402, 403.

51. Madison gave indication in early 1802 that the American threat of alliance with Britain was to have no substantive basis. Robert Livingston, on December 30, 1801, had cautiously suggested to Rufus King that he might induce the British to throw obstacles before the French plan, a suggestion which Jefferson and Madison considered highly "delicate." Because Britain might abuse the American situation, "sowing jealousies in France," Livingston and King were told that "too much circumspection cannot be employed." Livingston to King, Dec. 30, 1801 and Madison to Livingston, March 16, 1802, *A.S.P.F.R.,* II, 512, 514.

52. Though Livingston later took up the refrain to a limited extent, the American threats of alliance with Britain were delivered primarily through the French envoy to the United States, Pichon. In late December of 1801, Jefferson warned the French Chargé that a French possession of Louisiana would result, when war again broke out in Europe, in a rupture with the United States and an alliance between the United States and Britain. This prospect would be raised again by Madison and Jefferson in conversations with Pichon, though always with declarations, no doubt sincerely felt, of how distasteful such an alliance appeared to the administration. Pichon to Talleyrand, Jan. 2, July 7, and Dec. 23, 1802, *Affaires Etrangères,* LIV, 17–18, 413; LV, 126. This British–American alliance was always portrayed as a distant but inevitable event, and Paris was given no evidence of substantive dealings between the two countries, save for increasing friendliness between the Americans and the British envoy to Washington. James Monroe, before departing for Paris, delivered the strongest warning to the French, telling Pichon that the United States would "receive the overtures which England was incessantly making" should his negotiations fail. Pichon to Talleyrand, Feb. 17, 1803, quoted in Adams, *History,* I, 440. There was no basis for Monroe's statement, as Adams long ago pointed out.

53. Franklin B. Sawvel, ed., *The Complete ANAS of Thomas Jefferson* (New York, 1903), 219.

54. Madison to Livingston and Monroe, April 18, 1803, Hunt, ed., *Writings of Madison,* VII, 37–43.

55. The secret instructions to the Captain-General of Louisiana

stipulated that he make no "innovation" in the commercial policy he found established on his arrival in New Orleans, and the dictates of self-interest recommended a prudent recognition of the interests of the western states. Robertson, ed., *Louisiana,* I, 366–68. Indeed, as the prospect of detaching the West from the Union was viewed as a possible windfall to the French occupation, navigation would have served as a seductive carrot in French hands.

As with so many of the designs of Napoleon, the ultimate French intentions regarding the deposit remain somewhat inscrutable. To the Americans, France displayed a tendency toward dissimulation, while behind the scenes it gave indications of support for the Spanish policy of closing the deposit. In Louisiana, the newly arrived French prefect, Laussat, lent staunch encouragement to Morales as the Intendant fended off demands by various Spanish colonial officials that he reverse his action. And in Paris, Talleyrand conveyed Napoleon's approbation of the policy to the Spanish minister Azara. Lyon, *Louisiana in French Diplomacy,* 174. At the same time, Jefferson did receive indications that American rights would be observed. From his trusted intermediary Du Pont came word in October 1802 that there "can be no doubts that your treaties with Spain . . . will be respected, confirmed, and renewed." Du Pont to Jefferson, Oct. 4, 1802, Malone, ed., *Correspondence Between Jefferson and Du Pont,* 68. Although Livingston reported on November 11, 1802, that Talleyrand had given a verbal promise that American treaty rights on the Mississippi would be "strictly observed," he apparently considered the issue unsettled during the winter and spring of 1803. *A.S.P.F.R.,* II, 526, 528–55. Immediately following the issuance of the April 18 instructions, Jefferson and Madison received what they took to be a confirmation of American rights, rights "expressly saved by the instrument of [Spain's] cession of Louisiana to France." Jefferson to Dr. Hugh Williamson, April 30, 1803, Andrew A. Lipscomb and Albert Bergh, eds., *The Writings of Thomas Jefferson* (Washington, D.C., 1904–5), X, 385–86.

56. If we may judge by the persistent French preparations made throughout 1802 toward the occupation of Louisiana, it would seem that Napoleon was unshaken by the alarms sounded by Pichon. Considering the nature of the American threats, Napoleon's response was hardly surprising. Whether predicated on alliance with Britain or on a renewal of the conflict in Europe, the prospect that force would be employed by the United States was ultimately dependent on British intentions. The First Consul had good reason to believe that London would not break the peace over Louisiana; see above, note 50; and below, note 57. Noticeably absent from Pichon's dispatches is any evidence of actual U.S. military preparations; indeed, even Pichon's

remarks tend to betray the hollowness of the American threats. See, for example, Pichon to Talleyrand, July 7, 1802, *Affaires Etrangères,* LIV, 414. Undoubtedly, the windfall of an aroused West produced by Spanish policy at New Orleans did eventually endow the American position with a certain degree of strength, though Jefferson quickly acted to restrain western sentiment; see below, note 74. If the substance of American threats was suspect, the channel through which they were delivered was dubious as well. Beyond opposing Napoleon's colonial scheme from its inception, Pichon occasionally made himself appear a more zealous advocate of American than of French interests. Leclerc, commander of the French expedition on Santo Domingo and brother-in-law to Napoleon, had complained bitterly over the Chargé's conduct, eventually breaking off communications with him and recommending his replacement. Leclerc to Decrès, May 6, 1802 and Leclerc to Napoleon, June 6, 1802, Henry Adams, ed., *French State Papers* (Transcripts in Library of Congress), I. On July 4, 1802, Napoleon directed Talleyrand to rebuke Pichon and to admonish the Chargé against engaging in the extended general discussions "which will only diminish the French government." After a long lesson in diplomatic tact, Napoleon ordered that Pichon be reminded "that the first quality required of a diplomat is that he thinks not as those of the country where he is serving, but as those of the country which he represents." Albert Du Casse, ed., *Correspondance de Napoleon Ier* (Paris, 1858–70), VII, 508–9.

57. This impression of British policy was reported to Napoleon through his Ambassador to London, Andreossy, who detailed a profound British inattention to Louisiana and a fixation on the continued occupation of Holland by France. P. Coquelle, *Napoleon and England, 1803–1813: A Study from Unprinted Documents,* trans. Gordon D. Knox (London, 1904), 13–36.

58. Barbé-Marbois, *History of Louisiana,* 312. Napoleon's motives are discussed on pages 207–10.

59. See Alexander DeConde, *This Affair of Louisiana* (New York, 1976), 115, 135; Malone, *Jefferson and His Time,* IV, 285–86; Gilbert Chinard, *Thomas Jefferson: Apostle of Americanism,* 2nd ed. (Ann Arbor, Mich., 1957 [1st ed., 1929]), 406; Brant, *Madison: Secretary of State,* 80; and Bradford Perkins, *The First Rapprochement: England and the United States, 1795–1805* (Berkeley and Los Angeles, 1967), 167.

60. Madison to Livingston, May 1, 1802, Hunt, ed., *Writings of Madison,* VI, 452.

61. Pichon to Talleyrand, Jan. 2, 1802, *Affaires Etrangères,* LIV, 17–18.

62. Livingston to Madison, Jan. 13, 1802, *State Papers and Cor-*

respondence Bearing Upon the Purchase of the Territory of Louisiana (Washington, D.C., 1903), 11–12.

63. Jefferson to Livingston, Oct. 10, 1802, Paul Leicester Ford, ed., *The Works of Thomas Jefferson* (New York, 1904–5), IX, 397.

64. See above, note 59. Of the writers mentioned, Malone is more circumspect in stipulating a particular *casus belli*, stating that "if Jefferson's own words are to be believed, he was ready to use force eventually if his peaceful efforts to meet the imperative needs of his country should fail." Malone, *Jefferson and His Time*, IV, 285–86.

65. See Livingston to Madison, Feb. 26, 1802, *A.S.P.F.R.*, II, 513.

66. Livingston to Madison, April 24, 1802, *A.S.P.F.R.*, II, 515–16.

67. May 28, Livingston to Madison, 1802, *A.S.P.F.R.*, II, 518.

68. Livingston to Madison, Aug. 16, Nov. 11, 1802, *A.S.P.F.R.*, II, 524–26.

69. Madison to Livingston and Monroe, March 2, 1803, Hunt, ed., *Writings of Madison*, VII, 9–30. Also Madison to Monroe, March 2, 1803, ibid., 30–34.

70. Adams, *History*, I, 442.

71. Madison to Livingston and Monroe, April 18, 1803, Hunt, ed., *Writings of Madison*, VII, 37–43.

72. See above, note 55.

73. See Whitaker, *The Mississippi Question*, 207–9; Jefferson to Williamson, April 30, 1803, Lipscomb and Bergh, eds., *Writings*, X, 386.

74. Jefferson to Monroe, Jan. 13, 1803, Lipscomb and Bergh, eds., *Writings*, X, 343–46. Jefferson revealed his political and pacific intentions immediately to the French, telling Pichon "that Mr. Monroe was so well known to be a friend of the Western people that his mission would contribute more than anything else to tranquillize them and prevent unfortunate incidents." Pichon to Talleyrand, Jan. 21, 24, 1803, quoted in Brant, *Madison: Secretary of State*, 115. Brant discounts the notion that this remark had any significant negative effect, concentrating on the relatively stronger stance assumed by Madison at the same time (pp. 116–18). Nevertheless, reports of this statement and of the pacified West were not without effect in Paris, where Talleyrand reported his pleasure that the wise policies of the American government had defused the crisis. Talleyrand to Livingston, March 21, 1803, *A.S.P.F.R.*, II, 550.

75. Livingston to Madison, Nov. 11, 1802, *A.S.P.F.R.*, II, 526.

76. Lawrence Delbert Cress, *Citizens in Arms: The Army and the Militia in American Society to the War of 1812* (Chapel Hill, N.C., 1982), 153. Also Theodore J. Crackel, *Mr. Jefferson's Army: Political*

and Social Reform of the Military Establishment, 1801–1809 (New York, 1987), who argues that the actual reductions in military strength from the force existing at the close of the Adams administration was relatively minor, comprising a cut of about 300 regulars.

77. Second Annual Message, Dec. 15, 1802, Ford, ed., *Writings of Jefferson*, VIII, 181.

78. The measures taken on the Mississippi frontier in the winter and spring of 1803 are given in detail by Mary P. Adams, "Jefferson's Reaction to the Treaty of San Ildefonso," *Journal of Southern History*, 21 (May 1955), 173–88.

79. The point is simple yet critical. It was not the French occupation of New Orleans that was regarded as a *casus belli* but the denial of the free use of the Mississippi and a place of deposit. There ought to be no confusion on this point since the President and the Secretary of State were quite clear in setting it out. Nevertheless, historians persist in going beyond the record. Thus DeConde, intent on portraying a bellicose Jefferson, insists that the President was prepared to seize New Orleans by force if he could not get it by purchase. *This Affair of Louisiana*, 115–46. But the evidence he offers in support is unpersuasive. The American force at Fort Adams does not establish the intent to assault New Orleans. Nor does the January 1803 statement of the Governor of the Mississippi Territory, William C. C. Claiborne, that 600 of the 2,000 militia he had could take New Orleans provided that only Spanish troops defended the city.

In making her case that Jefferson prepared for every contingency and that "he was ready for a showdown" in the spring of 1803, Mary Adams, "Jefferson's Reaction to the Treaty of San Ildefonso," 177, cities the "long and important letter" of March 7, 1803, written by the Secretary of War, Henry Dearborn, to James Wilkinson. From this letter, Adams writes, "the aims of the government and the action it planned to take, if James Monroe's mission to France failed, become clear." What becomes clear from the letter, however, was not a plan of action in the event Monroe's mission failed, but the expectation that there would be no serious trouble on the Mississippi. Although ordering Wilkinson to take measures of precaution "for a different state of things," Dearborn wrote that "we have the most explicit assurance of the friendly disposition of the French and Spanish courts, and have good reason to believe that all difficulties relating to the navigation of the Mississippi will be amicably adjusted." Dearborn went on to note: "From the latest accounts from France it appears that no troops are to accompany General Victor . . . that Spain is to continue to garrison the country until the St. Domingo business is settled, that the troops which were destined for New Orleans are to go to the West Indies." The awaited results of Monroe's mission are brought up in the context of calling up, if necessary, the militia force of 80,000 authorized by Congress. But Dearborn then observed: "It is reasonable to expect that we under all circumstances shall be able to make such arrangements with France and Spain as will place our affairs relative to the Mississippi on a satisfactory footing." *Military Book, 1800–1803* (National Archives), 380–82.

80. The commitment to peace that dominated Jefferson's diplomacy may appear to be contradicted by the action undertaken against the Barbary powers during the first administration. One of Jefferson's first acts as President was to send a squadron of four vessels to the Mediterranean. He did so in response to persistent rumors of impending trouble in the form of attack by the Barbary powers on American commerce. In particular, the pasha of Tripoli was reported as dissatisfied with the terms of the 1796 treaty between Tripoli and the United States, a treaty providing for annual tribute to the pasha on condition that he remain at peace with the United States and refrain from the plunder of American commerce. To prevent any of the Barbary states, but especially Tripoli, from breaking the peace, the squadron was sent with orders to determine the disposition of these states. If any were found to be at war with the United States, the commander was directed to act "so as best to protect our commerce and chastise their insolence—by sinking, burning or destroying their ships and vessels wherever you shall find them." In his orders the commander was pointedly reminded that the "force of Tunis and Tripoli are contemptible, and might be crushed with any of the Frigates under your command." Acting Secretary of the Navy Samuel Smith to Captain Richard Dale, May 20, 1801, Captain Dudley W. Knox, ed., *Naval Documents Related to the United States Wars with the Barbary Powers* (Washington, D.C., 1939–44), I, 467.

In taking this measure, Jefferson appeared to act in accordance with earlier conviction. From an early date, he had disapproved of dealing with the petty states of North Africa by a system that, in effect, purchased peace. This was the system practiced by the European states, the reasoning for it being that the payment of tribute and ransom was less costly than war. Jefferson rejected such reasoning. As he wrote to John Adams in 1786: "I acknowledge, I very early thought it would be best to effect a peace through the medium of war." Jefferson to Adams, July 11, 1786, ibid., 10. Justice and honor favored this course, he then believed, as did interest, since in time it would prove to be the least expensive policy. While European cooperation in establishing a "perpetual cruise" against Barbary could make the American burden "a very light thing," Jefferson was plainly drawn to the experiment in coercion even if the action was to be unilateral. Jefferson to Edward Rutledge, July 4, 1790, Lipscomb and Bergh, eds., *Writings*, VIII, 59; "Report relative to the Mediterranean trade," Dec. 28, 1790, ibid., III, 96–103. And if American resources were not yet sufficient for the attempt at rejecting the tributary system, then at least the country should avoid the European practice of ransoming hostages, for doing so "would occasion the capture of greater numbers by increasing the incitements to cruise against us." Jefferson to William Carmichael, Sept. 25, 1787, Ford, ed., *Works of Jefferson*, V, 345. Although these early views were moderated during the 1790s, they were not abandoned.

In sending a naval force to the Mediterranean under orders to protect American commerce from attacks by the Barbary satrapies, Jefferson plainly demonstrated that he had no compunction in principle against the use of force. But the use of force was not necessarily synonymous with war in any but a formal sense. Even if one or more of these petty principalities declared war against this country—which Tripoli did in fact do—there was a world of difference between war with these ministates and war with a European power. The former was no more than what we would term today a police action. In its scope and effects, it was not to be confused with a war fought against one of the European states.

Certainly Jefferson did not indulge in any such confusion. He appreciated from the outset the limited nature of the action he had determined to carry out, and the rules of engagement given the com mander of the force reflected this. So did the objective the administration sought to achieve by undertaking the action, which was not to reform, let alone to abolish, the tributary system but to encourage the rulers of the North African states to behave more reasonably and, in particular, to abide by existing treaties they had concluded with the United States. These treaties, concluded in 1795 and 1796, had provided for annual tribute. But the Pasha of Tripoli, Yusuf Caramanli, aware that Algiers had managed to secure an American tribute substantially greater than his own, and facing the costs entailed by a rule recently usurped, had become dissatisfied and, in consequence, had broken the treaty and declared war on the United States. In deciding to employ force in the Mediterranean, Jefferson's consistent goal was simply to restrain the demands of Tripoli and the other Barbary states. (Christopher McKee provides a valuable account of the administration's formulation of policy in *Edward Preble: A Naval Biography, 1761–1807* [Annapolis, Md., 1972], particularly 86–88, 102–104. See also Ray W. Irvin, *The Diplomatic Relations of the United States with the Barbary Powers, 1776–1816* [Chapel Hill, N.C., 1931], Gardner W. Allen, *Our Navy and the Barbary Corsairs* [Hamden, Conn., 1965], Harold Sprout and Margaret Sprout, *The Rise of American Naval Power, 1776–1918* [Princeton, N.J., 1944], 55–61, and Louis B. Wright and Julia H. Macleod, *The First Americans in North Africa: William Eaton's Struggle for a Vigorous Policy Against the Barbary Pirates, 1799–1805* [Princeton, N.J., 1945].)

The squadron sent out to induce such moderation consisted initially of just four vessels. Its commander was directed by the administration to sink or destroy the vessels of any Barbary power at war with the United States, to interdict such forces in the enemy's home waters and at Gibraltar (to keep the Atlantic free of threat), and if resources permitted, to convoy American merchant vessels in the Mediterranean. The operational requirements of such a strategy proved greater than the means provided. Hampered by logistical problems of supply and distance, with their forces spread thin and from time to time employed in coercive calls at the ports of other restless Barbary states, a succession of American commanders proved unable to maintain an unbroken blockade of Tripoli or to bring the contest to decision. The consequent "temporizing mode of warfare," complained one advocate of a more energetic policy, would certainly fail to humble Tripoli (whose cruisers were able to avoid the porous blockade even on the rare occasions when it was applied), and would in turn "encourage the other Barbary powers to become insolent." William Eaton to Madison, June 8, 1802, Knox, ed., *Naval Documents*, II, 166–69.

Once this initial optimism attending the Mediterranean policy had dampened, strains within the administration over the proper course of action appeared. Jefferson and his Secretary of State had come to embrace the expedition on the assumption that it would entail quite limited costs; law required the maintenance of six frigates in any case, and these vessels could be employed with greater advantage and "at nearly the same expense" on the Barbary coast. Madison to William Eaton and Richard O'Brien, May 20, 1801, Knox, ed., *Naval Documents*, I, 460. Albert Gallatin, charged with carrying out the administration's ambitious plans for fiscal restraint and debt retirement, held a less sanguine view of the costs entailed. Writing to Jefferson after a year of inconclusive action in the Mediterranean, Gallatin rejected the pretensions of the policy and urged payment of an annuity to Tripoli: "I consider it no greater disgrace to pay them than Algiers. And indeed we share the dishonor of paying those barbarians with so many nations as powerful and interested as ourselves, that, in our present situation, I consider it a mere matter of calculation whether the purchase of peace is not cheaper than war, which shall not even give us the free use of the Mediterranean trade." Gallatin to Jefferson, Aug. 16, 1802, Henry Adams, ed., *The Writings of Albert Gallatin* (New York, 1960 [Philadelphia, 1879]), I, 88.

In large measure, Gallatin's worst fears would be realized as the administration's deepening commitment required increasing augmentations of force and the costs of a new building program. Conversely, the Secretary of the Navy, Robert Smith, who held that force would prove the cheaper means, offered an equally prescient criticism. Only an increase in the Mediterranean squadron could bring the war to a favorable conclusion, he argued in September 1802, and with "less force the war may continue for years, which would be playing a hazardous game." Quoted in McKee, *Preble,* 103. The President, torn between his commitment to economy and his insistence that the Pasha of Tripoli, at least, receive no additional tribute, continued to ratchet up the naval force and to widen its latitude for action.

Jefferson's conviction that the frustration of his policy owed in large part to the manner of its implementation—most notably in the tendency of Commodore Morris to favor convoy over blockade—was reflected in the instructions given to Captain Preble, the new commander of the force, in July 1803. Robert Smith to Edward Preble, July 13, 1803, Knox, ed., *Naval Documents,* II, 474–77; see also Jefferson to Madison, March 22, 1803, Ford, ed., *Works of Jefferson,* IX, 455. In addition to a tight blockade of Tripoli, Preble was admonished "you will by all the means in your power annoy the Enemy." The Commodore proved up to his task. Using smaller gunboats borrowed from the Kingdom of the Two Sicilies, the men under Preble's command engaged the Tripolitan cruisers in the harbor, fighting toe to toe with great success. Eluding the fire from shore batteries, the forty-four–gun *Constitution* sent repeated barrages into the fortress and town. This energetic policy was made more daring by the presence of over 300 American prisoners—the crew of the captured frigate *Philadelphia*—now in the custody of Pasha Yusuf. Warning that the heavy bombardment might drive the Pasha into taking retribution against the prisoners, the French Consul Beaussier thought that the hostage crew "must shudder at every discharge of your cannon." Bonaventure Beaussier to Preble, Aug. 6, 1804, Knox, ed., *Naval Documents,* IV, 369. The imprisonment of the crew also rendered Preble's campaign less effective, strengthening the Pasha's determination to ride out the attacks and await better terms.

Although Tripolitan possession of the *Philadelphia* proved fleeting—Stephen Decatur having set the ship ablaze in the bold raid which began Preble's summer campaign—its loss was an embarrassment to the Jefferson administration. In response to the apparently worsening military situation, the administration urged on Congress a new, if restrained, naval building program. The Mediterranean Fund, generated by enhanced import duties, authorized new construction and provided funding for stepped-up operations against Tripoli.

The action against the Barbary states had begun in early 1801. By the spring of 1804, the administration was eager for a settlement. A new commander, Commodore Samuel Barron, was dispatched with a still larger force and authorized to take measures necessary to bring the conflict to a successful conclusion. Barron was authorized, at his discretion, to unleash Hamet Caramanli, who had been displaced by his brother as Pasha. Robert Smith to Captain Samuel Barron, June 6, 1804, Knox, ed., *Naval Documents*, IV, 153. William Eaton, the former American Consul at Tunis, had long advocated using the ex-Pasha against his brother Yusuf. Marching from Egypt into eastern Tripoli, in the spring of 1805, Hamet Caramanli's "army" of loyal followers, mercenaries, and nine American marines captured and then held (with timely naval support) the town of Derne. Whether or not this force, in effect led by the colorful Eaton, could have made the trek to Tripoli and actually challenged the rule of Yusuf must remain an open question. For soon after the capture of Derne, in part as a consequence of this success, the Pasha of Tripoli was brought to terms. Derne was surrendered; the Americans, Hamet, and his entourage were spirited away in the night on board American vessels.

Thus the "war" against the Barbary states came to an end. Despite the capture of Derne and the large American squadron standing off Tripoli, Commodore Barron and Tobias Lear, the American negotiator, had continued to search for a settlement. In the terms finally accepted by Yusuf, $60,000 was offered for the balance of the prisoners (the Americans held a smaller number of Tripolitans), and $5,000 was to be paid on establishing a consulate at Tripoli. Although the terms were highly advantageous to the United States, the treaty nevertheless proved extremely unpopular at home. Lear's reputation was never to recover. Eaton, unrestrained in his criticism of the treaty, lambasted the "Machiavelian Commissioner" for the "purchased" peace, holding that Hamet, if supported more energetically by American forces, could have taken Tripoli and established an honorable and permanent peace with the United States. William Eaton to Secretary of the Navy, Aug. 9, 1805, Knox, ed., *Naval Documents*, VI, 213–19. Edward Preble called the settlement a "sacrifice of national honor." Preble to William Eaton, Feb. 8, 1806, *Naval Documents*, VI, 364.

Although the criticisms of Eaton and Preble were too harsh, it is the case that the settlement was no triumph of principle. Its unpopularity was due in no small measure to the feeling that it was instead of betrayal of principle. In the end, peace had been effected, to use Jefferson's earlier words, only in part "through the medium of war." It had also been effected in part by the very methods that Jefferson had always set his face against. Of greater relevance here, though, is recognition that the action against the Barbary states set no precedent for the Jefferson administration. It did not reflect a willingness to use force, even on behalf of vital interests, when such use went beyond the demonstrative bounds of a police action. If anything, the difficulties encountered in challenging such a minor state as Tripoli added a note of caution to the administration's view of military power. The Barbary War is noteworthy not as an exception to an otherwise consistent pattern of behavior respecting force but as an example of the difficulties of adhering to principle even in matters of less than vital interest to the state.

81. Jefferson to Joseph Priestley, Jan. 29, 1804, Ford, ed., *Writings of Jefferson*, VIII, 295. "I did not expect he [Napoleon] would yield till a war took place between France and England, and my hope was to palliate and endure . . . until that event. I believed the event not very distant, but acknowledge it came on sooner than I had expected."

82. Pichon to Talleyrand, July 22, 1801, *Affaires Etrangères*, LIII, 177–79.

83. By the time of Jefferson's conversation with Pichon, the administration had reason to believe that the rumors of a retrocession of Louisiana to France were based on fact. It has often been assumed that given this knowledge, the President saw from the outset the relationship between Santo Domingo and Louisiana in French plans. Why then, had the President offered to assist France in its effort to regain control of the island? One possible explanation is that the administration was suggesting or implying a deal between this country and France: the American government would cooperate in helping Napoleon suppress the rebellion on Santo Domingo and in return the French would abandon any ambitions they might have in Louisiana. Logan, *Relations of the United States with Haiti*, 119, and Merrill D. Peterson, *Thomas Jefferson and the New Nation*, 749. This explanation presumably receives support from an earlier exchange between Pichon and Madison. The French representative had come to Madison to inquire about the information the American government had from its Consul in Santo Domingo. The French feared that the rebel leadership on the island might soon declare independence from France and desired to know what the disposition of the American government would be in that event. Although Madison had allowed that the United States would not favor such a declaration by Toussaint, he offered the French Chargé little satisfaction by asserting the American desire to continue trading with the island and in accepting things "in the colony such as they were without intending to judge them." Moreover, before Pichon could withdraw, Madison raised the issue of the rumored retrocession of Louisiana, warning that a French possession must result in "daily collisions" on the Mississippi, "which

the distance between the two governments would render fatal to peace."
Pichon to Talleyrand, July 20, 1801, *Affaires Etrangères*, LIII, 169–
71. The Secretary's response was the first of a long series of exchanges
in which the administration warned that a French return to Louisiana
must eventually result in hostilities. But whether Madison intimated a
modus vivendi in which American assistance on Santo Domingo would
be contingent on a French promise to forgo a return to Louisiana may
be doubted. The argument supporting this interpretation must rest, in
the last analysis, simply on the fact that Madison raised Louisiana in
the context of a discussion of Santo Domingo. That he was resistant
to French pressure for cutting off American trade with Santo Domingo reflected his reluctance to concede what were, after all, important American commercial interests. In any case, Madison's remarks
were not further clarified, let alone strengthened, by Pichon's subsequent conversation with Jefferson. Had the administration intended
to suggest some sort of deal, its diplomatic method clearly misfired.
Napoleon, obviously encouraged, responded to Jefferson's promises
rather than the Secretary of State's discouraging tone, writing in the
secret instructions to General Leclerc: "Jefferson has promised that
the instant the French army arrives all measures will be taken to reduce Toussaint to starvation and to aid the army." Lokke, "The Leclerc Instructions," 93. If we may judge from the subsequent French
preparations for an occupation of Louisiana, it is apparent that Napoleon did not view this expected aid to be contingent on the abandonment of Louisiana. Indeed, if this is what Jefferson had in mind
when he made his promise of help to Pichon, it was the President as
much as his interlocutor who was deceived. Napoleon could not be
expected to enter into any such arrangement or, if entering into such
an arrangement, to observe it. His plans for Louisiana were an inte-
gral part of his general scheme of empire in North America. To give
them up would jeopardize the larger undertaking. Besides, there was
no reliable way to enforce an agreement of this kind, short of both
sides taking physical possession of their respective objectives at roughly
the same time, a prospect that was virtually excluded on the American
side. Napoleon might even have promised abstention with respect to
Louisiana, knowing that the French occupation of the colony would
have to be put off in any event until such time as he had sufficient
resources to undertake it. The idea that we might make an agreement of this kind with Napoleon, an agreement that would be self-
enforcing, betrayed a poor understanding of the effects that peace
would have on our position. In war, Louisiana was protected—at least
to a point—by British sea power against a French attempt to occupy
it. In peace, it was unprotected from a French occupation unless the
United States chose to mount a defensive effort it showed no disposition to make in 1801.

It may be, however, that Jefferson made his promise to Pichon in the expectation, or hope, that we would earn France's good will as a result and that this might be counted on with respect to Louisiana. Dumas Malone comes close to suggesting this by arguing that, having displayed his "excellent dispositions" toward France, Jefferson was in a position to deliver his warnings over Louisiana candidly. *Jefferson and His Time*, IV, 253. Was this an instance in which the President intended to put his well-known views about the gratitude of nations to a test? If so, he was even more deceived than in thinking that a deal involving Santo Domingo and Louisiana might be struck with the First Consul. Napoleon did not know of gratitude in the affairs of states. He would only have concluded that Jefferson was simple-minded or that the offer hid motives he could not discern. But what those motives might have been remains unclear so long as it is assumed that the President saw the relationship between Santo Domingo and Louisiana. They remain unclear for the reason that it is ludicrous to suggest that any conceivable interest might have prompted Jefferson to jeopardize the security of the West.

The mystery of Jefferson's motivations largely disappears, however, if it is once assumed that he did not see the relationship between Santo Domingo and Louisiana in the summer of 1801. On this assumption, it is reasonable to believe that the President made his promise to Pichon because—as noted in the text—he wanted to preserve the relationship with France set out in the convention of Mortfontaine and because he felt obliged to respond to the mounting fears of the southern slaveholders over the spread of the black rebellion on Santo Domingo. If the reimposition of French power on Santo Domingo carried little or no adverse implications for American security interests in the West, Jefferson's offer to Pichon made sense.

84. Pichon to Talleyrand, Feb. 23, 1802, *Affaires Etrangères*, LIV, 162–63.

85. Leclerc closed all of the island's ports save the two—Cap Francois and Port-au-Prince—through which trade with French forces was permitted. Leclerc decreed the confiscation of any ship violating the port closure and, presumably, trading with Toussaint. These measures the French had, as sovereign over Santo Domingo, every right to take and the American government did not contest the exercise of that right. It did reject the French insistence that it take measures to forbid all trade between its citizens and Toussaint, contending instead, as Jefferson did to the French Chargé, that in standing aside while American merchants suffered the consequences of violating the decrees of Leclerc, this government behaved altogether properly. Logan, *Relations of the United States with Haiti*, 130. At the same time, the administration contended that there was virtually no trade going on between its citizens and Toussaint's forces, by virtue of the effec-

tiveness of the French measures. Madison to Livingston, July 6, 1802, Hunt, ed., *Writings of Madison*, VI, 458.

86. King to Madison, Nov. 20, 1801, King, ed., *Correspondence of King*, IV, 15–16.

87. Livingston to Madison, Dec. 10, 1801, Jan. 13, 1802, and Livingston to King, Dec. 30, 1801, *A.S.P.F.R.*, II, 512–13.

88. Livingston to Madison, Dec. 12 and 31, 1801, and Jan. 13, 1802, *A.S.P.F.R.*, II, 512–13.

89. Lear's dismissal actually conformed to French policy, as American consuls and commercial agents were not recognized in French colonies. Leclerc's initial desperation for American supplies had forced him to tolerate the presence of Lear for a brief period. Treudley, "The United States and Santo Domingo," 139–40.

90. Lear to Madison, March 29, 1802, *Consular Dispatches* (Department of State Archives, in National Archives).

91. Jefferson to Livingston, April 18, 1802, Ford, ed., *Writings of Jefferson*, IX, 363–68. In a dispatch of March 29, Tobias Lear reached a similar conclusion. As previously translated from code, Lear affirmed "that part of the forces sent out are intended that country [Louisiana] there is no doubt, but I [am] still of the opinion that if they touch here there will be employment for the troops in the island for some time to come." Lear to Madison, March 29, 1802, *Consular Dispatches*. See also Lear to Madison, Feb. 12, 1802, ibid.

92. Lear to Madison, Feb. 12, 1802, *Consular Dispatches*.

93. Thomas O. Ott, *The Haitian Revolution, 1764–1804* (Knoxville, Tenn., 1973), 144, concludes that "the First Consul's misconception about the Negroes was the greatest flop in his plans for Western empire." Ott further argues that Napoleon's plans for his colonial scheme generally reveal poor preparation. To the First Consul's illusions about the inferior fighting ability of the blacks must be added, Ott writes, the neglect of supply and logistics and an expedition that needed to be at least twice as large as the one Napoleon sent.

94. Bonaparte to Decrès, June 4, 1802, Du Casse, ed., *Correspondance de Napoleon*, VII, 485.

95. Despite the devastating impact of yellow fever on the French army, Leclerc's position in midsummer was not bad. The powerful black generals on the island, among them Dessalines, Christophe, and Laplume, with their native troops, seemed under Leclerc's sway and were brutally executing his orders to disarm and annihilate the rebels. The general's optimism was shattered as word of Napoleon's reversal on slavery spread across the island. On August 6, 1802, he despairingly wrote to Bonaparte: "I entreated you, Citizen Consul, to do nothing which might make them anxious about their liberty until I was ready. . . . Suddenly the law arrived here which authorises the

slave-trade in the colonies. . . . More than all that, General Riche-
panse has just taken a decision to re-establish slavery in Guadeloupe.
In this state of affairs, Citizen Consul, the moral force I had obtained
here is destroyed. I can do nothing more by persuasion. I can depend
only on force and I have no troops." Leclerc to the Minister of Marine, Aug. 6, 1802, in C. L. R. James, *The Black Jacobins: Toussaint
L'Ouverture and the San Domingo Revolution* (New York, 1963),
343.

96. Leclerc to Decrès, May 8, June 6, Aug. 25, 1802, and Leclerc
to Napoleon, July 6, 1802, in *French State Papers,* I (Henry Adams
Transcripts, Library of Congress).

97. Criticizing the earlier work of Sir John Fortescue (*History of
the British Army* [London, 1910–30], IV), David P. Geggus's recent
account of the British effort to subdue Santo Domingo in the 1790s
gives evidence that British losses on the island were significantly more
modest than previously contended. In Santo Domingo, he writes, of a
total of some 20,000 British troops about 12,700 lost their lives, the
great majority from disease (mainly yellow fever). Only about 1,000
were battle casualties. If these figures are roughly correct, they show
that British losses on the island were quite modest compared with
French losses in 1802–3. David P. Geggus, *Slaves, War and Revolution: The British Occupation of Saint Domingue, 1793–1798* (Oxford, 1982).

98. For an account of the various abortive expeditions during this
period, see Lyon, *Louisiana in French Diplomacy,* 129–44. In the
case of the last expedition, the commander had received orders from
Napoleon in late December to depart immediately. A further delay to
take on food and wood, however, immobilized the soon icebound
expedition.

99. Events in Europe, destabilized by Bonaparte's encroachments
on the Amiens peace, finally led Britain to abandon the treaty and
decide on war. By the treaty, Britain was to abandon the island of
Malta, but with indications that France might make a new attempt at
Egypt the British refused to abandon the islands. In addition, the British had expected France to evacuate the Low Countries, an action not
specifically required by the terms of Amiens but which would have
been in accord with other continental treaties. But Napoleon showed
no disposition to leave. From the British came proposals that Britain
be allowed a ten-year occupation of Malta as a quid pro quo for the
French occupation of Holland, or that both countries abandon their
positions. So eager were the British to lead Napoleon to a reasonable
compromise, that bribes were employed among his inner circle in an
attempt to facilitate negotiations. Carl L. Lokke, "Secret Negotiations
to Maintain the Peace of Amiens," *American Historical Review,* 49

(October 1943), 55–64, and P. Coquelle, *Napoleon and England, 1803–1813. A Study from Unprinted Documents,* trans. Gordon D. Knox (London, 1904), 13–36.

100. Jefferson to Horatio Gates, July 11, 1803, Ford, ed., *Writings of Jefferson,* VIII, 250.

101. See Ott, *The Haitian Revolution,* 179–82.

102. Jefferson to William H. Harrison, Feb. 27, 1803, Lipscomb and Bergh, eds., *Writings,* X, 368–73. The same method of gaining Indian lands is put forward in a letter to the Secretary of War. Jefferson to Henry Dearborn, Aug. 12, 1802, Clarence Edwin Carter, ed., *The Territorial Papers of the United States* (Washington, D.C., 1939), VII, 68–70. Bernard W. Sheehan views this resort to "reprehensible coercion" as an exception to Jefferson's more humanely motivated Indian policy. *Seeds of Extinction: Jeffersonian Philanthropy and the American Indian* (Chapel Hill, N.C., 1973), 171. To the extent to which this is true, it indicates the weight of necessity dictated by Jefferson's reason of state. Though the President was unwilling to fundamentally alter his methods of diplomacy during the Louisiana crisis, less imperative principles, such as his philanthropic attitude toward the Indians, were easily shunted aside when set against the demands of territorial expansion. The manipulative use of debt to gain Indian lands, while surely an extreme manifestation of Jeffersonian policy, was not in practice inconsistent with its overall aims and methods. For it was a given under Jefferson, as it had been under his predecessors, that the land hunger of the westerners would have to be satisfied. At the same time, the whites would have to be restrained. As conceived during the first Washington administration, the policy which Jefferson inherited entailed a recognition of the Indians' rights to their lands and stipulated that at least the form of fair purchase be followed in appropriating these lands. Although the policy was in part a product of the genuine concern over the future of the Indians as expressed by the likes of Henry Knox, Washington, and Jefferson, it was also seen as the most cost-effective means of acquiring these lands (the expensive Indian wars of the late 1780s and early 1790s were seen as the result of unchecked white encroachment). Reginald Horsman, "American Indian Policy in the Old Northwest, 1783–1812," *William and Mary Quarterly,* 18 (January 1961), 35–53. In Jefferson's specific formulation of this policy, the best means of securing lands and the greatest good for the Indians was to encourage their assimilation into the white, agricultural society. A proliferation of government trading houses would display to the Indians the better lifestyle to be obtained through the abandonment of hunting and adoption of agriculture. The government would aid the Indians by teaching agricultural methods and provide them with needed capital through the pur-

chase of lands made extraneous by the abandonment of hunting. Jefferson to Benjamin Hawkins, Feb. 18, 1803, Lipscomb and Bergh, *Writings,* X, 360–65, and a confidential message to Congress, Jan. 18, 1803, ibid., 489–94. For its good efforts to "civilize" the Indians, however, the government could afford only a small sum annually, the bulk of which went into gift-giving, in essence a form of bribery meant to facilitate the purchase of lands. Despite the noble sentiments expressed by Jefferson, the reality of Indian policy remained what it had been under his predecessors: the cheap purchase of land. "In keeping agents among the Indians," Jefferson wrote to Andrew Jackson, "two objects are principally in view: 1. The preservation of peace; 2. The obtaining lands. Towards effecting the latter object, we consider leading the Indians to agriculture as the principal means from which we can expect much effect in the future." Jefferson to General Andrew Jackson, Feb. 16, 1803, ibid., X, 357–58.

The identification of Jefferson as an enlightened defender of Indian interests is often predicated on his statements of high regard for the race, particularly when this regard is contrasted with his low opinion of blacks. Thomas Jefferson, *Notes on the State of Virginia* (New York, 1954 [1782]), William Peden, ed., 58–63, 138–43. Jefferson regarded the Indians as physiologically equal to whites, advocating their assimilation into white society and even the intermixing of the two races. Ralph Lerner has noted the irony that "the black slave who took over the whites' language, religion, and work, was to remain beyond the pale—or be returned to Africa—while the red native, who rejected all these out of hand, was to be assimilated—or removed beyond the frontier." Ralph Lerner, *The Thinking Revolutionary: Principle and Practice in the New Republic* (Ithaca, N.Y., 1987), 164. Though Jefferson was clearly concerned with the fate of the Indians, his solution, assimilation, required in his own words "the termination of their history." Jefferson to Harrison, Feb. 27, 1803, Lipscomb and Bergh, eds., *Writings,* X, 370. As Sheehan, whose portrait of Jefferson is generally sympathetic, concludes: "Ultimately, hating Indians could not be differentiated from hating Indianness. If the frontiersman adopted the direct method of murdering Indians, humanitarians were only more circumspect in demanding cultural suicide of the tribes" (p. 277).

Except for the overriding fact that white settlers did obtain the lands, the ends of Jefferson's policy were not achieved. The Indians, hanging on to their culture, were simply divested of their lands, not assimilated. In the War of 1812, the British not surprisingly found easy recruitment among the Indians, a fact that Jefferson the philanthropist could not fathom. Undoubtedly swept up in the spirit of the conflict, he blamed the "unprincipled policy of England" for seducing the Indians to again "take up the hatchet against us." Although the

United States had pursued a "benevolent plan" toward the Indians, under which they "would have mixed their blood with ours, and been amalgamated and identified with us within no distant period of time," the despondent Jefferson of 1813 lamented that Indian atrocities "will oblige us now to pursue them to extermination, or drive them to new seats beyond our reach." Jefferson to Baron Alexander von Humboldt, Dec. 6, 1813, Lipscomb and Bergh, eds., *Writings*, XIV, 23.

103. Jefferson to John Bacon, April 30, 1803, Ford, ed., *Writings of Jefferson*, VIII, 228.

104. On the same day, Jefferson expressed his willingness to do without the right of deposit at New Orleans. "I am persuaded that had not the deposit been so quickly rendered we should have found soon that it would be better now to ascend the river to Natchez, in order to be clear of the embarrassments, plunderings, and irritations at New Orleans, and to fatten by the benefits of the depot a city and citizens of our own, rather than those of a foreign nation." Jefferson to Williamson, April 30, 1803, Lipscomb and Bergh, eds., *Writings*, X, 386–87.

105. Du Pont initially warned Jefferson against the impact of a French occupation, writing on April 30, 1802: "Offer her enough to make her make up her mind before she takes possession. For the interests of governors, of prefects, and of business companies would become powerful obstacles." To this list Du Pont might easily have added the obstacle of Napoleon's prestige once committed. Du Pont to Jefferson, April 30, 1802, Malone, ed., *Correspondence Between Jefferson and Du Pont*, 60.

106. See above, note 14.

107. "Pericles," in Syrett, ed., *Hamilton Papers*, XXVI, 82–85.

108. The indictment of Hamilton is set out most forcefully by Whitaker, *The Mississippi Question*, 211–14. Although the argument is in part vulnerable to the points subsequently made in the text, it remains a strong one given Whitaker's assumption that "the use of force was out of the question." This being so, he argued, procrastination was of necessity Jefferson's policy. Malone, *Jefferson and His Time*, IV, 277–78, makes much the same criticism of Hamilton as Whitaker. But unlike Whitaker, he does not say that force was out of the question. Instead, Malone writes that, if necessary, Jefferson would have used force in the spring of 1803 to take New Orleans. "That Jefferson was determined to get hold of New Orleans by one means or another in the near future, and would settle for no less, could hardly have been doubted—except by foes who completely mistrusted him" (p. 287). But if this were indeed the case, how then did Jefferson's position differ, in the last analysis, from Hamilton's, of whom Malone writes: "It is hard to believe that a statesman of his intelligence

could have regarded immediate recourse to war as either justifiable or practicable under existing circumstances" (p. 277). Hamilton is criticized because he advocated seizing New Orleans before a European war broke out, whereas Jefferson was determined to wait until the war occurred and we could be assured of Britain's cooperation. But Jefferson did not know, as he later admitted, when a war would begin. Had he been as "determined" to get hold of New Orleans "in the near future" as Malone insists, then he too was intent on a course of action subject to the same criticism as that made against Hamilton. One cannot have it both ways. Either Jefferson was determined to wait until war broke out between France and England, despite the interim risk of a French military presence in New Orleans, or he was not. One cannot reasonably argue, however, that what was terribly misguided in February was sound policy in March or April.

109. Madison to Livingston and Monroe, March 2, 1803, Hunt, ed., *Writings of Madison*, VII, 9.

110. The territorial extent of the cession is discussed below. The treaty of cession was signed on April 30, along with two other conventions that dealt with payment for the ceded territory and claims of American citizens against France arising out of the quasi war, 1798–1800. In return for the cession of all of Louisiana, the American ministers agreed to pay a sum of fifteen million dollars. Of this, sixty million francs ($11,250,000) was to go to France, the form of payment being in 6 percent bonds redeemable only after fifteen years. The remaining sum of twenty million francs ($3,750,000) was to be paid to American creditors for claims against France. The treaties, and particularly the claims convention, were drawn up in considerable haste, the parties acting with an awareness of the impending war between France and Great Britain (a declaration of war by England came on May 18). The claims convention was marked by several defects and subsequently led to considerable difficulty.

111. Livingston to Madison, April 11, 1803, *A.S.P.F.R.*, II, 552.

112. Livingston to Rufus King, April 27, 1803, Edward Alexander Parsons, ed., *The Original Letters of Robert R. Livingston, 1801–1803* (New Orleans, 1953), 57.

113. Jefferson to John C. Breckenridge, Aug. 12, 1803, Ford, ed., *Writings of Jefferson*, VIII, 243.

114. An undated draft cited in George Dangerfield, *Chancellor Robert R. Livingston of New York, 1746–1813* (New York, 1960), 368.

115. Livingston to Madison, April 13, 1803, *A.S.P.F.R.*, II, 553.

116. Dangerfield, *Livingston*, 368, writes that the minister had become persuaded by the end of April that the wording of San Ildefonso would indeed support a claim to West Florida, whereas Monroe

had continued to believe a specific provision to this effect was desirable. But time pressed the American negotiators and the matter was dropped.

117. Livingston to Madison, May 20, 1803, *A.S.P.F.R.*, II, 561.

118. The best sources on the West Florida controversy remain the accounts of Henry Adams, *History*, II, III, and Isaac Joslin Cox, *The West Florida Controversy, 1798–1813: A Study in American Diplomacy* (Baltimore, 1918). In more recent literature, see Clifford L. Egan, "The United States, France, and West Florida, 1803–1807," *Florida Historical Quarterly*, 47 (January 1969), 227–52, and *Neither Peace Nor War: Franco-American Relations, 1803–1812* (Baton Rouge and London, 1983), 46–66.

119. Monroe to Livingston, undated, Stanislaus Murray Hamilton, ed., *The Writings of James Monroe* (New York, 1898–1903), IV, 42. See also Livingston and Monroe to Madison, June 7, 1803, *A.S.P.F.R.*, II, 563–65.

120. Monroe to Madison, June 19, 1803, Hamilton, ed., *Writings of Monroe*, IV, 39. The "Opinion Respecting West Florida" was enclosed in the June 19 letter and is reprinted in IV, 503–9.

121. Madison to Livingston, July 29, 1803, *A.S.P.F.R.*, II, 566. Brant, *Madison: Secretary of State*, 150, cites the letter of the same day from Madison to Charles Pinckney in which Madison declared: "The Floridas are not included in the treaty being it appears still held by Spain." Brant writes that this was no rejection of the Livingston–Monroe claim to the Perdido but was intended "to prevent the distrusted minister [Pinckney] from raising the boundary issue before Monroe joined him." No evidence is given, however, in support of the contention.

122. Madison to Livingston, Oct. 6, 1803, Hunt, ed., *Writings of Madison*, VII, 70.

123. Madison to Livingston, March 31, 1804, Hunt, ed., *Writings of Madison*, VII, 123–40.

124. Jefferson to Gallatin, August 23, 1803, Adams, ed., *Writings of Gallatin*, I, 144. Here and in a letter to John Dickinson (Aug. 9, 1803, Ford, ed., *Writings of Jefferson*, VIII, 261), Jefferson expressed the opinion that while the claim in the east to the Perdido was "solid," "strong," etc., the claim in the west to the Rio Bravo del Norte (Rio Grande) was considerably weaker. The claim in the west involved most of the province of Texas. In fact, it was the other way around. The claim in the east was "speculative" or "argumentative," whereas the claim in the west was "solid," at least to the point that one party to the original treaty, France, had interpreted Louisiana to include Texas. But the desire for Texas was nowhere near as great as the desire for the Floridas—particularly West Florida.

125. Thomas Jefferson, "The Limits and Bounds of Louisiana," in American Philosophical Society, *Documents Relating to the Purchase and Exploration of Lousiana* (Boston and New York, 1904), 37–39. Jefferson simply asserted that "Louisiana with the same extent that it now has in the hands of Spain" was a reference to the ancient *"thing,* as it is in her hands, unaffected by new names," rather than a reference to the then current understanding of Louisiana bounded by the Mississippi: "when the portion newly called West Florida came back to *the hands of Spain,* it was still a part of ancient Louisiana, as possessed by France, *as now in the hands of Spain."*

126. Earlier, while still presumably of an open mind on the limits of Louisiana, Jefferson had asked for the opinion of persons whose views he respected. Jefferson to William Dunbar, July 17, 1803, Ford, ed., *Writings of Jefferson,* VIII, 254. The response was disappointing. Although Governor Claiborne reversed his opinion in September after seeing the administration's claim to West Florida put forward in the *National Intelligencer,* in August he reported the understanding that "the Island of N. Orleans was the only Tract of Country east of the Mississippi" ceded by France to Spain by the treaty of 1763. Claiborne to Jefferson, Aug. 24 and Sept. 29, 1803, Carter, ed., *Territorial Papers,* IX, 16, 59. Daniel Clark, the American Consul at New Orleans, first gave without comment the Spanish understanding that West Florida was excluded from the cession (Clark to Madison, Sept. 8, 1803, ibid., 30), later putting it more forcefully, "that [West Florida] has never been reannexed to Louisiana since its conquest from Great Britain" and that neither France nor Spain believed "that any part of West Florida or ancient Louisiana to the East of the Mississippi except the Island of Orleans has been ceded to us by the late Treaty." Clark to Madison, Dec. 13, 1803, "Despatches from the United States Consulate in New Orleans, 1801–1803, II," *American Historical Review,* 33 (January 1928), 356–57. Only William Dunbar could make a tentative case "which might induce an idea that a portion of West Florida had been consolidated with Louisiana." Dunbar to Jefferson, Oct. 21, 1803, Carter, ed., *Territorial Papers,* IX, 85. Asked to indicate his understanding of the boundaries of Louisiana, the French commissioner sent by Napoleon in January 1803 to receive possession of New Orleans from the Spanish authorities had also given evidence against the American claim. Laussat, to the chagrin of his superiors, had responded truthfully—that is, he had answered by giving the French understanding of the boundaries—by defining them as the Rio Bravo (or Rio Grande) in the west and the Mississippi in the east. William C. C. Claiborne and James Wilkinson to Madison, Dec. 27, 1803, Robertson, ed., *Louisiana,* II, 290–91.

127. Jefferson to Breckenridge, Aug. 12, 1803 and to Monroe,

Jan. 8, 1804, Ford, ed., *Writings of Jefferson,* II, 242, 289. It was the case, however, that on August 18, 1803, Jefferson declared that "our right to the Perdido is substantial and can be opposed by a quibble on form only." Jefferson to Madison, Aug. 18, 1803, ibid., 245.

128. Interest apart, Livingston seems to have been persuaded that Louisiana did not comprise the Floridas. "Since the possession of the Floridas by Britain, and the treaty of 1763," he wrote to Madison on July 30, 1802, "I think there can be no doubt about the precise meaning of the terms." The eastern boundary of Louisiana was the Mississippi and not the Perdido. Even during the course of the negotiations over the purchase, Livingston did not contend the contrary. Instead, he asked that the French help the United States obtain the Floridas, a very different thing. In another letter to Madison, written after the cession and eight days before pledging that the right to West Florida was good, Livingston had this comment on the attempts by France to gain the Floridas: "I endeavored, as far as possible, to obstruct that negotiation, and, at the same time, urged that absurdity of attempting to colonize Louisiana without ports in the Gulf." Livingston then added that "if [France] could have concluded with Spain, we should also have had West Florida." Livingston to Madison, July 30, 1802 and May 12, 1803, *A.S.P.F.R.,* II, 520, 557–58.

129. The instructions from Decrès to Victor confirm France's understanding, the boundary being defined as a line running along the Mississippi and Iberville rivers, that "all the territory east and north of that boundary forms part of the United States or of West Florida." Robertson, ed., *Louisiana,* I, 363. Laussat repeated this understanding to Claiborne and Wilkinson, ibid., II, 289–91. See also Lyon, *Louisiana in French Diplomacy,* 125–26, 191–95.

130. Adams, *History,* II, 246–47.

131. Livingston to Madison, May 20, 1803, *A.S.P.F.R.,* II, 561.

132. Talleyrand to General John Armstrong, Dec. 21, 1804, ibid., 635.

133. Madison to Livingston, March 31, 1804, Hunt, ed., *Writings of Madison,* VII, 126–27.

134. Don Carlos Martinez Irujo to Madison, Sept. 4 and 27, 1803, Robertson, ed., *Louisiana,* II, 77–81. Madison to Irujo, Oct. 4, 1803, *A.S.P.F.R.,* II, 569.

135. Irujo to Don Pedro de Cevallos, Nov. 5, 1803, Robertson, ed., *Louisiana,* II, 121.

136. Madison to Monroe, July 29, 1803, Hunt, ed., *Writings of Madison,* VII, 54.

137. Monroe was authorized to offer as much as two and a quarter million dollars for the Floridas. But this sum, Madison instructed Monroe, was to "be made applicable to the discharge of debts and

damages claimed from Spain, as well as those not yet admitted by the Spanish government. . . . On the subject of these claims you will hold a strong language." The spoliation claims referred to were in the main those that had arisen during the period of the quasi war with France. Monroe was further instructed not to entertain any proposals from Spain looking to an exchange of the Floridas for "Louisiana beyond the Mississippi." Thus the exchange that both Livingston and Monroe had considered before and during the negotiations over the Louisiana cession was ruled out, in part for the reason that "in intrinsic value there is no equality" between the two and, in part because "their [Floridas] position and the manifest course of events guarantee an early acquisition of them." Madison to Monroe, July 29, 1803, Hunt, ed., *Writings of Madison,* VII, 53–59.

138. Madison to Pinckney, Oct. 12, 1803, Hunt, ed., *Writings of Madison,* VII, 74.

139. George W. Erving to Madison, Dec. 7, 1805, State Department Archives.

140. Madison to Monroe, April 15, 1804, Hunt, ed., *Writings of Madison,* VII, 141–53.

141. Jefferson to Madison, July 5, 1804, Ford, ed., *Writings of Jefferson,* VIII, 309–12.

142. Madison to Monroe, July 29, 1803, Hunt, ed., *Writings of Madison,* VII, 56.

143. See *Annals of Congress,* 8 Cong., 1st sess., Feb. 24, 1804, 1257–58, and Gallatin to Hore Browse Trist, Feb. 27, 1804, Carter, ed., *Territorial Papers,* IX, 192–97. For interpretations of this episode see Adams, *History,* II, 257–63; Cox, *The West Florida Controversy,* 99–100; Malone, *Jefferson and His Time,* IV, 346–47, and Brant, *Madison: Secretary of State,* 188–99.

144. Pinckney to Cevallos, June 1 and 22, July 5 and 14, 1804, and Cevallos to Pinckney, July 2 and 8, 1804, *A.S.P.F.R.,* II, 618–24.

145. Monroe to Talleyrand, Nov. 8, 1804, and Monroe to Livingston, Nov. 13, 1804, Hamilton, ed., *Writings of Monroe,* IV, 266–77.

146. Talleyrand had condemned the American claim and voiced disapproval over U.S. conduct toward Spain in a note to Armstrong in late December of the previous year. Talleyrand to Armstrong, Dec. 21, 1804, *A.S.P.F.R.,* II, 635. Although Monroe had not heard such an explicit statement of French policy when he began the negotiations at Aranjuez, a letter to Madison on December 16, 1804, reveals that he had reason to suspect the imminent declaration by Talleyrand. Hamilton, ed., *Writings of Monroe,* IV, 293–94.

147. Monroe to Jefferson, March 15 and September 25, 1804, Hamilton, ed., *Writings of Monroe,* IV, 155, 254.

148. Armstrong to Monroe, March 12, 1805, *A.S.P.F.R.*, II, 636.

149. Jefferson to James Bowdoin, April 27, 1805, Ford, ed., *Works of Jefferson*, X, 140.

150. Monroe and Pinckney to Madison, May 23, 1805, *A.S.P.F.R.*, II, 667.

151. Madison to Armstrong, June 6, 1805, Hunt, ed., *Writings of Madison*, VII, 185.

152. Later historians have shared Madison's view. See Egan, *Neither Peace Nor War*, 46.

153. Monroe to Armstrong, March 1, 1805, "Despatches from U.S. Ministers to Spain, 1792–1906," National Archives, No. 31, Roll 9, 123.

154. Irujo to Cevallos, Aug. 3, 1803, Robertson, ed., *Louisiana*, II, 69–70.

155. As, for example, Egan, *Neither Peace Nor War*, 66, argues.

156. Adams, *History*, II, 245–49.

157. Adams, *History*, III, 22–23.

158. Henry Adams simply asserts the importance of these pressures. But in a letter from John Breckenridge to Jefferson, dated September 10, 1803, the Kentucky Senator wrote: "As to the Florida's, I really consider their acquisition as of no consequence for the present. We can obtain them long before we shall want them, and upon our own terms." Carter ed., *Territorial Papers*, IX, 48.

159. Egan, *Neither Peace Nor War*, 48.

160. In a letter to Talleyrand of July 8, 1805, the French minister to Washington, Turreau, suggested that at the end of the war Spain cede Cuba and the Floridas to France, in order that France might dominate the Gulf. Robertson, ed., *Louisiana*, II, 142. Turreau's suggestion assumed, of course, France would be victorious.

161. Malone, *Jefferson and His Time*, IV, 317.

162. Peterson, *Jefferson and the New Nation*, 772–73.

163. Julian P. Boyd, "Thomas Jefferson's 'Empire of Liberty,' " *Virginia Quarterly Review*, 24 (1948), 548.

164. Jefferson to Nathaniel Niles, March 22, 1801, Ford, ed., *Writings of Jefferson*, VIII, 24.

165. Jefferson to Breckenridge, Aug. 12, 1803, Ford, ed., *Writings of Jefferson*, VIII, 243–44.

166. Jefferson to Monroe, Nov. 24, 1801, Ford, ed., *Writings of Jefferson*, VIII, 105. It was this essential similarity that apparently provided the basis for the concept of sister republics. The same sense of commonality that marked the nation and the sense of nationhood would also characterize a hemisphere made up of sister republics. Alluded to on several occasions during his life, Jefferson's idea of sister republics was never articulated with any clarity or detail. It was one

of his many ideas that tantalized, in part because it remained largely unexplained. Certainly it raised far more questions than it answered. On the one hand, if the Empire of Liberty was to be made up of one people that was alike in language, religion, customs, and beliefs, why the need of sister republics? Of course, one answer is that the difficulties imposed by distance and the inadequacy of communications would necessitate sister republics. But if these barriers might in time be overcome, at least in part, then the question persists: why the need? On the other hand, if the Empire of Liberty was to be composed of sister republics, how were wars among them to be prevented and Europe's experience to be avoided? Was peace to be guaranteed by virtue of their essential similarity and, particularly, because being republics they would be peaceful? Or was peace to be guaranteed because the parent republic—Jefferson's "nest"—would stand as the guarantor? Jefferson provided no clear answer, though it is not unreasonable to assume that he intended peace to be guaranteed by the Union, that is, the "nest." Peterson, *Jefferson and the New Nation,* 746, writes: "He did not think it necessary that this 'empire' be united under one government, but it must be one people united in the enterprise of freedom, his own country serving as common parent, midwife, and nurse of the entire 'American system.' " How this role could be made compatible with the concept of "sister republics," which implies after all a condition of rough equality, Peterson leaves unexplained.

167. Frederick Merk, *Manifest Destiny and Mission in American History: A Reinterpretation* (New York, 1963), 74, writes that although Jefferson saw the American people as occupying the entirety of the hemisphere, the "nest," i.e., the Union, had limits and they were bounded by the Rio Grande and the Rocky Mountains. He further writes: "As for any concept of hegemony by the United States over all the continent, Federalists and Jeffersonians would have been at one in repudiating it as unthinkable." Merk does not explain how he reconciles the limits he lays down and Jefferson's desire—to take one example—for Cuba. His insistence that the idea of hegemony was unthinkable would come as a surprise to those Federalists who argued that Louisiana must be held as a colonial dependency. Nor is it at all clear how peace would have been maintained in the hemisphere in the absence of "any concept of hegemony."

168. Historians do not deny the great significance Jefferson attached to the Union as a means to the realization of the great end—Liberty. Quite the contrary, they have been at pains to emphasize the critical importance Jefferson attached to this means. "Nothing could be more mistaken," Merrill Peterson writes of Louisiana, "than to suppose that he resigned himself to separation and disunion," noting Jefferson's argument that by "enlarging the empire of liberty, we mul-

tiply its auxiliaries, and provide new sources of renovation, should its principles at anytime degenerate, in those portions of our country which gave them birth." *Jefferson and the New Nation,* 773. Julian Boyd, while insisting that the Empire of Liberty is the "realm of the mind and spirit of man," and not to be confused with coercion, political boundaries, and the like, is equally insistent in pointing out that this empire is, after all, the realm of coercion as well as the realm of freedom, that Jefferson intended it to be of such a nature as to make national dissolution impossible. Thus Boyd points to the Northwest Ordinance of 1784, which, under Jefferson's hand, would have riveted the bonds of union so firmly as to make national dissolution impossible, since it provided that the new governments should "forever remain a part of the United States of America." "Jefferson's 'Empire of Liberty,' " 548. The impact of these statements is clear enough. The Union was not only a means but very nearly an indispensable means to the great end of liberty.

169. These themes are brilliantly developed by Drew R. McCoy, *The Elusive Republic: Political Economy in Jeffersonian America* (New York, 1980), 185–235.

170. And the Union, of course, was the perfect instrument of that expansion. In a letter written soon after his retirement from office, Jefferson declared: "I am persuaded no constitution was ever before so well calculated as ours for extensive empire and self government." Jefferson to Madison, April 27, 1809, *The Papers of James Madison: Presidential Series* (Charlottesville, Va., 1984), I, 140. See also Jefferson's second inaugural, Ford, ed., *Writings of Jefferson,* VIII, 344.

171. Jefferson to John B. Colvin, Sept. 20, 1810, Ford, ed., *Writings of Jefferson,* IX, 279. The letter was occasioned by the question Colvin posed to Jefferson: "Whether circumstances do not sometimes occur, which make it a duty in officers of high trust, to assume authorities beyond the law." Jefferson responded that the question "is easy of solution in principle, but sometimes embarrassing in practice. A strict observance of the written laws is doubtless *one* of the high duties of a good citizen, but it is not the *highest.* The laws of necessity, of self-preservation, of saving our country when in danger, are of higher obligation." It is not Jefferson's general response that occasions comment, but the examples he gives—that is, his application of the principle. Jefferson gives one "hypothetical case." The case and his response merit scrutiny.

> Suppose that it had been made known to the Executive of the Union in the autumn of 1805, that we might have the Floridas for a reasonable sum, that that sum had not indeed been appropriated by law, but that Congress were to meet within three weeks, and might appropriate it on the first or second day of

their session. Ought he, for so great an advantage to his country, to have risked himself by transcending the law and making the purchase? The public advantage offered, in this supposed case, was indeed immense; but a reverence for law, and the probability that the advantage might still be *legally* accomplished by a delay of only three weeks, were powerful reasons against hazarding the act. But suppose it foreseen that a John Randolph would find means to protract the proceeding on it by Congress, until the ensuing spring, by which time new circumstances would change the mind of the other party. Ought the Executive, in that case, and with that foreknowledge, to have secured the good to his country, and to have trusted to their justice for the transgression of the law? I think he ought, and that the act would have been approved.

This, however, is not a case of necessity. At least, Jefferson does not so represent it here. He himself speaks only of "advantage," though he describes it as "so great an advantage." The example illustrates the danger that must always attend the plea of necessity: the confusion of what is necessary with what is merely desirable (or advantageous). The example also suggests how Jefferson viewed the Floridas. They were an advantage he wished to represent as a necessity.

172. Gallatin to Jefferson, Jan. 13, 1803, Adams, ed., *Writings of Gallatin*, I, 111.

173. Ford, ed., *Writings of Jefferson*, VIII, 241 (specific date not given).

174. Jefferson to Dickinson, Aug. 9, 1803, Ford, ed., *Writings of Jefferson*, VIII, 261.

175. Jefferson to Breckenridge, Aug. 12, 1803, Ford, ed., *Writings of Jefferson*, VIII, 244.

176. Jefferson to Gallatin, Aug. 23, 1803, Adams, ed., *Writings of Gallatin*, I, 145.

177. Livingston and Monroe to Madison, June 7, 1803, A.S.P.F.R., II, 564.

178. Livingston to Madison, June 25, 1803, A.S.P.F.R., II, 566.

179. Gallatin to Jefferson, Aug. 31, 1803, Adams, ed., *Writings of Gallatin*, I, 147. Madison to Jefferson, Aug. 20, 1803, cited in Brant, *Madison: Secretary of State*, 143.

180. Cited in Everett S. Brown, *The Constitutional History of the United States, 1803–1812* (Berkeley, Calif., 1920), 26–27.

181. Jefferson to Wilson Cary Nicholas, Sept. 7, 1803, Ford, ed., *Writings of Jefferson*, VIII, 247–48.

182. Malone, *Jefferson and His Time*, IV, 319.

183. Adams, *History*, II, 89.

184. Adams, *History*, II, 91.

185. Adams, *History*, II, 92–93.

186. Brown, *Constitutional History,* 29.

187. Charles Francis Adams, ed., *Memoirs of John Quincy Adams* (Philadelphia, 1874–77), V, 364–65 (Oct. 20, 1821).

188. Henry Adams, ed., *Documents Relating to New England Federalism, 1800–1815* (Boston, 1877), 157–58.

Part IV. The Maritime Crisis (1805–9)

1. First Annual Message, Dec. 8, 1801, Andrew A. Lipscomb and Albert Bergh, eds., *The Writings of Thomas Jefferson* (Washington, D.C., 1904–5), III, 331.

2. See Dumas Malone, *Jefferson and His Time,* (Boston, 1948–81), V, xix. See also Merrill D. Peterson, *Thomas Jefferson and the New Nation: A Biography* (Oxford, 1970), 882: "Between the tightening Continental System and the expected British orders, American commerce was caught in the jaws of a vise, a maniacal war of blockades, from which there seemed to be no appeal to reason or justice or right. The ruthless had combined to slaughter the innocent." But see also pp. 916–18, where Peterson sharply criticizes Jefferson's conduct in 1808 and 1809.

3. See Henry Adams, *History of the United States During the Administrations of Jefferson and Madison* (New York, 1903 [1889–91]), IV, 289. On the leading points, Adams's interpretation is quite close to that set forth by Alfred Thayer Mahan, *Sea Power in Its Relations to the War of 1812* (Boston, 1905). For both, the most grievous flaw of Jeffersonian statecraft lay in its absence of military preparedness and in its inability to understand that its weakness would breed the contempt of other powers. Adams, like Mahan, did not dispute the legitimacy of American claims on the high seas; on the contrary, he believed that America was the innocent victim of English rapacity and thought that the restrictions on American trade and the continuing practice of impressment called for war in 1806. These were injuries which no self-respecting nation could admit. For a similar view, see also Clifford L. Egan, *Neither Peace Nor War: Franco-American Relations, 1803–1812* (Baton Rouge and London, 1983).

4. The brief summaries of these rival explanations of Jefferson's failure do not do complete justice to the complexity of historical interpretation of the episode. There are many ways into the subject, and many variations on the three distinct interpretations identified in the text. Jefferson's failure to order properly the two great desiderata— neutral rights and the balance of power—is the principal theme of Lawrence Kaplan, "Jefferson, The Napoleonic Wars, and the Balance of Power," in Kaplan, *Entangling Alliances with None: American*

Foreign Policy in the Age of Jefferson (Kent, Ohio, 1987), 111–26. The theme is also intelligently handled in Daniel G. Lang, *Foreign Policy in the Early Republic: The Law of Nations and the Balance of Power* (Baton Rouge, 1985). A generation earlier, the essentials of the interpretation were set forth by W. Allison Phillips, though Phillips developed the theme more explicitly in relation to the issues arising in the 1790s. The classic study from this perspective remains the superb work of A. L. Burt, *The United States, Great Britain and British North America: From the Revolution to the Establishment of Peace After the War of 1812* (New Haven, Conn., 1940).

In the post–World War II period, the leading work in diplomatic history covering the second administration is Bradford Perkins, *Prologue to War, 1805–1812: England and the United States* (Berkeley, Calif., 1961). Perkins tends to be sympathetic to the British claim that Bonaparte constituted a genuine threat to the balance of power; like Adams, however, he is critical of the British conception of neutral rights and is very rough, in particular, on George Canning's conduct as British Foreign Secretary. From a neo-Hamiltonian point of view there is Forrest McDonald's vigorous critique, *The Presidency of Thomas Jefferson* (Lawrence, Kans., 1976), which stresses the disjunction between means and ends in Jefferson's foreign policy. Both Perkins and McDonald follow Henry Adams in holding that Jefferson's pusillanimity invited assault from foreign powers; unlike Adams, however, both authors tend to be more critical of American pretensions on the high seas.

The most provocative recent critique of Jefferson's maritime diplomacy is Bernard Spivak, *Jefferson's English Crisis: Commerce, Embargo, and the Republican Revolution* (Charlottesville, Va., 1979), wherein the story is rendered with a deftness of stroke that often rivals Adams in interpretive power. Here, and in "Thomas Jefferson, Republican Values, and Foreign Commerce" (in *Traditions and Values: American Diplomacy, 1790–1865* Norman A. Graebner, ed., [Lanham, Md., 1985]), Spivak explores Jefferson's deep-seated animus against Great Britain and his ironic encounter with American external commerce. He well develops the case against Jefferson's moralism and superbly recounts Randolph's bitter denunciation of the administration and Gallatin's frequent reservations over the course taken by the President and the Secretary of State. Spivak, however, is less persuasive on Jefferson's relationship to the embargo. See below, note 55.

5. On the fall of 1805 as a turning point, see Adams, *History*, III, 80–81. Adams's flight of fancy in describing the alternative that awaited Jefferson (which had the liability for the President of reverting to the precedents of a former administration) is sufficiently plausible to de-

serve mention here. Had Jefferson, Adams held, gone to war against Spain in August 1805, the act

> would probably have been supported, as the purchase of Louisiana had been approved, by the whole country, without regard to Constitutional theories; and indeed if Jefferson succeeded to the rights of Napoleon in Louisiana, such a step required no defence. Spain might then have declared war; but had Godoy taken this extreme measure, he could have had no other motive than to embarrass Napoleon by dragging France into a war with the United States, and had this policy succeeded, President Jefferson's difficulties would have vanished in an instant. He might then have seized Florida; his controversies with England about neutral trade, blockade, and impressment would have fallen to the ground; and had war with France continued two years, until Spain threw off the yoke of Napoleon and once more raised in Europe the standard of popular liberty, Jefferson might perhaps have effected some agreement with the Spanish patriots, and would then have stood at the head of the coming popular movement throughout the world,—the movement which he and his party were destined to resist. Godoy, Napoleon, Pitt, Monroe, Armstrong, John Randolph, and even the New England Federalists seemed combined to drag or drive him into this path. Its advantages were so plain, even at that early moment, as to overmaster for a whole summer his instinctive repugnance to acts of force.

6. Madison to John Armstrong, June 6, 1805, Gaillard Hunt, ed., *The Writings of James Madison* (New York, 1900–1910), VII, 183.

7. Monroe to Madison, Oct. 18, 1805, Stanislaus Murray Hamilton, ed., *The Writings of James Monroe* (New York, 1898–1903), VII, 362–65. Armstrong to Monroe, May 4, 1805, and to Madison, July 3, 1805, in Adams, *History,* III, 39–40.

8. Jefferson to James Bowdoin, April 27, 1805, Paul Leicester Ford, ed., *The Writings of Thomas Jefferson* (New York, 1892–99), VIII, 350–51.

9. Gallatin to Jefferson, Sept. 12, 1805, Henry Adams, ed., *The Writings of Albert Gallatin* (New York, 1960), I, 241–46; Irving Brant, *James Madison: Secretary of State, 1800–1809* (Indianapolis, Ind., and New York, 1953), 283. Gallatin's objections in 1805 to the justice of American territorial claims against Spain—dismissed by Forrest McDonald as a "tedious tract . . . filled with legal and moralistic philosophizing" (*Presidency of Thomas Jefferson,* p. 100)—deserve notice here. Gallatin considered the claim to the territory extending westward of New Orleans to be doubtful and attended with such uncertainty that "a resort to arms for that cause [would] appear unjustifiable in the opinion of mankind and even of America." The claim

to West Florida as far as the Perdido River he thought much better founded. The claim, however, was "not self-evident, but constructive"; the justice of a war in support of it was "extremely doubtful." He assigned three reasons:

—1st. Whether ascribed to policy, or to precipitancy, or to any other cause, it is not less the fact that the acquisition of Louisiana without any fixed boundaries was the act of the United States; for the act of their negotiators is theirs; if they intended at all events to obtain the now disputed territory between Mississippi and Perdido, if they then attached such value to it as to risk a war for securing it, they would not have signed the treaty without placing the subject beyond the possibility of dispute.
—2dly. Not only we neglected, when the treaty was made, to obtain from France, if not a guarantee, at least an official declaration of what she considered as the boundary of the territory ceded to her by the Treaty of St. Ildefonso, but Spain was not consulted on the subject. If, therefore, a previous explanation had taken place between Spain and France on that subject, however we may complain of the duplicity of France for having withheld such communication, Spain may justly oppose it to our demands. If A purchases from B a tract of land, and the boundaries are not precisely defined by the deed; if by subsequent articles the parties explain the meaning of the deed; if neither the deed nor articles have been made matter of public record; and if afterwards C shall purchase from A on the face of the first deed, and, notwithstanding its want of precision, shall neither ask from A a guarantee or even explanation of the boundaries, nor inquire from B what he had intended to convey; it is true that he may have recourse against A for the deception in not showing the articles, but it is very doubtful whether the disputed land can be recovered from B, who has in the mean while never given possession, and who had even, before C's purchase was ratified, warned him not to purchase.
—3dly. We cannot deny that we had before the ratification of the treaty a knowledge of the intention of the parties to the Treaty of St. Ildefonso, so far as related to the eastern boundaries. For we knew that Laussat was instructed to demand and the Spanish officers to deliver, east of the Mississippi, that part only which is in our possession.

These considerations, together with Gallatin's skepticism that Spain's refusal to ratify the claims convention constituted a just cause of war, led him to oppose the course that Jefferson had suggested to the cabinet. A war waged on such grounds could not "carry that conviction of the justice of our cause which is necessary to justify a war in the public opinion and to our own hearts. The high station which America and, I flatter myself, Mr. J's administration now occupy in the

eyes of other nations, is principally due to the opinion which is entertained of their wisdom, justice, and moderation; and I think it (exclu-
sively of every reason derived from duty) of primary importance that
nothing should be done to weaken those favorable impressions."
"Spanish Affairs," Sept. 12, 1805, Adams, ed., *Writings of Gallatin*,
I, 242–44.

Gallatin's observations in 1805 anticipate his later opposition to
the Mexican War. In a pamphlet written three years before his death
("Peace with Mexico" [1847], ibid., III, 555–91), Gallatin charged
that the war had been undertaken for an object (the left, or northern,
bank of the Rio Grande) which was "of no intrinsic value, to which
the United States have no legitimate right, which justice requires them
to yield, and which even expediency does not require." The pamphlet
is best remembered for its evocation of the mission of America, which
Gallatin said was "to improve the state of the world, to be the 'model
republic,' to show that men are capable of governing themselves, and
that this simple and natural form of government is that also which
confers most happiness on all, is productive of the greatest development of the intellectual faculties, above all, that which is attended
with the highest standard of private and political virtue and morality." The Founding Fathers, Gallatin held, had not deviated from these
principles. The sound sense, wisdom, probity, and respect for public
faith with which they had managed the internal concerns of the nation had "made our institutions an object of general admiration. Here,
for the first time, was the experiment attempted with any prospect of
success, and on a large scale, of a representative democratic republic.
If it failed, the last hope of the friends of mankind was lost or indefinitely postponed; and the eyes of the world were turned toward you.
Whenever real or pretended apprehensions of the imminent danger of
trusting the people at large with power were expressed, the answer
ever was, 'Look at America!' "

Until the Mexican War, Gallatin held, the United States had always acted, in its external relations, "in strict conformity with the
dictates of justice, and displayed the utmost moderation. They never
had voluntarily injured any other nation. Every acquisition of territory from foreign powers was honestly made, the result of treaties not
imposed, but freely assented to by the other party." Gallatin's examples were confined to the quasi war with France in 1798 and the
War of 1812 against Britain. He passed over in silence the American
claim to West Florida, occupied and annexed by the Madison administration from 1810 to 1813 despite the principled objections that
Gallatin himself had earlier raised. Nor did he mention the Adams–
Onís Treaty of 1819, which did not in the main conform to his image
of a treaty "not imposed, but freely assented to by the other party."

His purpose, of course, lay in recalling the country to a higher sense of duty, and it is understandable that an image of the "lofty position" occupied by the Founding Fathers should have been called in to sum up the account. His aim was to restrain the appeal to "cupidity" and to "the thirst of unjust aggrandizement by brutal force." The country had a purpose, he thought, and nothing was more serious than a betrayal of it: "Your mission was to be a model for all other governments and for all other less-favored nations, to adhere to the most elevated principles of political morality, to apply all your faculties to the gradual improvement of your own institutions and social state, and by your example to exert a moral influence most beneficial to mankind at large."

Gallatin no doubt went too far in holding that "the true law of nations is nothing else than the conformity to the sublime precepts of the gospel morality, precepts equally applicable to the relations between man and man and to the intercourse between nation and nation." This set a standard that was unrealistic and that might prove dangerous or irresponsible in certain circumstances. Though states might reasonably be expected to do unto others as they would themselves be done, a rule that has the merit of enshrining the principle of reciprocity (and the evident demerit of neglecting disparities in power), no state has ever loved thy neighbor as thyself. To make this principle controlling, as Hamilton once said, "is to misconceive or mistake what usually are and ought to be the springs of National Conduct." Hamilton's alternative formulation—that states ought to follow "a policy regulated by their own interest, as far as justice and good faith permit"—is much to be preferred.

But before the hard duty that Gallatin imposed is consigned to the verdict of a lack of realism, it must be acknowledged that his fidelity to justice was the genuine article. As such, it may be distinguished from the readiness with which Jefferson and Madison gave the claim to West Florida the status of a high moral principle, and from their zeal to clothe American commercial avarice in the same garment. Gallatin's conception of morality—similar, in this respect, to Hamilton's own outlook—lay in the recognition that duty (whether derived from the laws of nations, the dictates of natural justice, or the "eternal principles of morality and good faith") might be called upon to restrain desire. By contrast, the moralism of Jefferson and Madison, so far as it related to the justification of American diplomacy, lay in the insistence on rights. The one outlook sought to frustrate or contain avarice and ambition; the other operated in deadly combination with them.

10. *Annals of Congress* (Washington, D.C., 1834–56), 9th Cong., 1st sess., March 6, 1806, 594.

11. Fifth Annual Message, Dec. 3, 1805, Lipscomb and Bergh, eds., *Writings*, III, 388–89.

12. "Draft of Fifth Annual Message," Ford, ed., *Writings of Jefferson*, VII, 391.

13. Whether the administration supported the Logan bill and if so, its motives in doing so, remain a subject of dispute among historians. Both McDonald, *Presidency of Jefferson*, and Charles Callan Tansill, *The United States and Santo Domingo, 1798–1873* (Baltimore, 1938), 104–9, argue that the desire to conciliate the French was primary, whereas Adams was ambivalent on the point and Malone insists that "if this was a pro-French gesture it was a very slight one for which the administration itself bore little or no responsibility." Malone, *Jefferson and His Time*, V, 96–97. Yet there can be little question that the administration supported the measure; its reticence in doing so openly is attributable to its sensitivity to the charge that the French had demanded it, and that American foreign policy, as John Quincy Adams put it, was entirely prostrate to the wishes of Napoleon. Madison, typically, told the French that they had no right to demand such a measure, but he nevertheless insisted that the United States ought to gain credit with Napoleon for acquiescing in it. That credit was no doubt diminished by the perception that the bill would have little impact. William Plumer reports telling Logan that the bill would accomplish nothing, to which Logan replied that "it was not his intention to prohibit the trade—but only to pass a bill that would please the French." Plumer concluded that "the French cannot be deceived by such flimsy acts as these." *William Plumer's Memorandum of Proceedings in the United States Senate, 1803–1807*, ed. Everett Somerville Brown (New York, 1923), 387. See also the instructive account of Egan, *Neither Peace Nor War*, 57–58.

14. Jefferson to William C. C. Claiborne, April 27, 1806, Ford, ed., *Writings of Jefferson*, VII, 442.

15. See Malone, *Jefferson and His Time*, V, 96–97 and note 8 above. See also Adams, *History*, III, 148: "Jefferson disliked and dreaded the point in dispute with England. The Spanish policy was his own creation, and he looked upon it with such regard as men commonly bestow upon unappreciated inventions. . . . Where England was to be dealt with, Madison took the lead which Jefferson declined." To similar effect, see Spivak, *Jefferson's English Crisis*, 50. The carrying trade, Spivak believes, "troubled Jefferson the agrarian even as Jefferson the president struggled to justify and protect it. Like his southern congressional brethren, he would not risk the nation's peace and the South's prosperity for its sake."

16. *Annals of Congress*, 9th Cong.; 1st sess., March 5, 1806, 555–74.

17. Jefferson to William Duane, March 22, 1806, Lipscomb and Bergh, eds., *Writings,* XI, 96.

18. In this light, too, Randolph's complaint takes on a new meaning, and perhaps should be considered more seriously than is normally the case. It is a curious feature of Jefferson's reserve that while it at first sight appears as an instance of his willingness to defer to Congress and public opinion, it also had the additional advantage of repelling responsibility from himself—a consideration that he sought to make use of in his dealings with foreign powers and which increased rather than diminished his power over foreign policy. One of the reasons for Randolph's bitterness toward the administration was that it would not present a clear object of attack. However artful he was in making out the indictment—and even his harshest critics seldom deny Randolph's brilliance in this respect—the mud could not be made to stick on his shifting and uncertain target. His criticism, nevertheless, raised a serious constitutional issue. On Randolph's view, it was the administration's responsibility to provide a plan, and this it had not done. Only if it did so might it be held accountable. He cited the provision of the Constitution that recognized the President's duty of giving to Congress "information of the state of the Union" and recommending "to their consideration such measures as he shall judge necessary and expedient," and he believed that there was nothing in this open procedure that prevented a Republican congressman from exercising his own independent responsibility in approving or rejecting such presidential recommendations. Jefferson did not follow this open procedure, preferring a more circuitous and seemingly deferential route. And yet he had gotten his way! This was too much for Randolph to bear, and so he doubled the ferocity of his attacks when it became apparent that his prey would slip beyond his grasp. Randolph, as Adams said, "took rare pleasure in making enemies, while Jefferson never made one enemy except to gain two friends." Adams, *History,* III, 159.

Jefferson's skills as party and presidential leader are examined in two modern studies that generally award him very high marks: Noble E. Cunningham, Jr., *The Process of Government Under Jefferson* (Princeton, N.J., 1978), and Robert M. Johnstone, Jr., *Jefferson and the Presidency: Leadership in the Young Republic* (Ithaca, N.Y., 1978). Both scholars note, as Adams did before them, Jefferson's uncanny ability to wield power while nevertheless appearing to yield.

19. Jefferson to Thomas Paine, March 25, 1806, Ford, ed., *Writings of Jefferson,* VII, 438.

20. Jefferson to Monroe, May 4, 1806, Ford, ed., *Writings of Jefferson,* VII, 449.

21. Perkins, *Prologue to War,* 29.

22. Paul S. Holbo, "Trade and Commerce," *Encyclopedia of American Diplomatic History*, 947.

23. James Stephen, *War in Disguise; or, The Frauds of the Neutral Flags*, 2nd. ed. (London, 1805), 5, 9–10.

24. Stephen, quoting the judgment of Sir William Scott, in the case of the *Immannuel*, at the Admiralty, Nov. 1799, *War in Disguise*, 13.

25. Fifth Annual Message, Dec. 3, 1805, Lipscomb and Bergh, eds., *Writings*, III, 386–87.

26. Jefferson to C. F. Comte De Volney, Feb. 11, 1806, Ford, ed., *Writings of Jefferson*, VII, 419.

27. Ford, ed., *Writings of Jefferson*, VII, 340.

28. *Annals of Congress*, 9th Cong., 1st sess., March 5, 1806, 557.

29. *Annals of Congress*, 9th Cong., 1st sess., March 5, 1806, 540–41 (Gregg).

30. Stephen, *War in Disguise*, 124.

31. *An Examination of the British Doctrine, Which Subjects to Capture A Neutral Trade, Not Open in Time of Peace*, Hunt, ed., *Writings of Madison*, VII, 332.

32. *Annals of Congress*, 9th Cong., 1st sess., Dec. 20, 1805, 33 (Mitchill).

33. See Spivak, *Jefferson's English Crisis*, 203–10.

34. Monroe to Jefferson, Jan. 11, 1807, Hamilton, ed., *Writings of Monroe*, V, 2.

35. The course of the negotiations is well summarized in Perkins, *Prologue to War*, 117–39.

36. Quoted in Eli F. Heckscher, *The Continental System: An Economic Interpretation* (Oxford, 1922), 99.

37. Madison to David Erskine, March 20, 1807, Walter Lowrie and Matthew St. Clair Clarke, *American State Papers: Foreign Relations* (Washington, D.C., 1832–61), III, 159. Hereafter cited as *A.S.P.F.R.*

38. Madison to Monroe, March 31, 1807, Hunt, ed., *Writings of Madison*, VII, 405.

39. Madison to Monroe and Pinkney, May 20, 1807, Hunt, ed., *Writings of Madison*, VII, 444–45.

40. Jefferson to Bowdoin, April 2, 1807, Ford, ed., *Writings of Jefferson*, VIII, 40–41.

41. See the discussion in Spivak, *Jefferson's English Crisis*, 65; and Perkins, Prologue to War, 88–94.

42. Jefferson to Madison, April 21, 1807, Ford, ed., *Writings of Jefferson*, VIII, 47.

43. A. L. Burt, *The United States, Great Britain, and British North America*, argues that "the train of circumstances which led to the at-

tack on the *Chesapeake* would certainly have been broken" if Washington had accepted the Monroe–Pinkney Treaty, with its "self denying ordinance" imposed upon the British Navy. "A different spirit on each side would have called a halt somewhere between the escape of the men and the order for their recovery" (pp. 241–42). Burt was certainly right in thinking that a different spirit on each side would have prevented an incident of this character, though it should not obscure the point that the settlement over impressment annexed to that treaty was defective in several particulars. Though the British pledged to practice the greatest restraint in the matter of impressment, the large number of British sailors in American employ and the rising exigencies of the Royal Navy meant that future incidents were virtually inevitable. The settlement of the issue required something more than empty American pledges to return deserters and empty British promises to show circumspection in impressment.

44. Jefferson to William H. Cabell (Governor of Virginia), June 29, 1807, Ford, ed., *Writings of Jefferson*, IX, 87–88.

45. Leonard W. Levy, *Jefferson and Civil Liberties: The Darker Side* (Cambridge, Mass., 1963), 138–39. It may also be noted that the measures to enforce the embargo were more draconian than anything attempted by British authorities throughout the years leading up to the American Revolution.

46. See the discussion in Perkins, *Prologue to War,* 162, and Malone, *Jefferson and His Time,* V, 475–79. See also Gallatin to Chairman Thomas Newton, Dec. 5, 1807, Walter Lowrie and Matthew St. Claire Clarke, eds., *American State Papers: Commerce and Navigation* (Washington, D.C., 1832), I, 699, and the witty account in Herbert Heaton, "Non-Importation, 1806–1812," *Journal of Economic History,* I (1941), 178–98. Since nonimportation received an extraordinary amount of attention in the 1806 congressional session, the failure to complement the embargo on shipping and exports with total nonimportation was peculiar. Though the surviving records are not particularly illuminating on the question, it is likely that the reluctance to embrace total nonimportation owed most to the following consideration: Given Britain's dependence on American exports, total nonimportation would be unnecessary for coercive purposes, and it would be highly inconvenient to American finances. The near total dependence of the government on custom duties—exacerbated by the administration's earlier repeal of internal taxes—made it dangerous to block imports entirely.

47. This is one of the principal themes of Levy, *Jefferson and Civil Liberties,* 94–96. Spivak, *Jefferson's English Crisis,* notes the resemblance between Jefferson's initial draft of his Seventh Annual Message, discussed below, and the Declaration of Independence. "Both

described differences that had passed the point of quiet negotiation. Both justified the legitimate use of force by an aggrieved people. The Declaration had secured the moral right of revolution by distinguishing between transient, accidental grievances and long-practiced conspiracies to deprive the colonists of their liberty. The first situation did not justify war; the second did. Jefferson's 1807 remarks made the same distinction," blending the attack on the *Chesapeake* "into the history of prior and contemporary British oppressions" (pp. 86–87). Yet the annual message was vitiated by Gallatin's objections, and hence "Jefferson's revised address did not involve Congress in his true perceptions of the English crisis" (p. 98). Once the embargo was enacted, the President scarcely made any effort at all to communicate to the nation a justification for the measure. Instead he presented what Levy calls "an imperturbable, almost sphinxlike silence." *Jefferson and Civil Liberties*, 96.

48. Jefferson to Madison, Aug. 16, 1807, Ford, ed., *Writings of Jefferson*, VIII, 125.

49. See in particular, Adams, *History*, IV, 33–37; and Peterson, *Jefferson and the New Nation*, 878: "War was an easy temptress," Peterson writes; "still he remained faithful to the demanding mistress of peace."

50. Gallatin to Jefferson, Oct. 21, 1807, Adams, ed., *Writings of Gallatin*, I, 358; Jefferson to Thomas Mann Randolph, Oct. 26, 1807, Jefferson Papers, Library of Congress.

51. Gallatin to Jefferson, Oct. 21, 1807, Ford, ed., *Writings of Jefferson*, VIII, 145–48.

52. Jefferson to General John Mason (n.d.), Lipscomb and Bergh, eds., *Writings*, XI, 402.

53. Gallatin to Jefferson, Dec. 18, 1807, Adams, ed., *Writings of Gallatin*, I, 368.

54. *National Intelligencer*, December 23, 1807. The authorship of this and other editorials on the same subject is discussed in Brant, *Madison: Secretary of State*, 397–403.

55. Adams, *History*, III, 400. Spivak, in *Jefferson's English Crisis*, argues that in imposing the embargo Jefferson's motives were at the outset wholly precautionary, and that only in the course of 1808 would coercion "challenge, equal, and finally overwhelm the prudent, defensive, and time-bound goals that lay behind Jefferson's original recommendation" (p. 111). Spivak portrays a President shifting from one expedient to another, incapable of pursuing any fixed design, experiencing a kaleidoscope of conflicting impulses and divergent passions in the course of his final eighteen months in office. In contrast to Henry Adams, who portrays the embargo as the natural culmination of Jefferson's faith in peaceable coercion (and who therefore also dis-

counts the bellicose noise emitted by Jefferson in the latter half of 1807 as "flighty talk, of no substance or importance"), Spivak argues that for the President the embargo was not initially an experiment in peaceable coercion, that its meaning changed radically over time, that it was taken at a confused and midway point between Jefferson's passion for war and his passion for peace; and that only in the end did it come to represent for him "a naive and heroic pacifism."

The ingenuity of this account cannot be gainsaid. Yet its central proposition is not wholly convincing. In sharply distinguishing between the two traditions of nonimportation and embargo (pp. x–xi, 68–70, 106–7), Spivak slights the role that nonexportation had always played in the Republican system. For Jefferson, as for Madison, the power of American supply—of naval stores and foodstuffs to Great Britain, of lumber and provisions to the West Indies—had always been recognized as potent. Expressions of the absolute dependence of the West Indies on American foodstuffs, and predictions of imminent starvation if they were withdrawn, constituted a persistent theme of Madison's writings on the subject, and the consideration was often raised in debate in both America and England. (A valuable survey of Madison's thinking on this point is provided by J. C. A. Stagg, *Mr. Madison's War: Politics, Diplomacy and Warfare in the Early American Republic, 1783–1830* (Princeton, N.J., 1983), esp. pp. 20–22.) A ban on American shipping and exports thus by no means "caricatured the Republican tradition of economic retaliation." If ineffective, it is true, the measure would destroy "much of the foreign market for American agriculture," and it would support "England's quest for monopoly of the world's carrying trade." The central belief of the Republicans, however, had long been that the commercial retaliation would not be ineffective, that it would expand markets and break down England's monopolistic pretensions.

It may also be remarked here that the administration appeared to believe that British imports would be sharply reduced as well—so, at least, Madison argued in the *National Intelligencer*. The 1806 nonimportation act would stifle the demand for many British products; so, too, would the collateral effects of the embargo. This, at least, appears to have been the expectation. That the coercive character of the embargo appears not to have figured in Jefferson's correspondence until March is indeed a curious fact, but it does not show that he was oblivious to the coercive considerations at the outset. Had he been so disposed, he would not have resolved the dispute in cabinet over temporary versus permanent embargo in the fashion that he did. This decision was of crucial significance and made the embargo at the outset something quite different from the "prudent, defensive, and time-bound" measure Spivak calls it. The decision to accept "permanent embargo" overshadows both Jefferson's curious faith in time in January and February and his skepticism over nonimportation in October.

It would be misleading to portray Jefferson as a man with a clear idea of how things would turn out, or as possessing, in late December 1807, a decided preference for peace or war. As in the Louisiana crisis, his great hope was that he might play for time: "Time prepares us for defense. Time may produce peace in Europe. That removes the ground of differences with England till another war and that may find our revenues liberated by the discharge of our national debt, our wealth and numbers increased, our friendship and our enmity more important to every nation." Since there was no prospect at all that Britain and France would soon come to terms and since his own favored course of action—reduced revenue caused by the embargo, increased spending because of military preparations—would increase the debt, it is difficult to understand why he thought that something would turn up to release him from his predicament. In all probability, there were many moments when he was simply confused and did not know what to think. The general contours of Spivak's account, indeed, sometimes makes this construction seem more plausible than any other.

56. Turreau to Champagny, June 28, 1808, cited in Adams, *History*, IV, 309.

57. Jefferson to Madison, March 11, 1808, Ford, ed., *Writings of Jefferson*, IX, 181–85.

58. Madison to Armstrong, May 2, 1808, Hunt, ed., *Writings of Madison*, VIII, 30.

59. Jefferson to John Langdon, Aug. 2, 1808, Ford, ed., *Writings of Jefferson*, IX, 201.

60. Spivak, *Jefferson's English Crisis*, 203–10.

61. Gallatin to Jefferson, Nov. 2, 1808, Adams, ed., *Writings of Gallatin*, I, 423.

62. Jefferson to Benjamin Stoddard, Feb. 18, 1809, Lipscomb and Bergh, eds., *Writings*, XII, 249.

63. Report on Manufactures, Hamilton's Final Version, Harold C. Syrett, ed., *The Papers of Alexander Hamilton* (New York, 1961–79), X, 262–64.

64. See Erskine To Madison, Feb. 23, 1808, *A.S.P.F.R.*, III, 209–10.

65. The requirements imposed by the 1807 Orders in Council, Captain Mahan wrote, were "literally, and in no metaphorical sense, the reimposition of colonial regulation, to increase the revenues of Great Britain by reconstituting her the *entrepot* of commerce between America and Europe." He then cited John Quincy Adams, who held that "the Orders in Council, if submitted to, would have degraded us to the condition of colonists." Mahan, *Sea Power*, I, 178.

66. Quoted in Adams, *History*, III, 417. Perceval was then in opposition but would come to power in March 1807 as Chancellor of the Exchequer in a Tory government headed by the Duke of Portland, whose Foreign Minister was George Canning.

67. Burt, *United States, Britain, and British North America*, 221.

68. The Stand No. 4, April 12, 1798, Syrett, ed., *Hamilton Papers*, XXI, 413.

69. *Annals of Congress*, 9th Cong., 1st sess., March 5, 1806, 559.

70. "The Successes of Bonaparte," March 1806, W. B. Allen, ed., *The Works of Fisher Ames* (Indianapolis, Ind., 1983 [1854]), I, 494–95.

71. Jefferson to Monroe, Jan. 8, 1804, Ford, ed., *Writings of Jefferson*, VIII, 292.

72. Jefferson to Benjamin Rush, Oct. 4, 1803, Lipscomb and Bergh, eds., *Writings*, X, 422.

73. Jefferson to Thomas Lomax, Jan. 11, 1806, quoted in Malone, *Jefferson and His Time*, V, 95. These themes continued to find frequent expression in Jefferson's correspondence after his retirement from the presidency. Toward the end of the Napoleonic wars, he readily acknowledged that "it cannot be to our interest that all Europe should be reduced to a single monarchy," and he hoped that "a salutary balance may be ever maintained among nations"; even at that late date, however, he hoped that Bonaparte would "be able to effect the complete exclusion of England from the whole continent of Europe." Jefferson to Thomas Leiper, Jan. 1, 1814, and June 12, 1815, Lipscomb and Bergh, eds., *Writings*, XIV, 44, 307. His hope and expectation that France would form a useful counterpoise to British ambition was joined to the conviction that French power was scarcely likely to grow so large and threatening as to pose a danger to the United States. A letter written a year after his retirement gave full vent to these views. "The fear that Bonaparte will come over to us and conquer us" he considered "too chimerical to be genuine." Too many obstacles—Spain and Portugal, England and Russia—loomed to block Napoleon's supposed design. And even were he to succeed, his attention would not at first turn to America:

> Ancient Greece and Macedonia, the cradle of Alexander, his prototype, and Constantinople, the sea of empire for the world, would glitter more in his eye than our bleak mountains and rugged forests. Egypt, too, and the golden apples of Mauritania, have for more than half a century fixed the longing eyes of France; and with Syria, you know, he has an old affront to wipe out. Then come "Pontus and Galatia, Cappodocia, Asia and Bithynia," the fine countries on the Euphrates and Tigris, the Oxus and Indus, and all beyond the Hyphasis, which bounded the glories of his Macedonian rival; with the invitations of his new

British subjects on the banks of the Ganges, whom, after receiving under his protection the mother country, he cannot refuse to
visit. When all this is done and settled, and nothing of the old
world remains unsubdued, he may turn to the new one. But will
he attack us first, from whom he will get but hard knocks and
no money? Or will he first lay hold of the gold and silver of
Mexico and Peru, and the diamonds of Brazil? A *republican* Emperor, from his affection to republics, independent of motives of
expediency, must grant to ours the Cyclop's boon of being the
last devoured. While all this is doing, we are to suppose the
chapter of accidents read out, and that nothing can happen to
cut short or to disturb his enterprises. (Jefferson to Langdon,
March 5, 1810, ibid., XII, 374–75)

Whether Jefferson was correct in dismissing any danger from Napoleon even after a conquest of Britain cannot be known. He appears,
in any case, not to have been troubled by the idea that the United
States might place its weight on the side of Bonaparte and aid him in
his gigantic enterprises. He never accepted, and often ridiculed, the
idea that Britain was "fighting for the liberties of mankind." Jefferson
to Duane, March 22, 1806, ibid., XI, 95. The nature of England's
government unfit that country, he thought, "for the observation of
moral duties," and nothing would prevent it from making a separate
peace with Napoleon and exposing the United States to his wrath
were this country to join forces with England. On what might the
United States rely to stop England from leaving the country in the
lurch? "Her good faith! The faith of a nation of merchants! The *Pun-
ica fides* of modern Carthage! Of the friend and protectress of Copen-
hagen! Of the nation who never admitted a chapter of morality into
her political code! And is now boldly avowing, that whatever power
can make hers, is hers of right." Jefferson to Langdon, March 5, 1810,
ibid., XII, 375–76. France and Britain, he thought, were morally
equivalent; both had gone beyond the pale. One was "a conquerer
roaming over the earth with havoc and destruction," the other "a
pirate spreading misery and ruin over the face of the ocean." Jefferson
to Walter Jones, March 5, 1810, ibid., XII, 372. Under these circumstances, the imperative of policy was to assert neutral rights against
England and otherwise trust to the chapter of accidents to preserve a
balance of power.

74. Jefferson to Colonel John Taylor, Aug. 1, 1807, Lipscomb
and Bergh, eds., *Writings*, XI, 305.

75. Jefferson to Thomas Leiper, Aug. 21, 1807, Ford, ed., *Writings of Jefferson*, IX, 130.

76. Gordon A. Craig, "The United States and the European Balance," in William P. Bundy, ed., *Two Hundred Years of American*

Foreign Policy (New York, 1977), writes that Jefferson's embargo may have been influenced by "the suspicion that the French Empire was potentially a greater threat to American liberties than Great Britain and that, despite their present differences, Britain and the United States were tacit partners against the possibility of a domination of the Western world by Napoleon" (p. 72). The reverse, however, is true: the Anglo-American quarrel over neutral rights was, in the end, of far greater moment for Jefferson than the possibility of Napoleonic domination.

77. Though we treat the issue historically here, the threat held out by Napoleon may also be understood in the context of a longstanding dispute in the science of international politics—a dispute focused on whether an inherent tendency toward equilibrium exists as an iron law of international politics, or, put differently, whether the tendency to "balance" is of such strength in human affairs that it inevitably outweighs the temptation to "bandwagon." The view taken here is that both questions must be answered in the negative. Then as now, the balance of power, as a human institution, requires human understanding and support; ironically, it may be most endangered precisely at times when men persuade themselves that the tendency toward equilibrium is an iron law, for in this way they free themselves from the onus of maintaining it.

78. Adams, *History,* IV, 112.

79. Armstrong to Madison, Sept. 14, 1805. Despatches from France, National Archives.

80. See, for example, Fisher Ames, "Dangerous Power of France IV," March 1808, Allen, ed., *Works of Fisher Ames,* I, 362–73.

81. On the relative weighting of the balance of power and neutral rights, see the discussion in Lang, *Foreign Policy in the Early Republic,* and W. Allison Phillips and Arthur H. Reede, *Neutrality: Its History, Economics and Law, II: The Napoleonic Period* (New York, 1936).

82. Adams, *History,* IV, 288.

83. Gallatin to Jefferson, July 29, 1808, Adams, *Writings of Gallatin,* I, 396–99; Jefferson to Gallatin, Aug. 11, 1808, Ford, ed., *Writings of Jefferson,* VIII, 202. See the discussion in Spivak, *Jefferson's English Crisis,* and Levy, *Jefferson and Civil Liberties.* In his generally excellent study, *Presidents Above Party: The First American Presidency, 1789–1829* (Chapel Hill, N.C., 1984), Ralph Ketcham insists that Jefferson's objectives in foreign policy were always undertaken "within republican guidelines," pointing to both the Louisiana Purchase and the embargo as evidence of the Republicans' careful observance of constitutional restraints. He contrasts not only the differing reactions of Hamilton and Jefferson toward the 1803 crisis at New

Orleans, but also their likely responses to the constitutional problems brought on by the embargo. In 1808, he writes, "when Gallatin reported it would take 'a little army' to uphold the Embargo on the Canadian border and 'arbitrary powers . . . equally dangerous and odious' to prevent smuggling along the seacoast, Jefferson and Madison decided to repeal the Embargo rather than be either oppressive or impotent to enforce the laws. Again, one can readily imagine Hamilton's likely different reaction in the face of such a dilemma" (p. 107). It may be observed here that Hamilton, "had he been in power between 1801 and 1809," was most unlikely to have faced such a dilemma, since he had warned repeatedly that the Republicans' strategy of economic coercion would fail and since he would most certainly have taken a different approach in the diplomatic preliminaries that led to the embargo. (Nor, for that matter, did Hamilton's belief that war was the alternative to a failure to settle amicably differences with foreign powers show that he was unmindful of the fact that it lay in the constitutional jurisdiction of Congress to declare it.) It may also be observed, as noted in the text, that Jefferson's response to Gallatin's warning was far different from that described by Ketcham. Rather than drawing back from a course of "oppression," he pushed ahead, valuing the embargo above all else. When the embargo finally was repealed, in the spring of 1809, the action was taken over the known objections of the President, who continued to believe even after leaving office that the measure would have been successful had it been given a fair trial and not been undercut by Republican malcontents. The embargo, in truth, was both oppressive against American citizens and impotent against foreign powers, yet Jefferson clung to it ever more obstinately even as the evidence on both points continued to mount.

84. On this point, see the valuable work of Lawrence Delbert Cress, *Citizens in Arms: The Army and the Militia in American Society to the War of 1812* (Chapel Hill, N.C., 1982). See also Harold Sprout and Margaret Sprout, *The Rise of American Naval Power, 1776–1918* (Princeton, N.J., 1944).

85. *Annals of Congress,* 9th Cong., 1st sess., March 5, 1806, 557 (Randolph).

86. See Robert W. Tucker and David C. Hendrickson, *The Fall of the First British Empire: Origins of the War of American Independence* (Baltimore, 1982), and David Potter, *The Impending Crisis: 1848–1861,* ed. Don E. Fehrenbacher (New York, 1976).

87. It may be argued that the conquest of Canada would have represented a potent equivalent that would have been successful in forcing the British to retreat from their maritime pretensions—and it was in this light, for the most part, that the Republicans viewed it. If

the constitutional objections were taken seriously, of course, it could form no equivalent at all, since it is not likely that the inhabitants of Canada would have voluntarily submitted to incorporation in the American union. That this objection did not sway the Jeffersonians indicates that, once acquired, Canada might have proven difficult to disgorge, and especially so if one presumes that the "War of 1812" had begun five years earlier and the conquest of Canada had been achieved at that time. Madison, for one, anticipated that Canada would one day be incorporated but had earlier believed that the "fruit" was not "ripe." It surely would have become so in the event of a successful conquest.

Part V. The Jeffersonian Legacy

1. Max Farrand, ed., *The Records of the Federal Convention of 1787* (New Haven, Conn., 1937), I, 466–67.

2. Alexander Hamilton et al., *The Federalist Papers,* ed. Jacob E. Cooke (Middletown, Conn., 1961 [1787–88]), No. 8, 45.

3. Jefferson to John B. Colvin, Sept. 20, 1810, Paul Leicester Ford, ed., *The Writing of Thomas Jefferson* (New York, 1892–99), IX, 279.

4. On Jefferson's indifference to the Dos de Mayo uprising, Henry Adams wrote: "While all Europe, except France, joined hands in active or passive support of Spanish freedom, America, the stronghold of free government, drew back and threw her weight on the opposite side." *History of the United States of America during the Administrations of Jefferson and Madison* (New York, 1903 [1889–91]), IV, 301. Jefferson's aspirations, in fact, reached beyond the Floridas to Cuba.

5. June 12, 1783, Worthington C. Ford et al., eds., *Journals of the Continental Congress* (Washington, D.C., 1904–37), XXIV, 394.

6. Hamilton et al., *The Federalist,* No. 11, 73.

7. "Farewell Address," Sept. 19, 1796, John C. Fitzpatrick, ed., *The Writings of George Washington, 1745–1799* (Washington, D.C., 1931–44), XXXV, 233–35.

8. The background of the Farewell Address may be found in Felix Gilbert's classic essay *To the Farewell Address: Ideas of Early American Foreign Policy* (Princeton, N.J., 1961).

9. Jefferson to Elbridge Gerry, Jan. 26, 1799, Ford, ed., *Writings of Jefferson,* VII, 328.

10. Third Annual Message to Congress, Oct. 17, 1803, Ford, ed., *Writings of Jefferson,* VII, 266.

11. Jefferson to Monroe, Oct. 24, 1823, Ford, ed., *Writings of Jefferson,* X, 277.

12. "Political Observations" (1795), *Letter and Other Writings of James Madison* (New York, 1884 [Philadelphia, 1865]), IV, 492.

13. Jefferson to Thomas Lomax, March 12, 1799, Ford, ed., *Writings of Jefferson,* VII, 373.

14. In later years Jefferson did pour out his execration of Robespierre. He made no objection, however, to the bloody measures of the Girondists.

15. Jefferson to Benjamin Rush, Oct. 4, 1803, Ford, ed., *Writings of Jefferson,* VII, 264.

16. Jefferson to William Short, Oct. 3, 1801. Ford, ed., *Writings of Jefferson,* VII, 95.

17. Gilbert Chinard, *Thomas Jefferson: The Apostle of Americanism,* 2nd ed. (Ann Arbor, Mich., 1957 [1st ed., 1929]).

18. Julian P. Boyd, "Thomas Jefferson's 'Empire of Liberty,' " *Virginia Quarterly Review,* 24 (1948).

19. Reginald C. Stuart, *The Half-way Pacifist: Thomas Jefferson's View of War* (Toronto, 1978), 58.

20. See for example Jefferson's letter to La Fayette, October 28, 1822, Ford, ed., *Writings of Jefferson,* X, 227. "I will hazard . . . but the single expression of assurance that this general insurrection of the world against its tyrants will ultimately prevail by pointing the object of government to the happiness of the people and not merely to that of their self-constituted governors."

21. Jefferson to Thomas Cooper, Nov. 29, 1802, Ford., ed., *Writings of Jefferson,* VIII, 177.

22. Jefferson to Monroe, June 11, 1823, Ford, ed., *Writings of Jefferson,* X, 256.

23. Jefferson to Baron Alexander von Humboldt, April 14, 1811, Andrew A. Lipscomb and Albert Bergh, eds., *The Writings of Thomas Jefferson* (Washington, D.C., 1904–5), XIII, 34.

24. Jefferson to Baron von Humboldt, December 6, 1813, Ford, ed., *Writings of Jefferson,* X, 430. "History, I believe, furnishes no example of a priest-ridden people maintaining a free civil government."

25. Jefferson to Adams, May 17, 1818, Ford, ed., *Writings of Jefferson,* X, 107.

26. Jefferson to Joseph Priestley, June 19, 1802, Ford, ed., *Writings of Jefferson,* IX, 158.

27. See, for example, Carl N. Degler, "The American Past: An Unsuspected Obstacle in Foreign Affairs," *American Scholar* (Spring 1963), 32, 192.

28. Jefferson to John Dickinson, March 6, 1801, Ford, ed. *Writings of Jefferson,* VIII, 7–8.

29. Jefferson to Adams, April 8, 1816. Lipscomb and Bergh, eds., *Writings,* XIV, 467.

30. Jefferson to David Bailey Warden, Dec. 26, 1820, Ford, ed., *Writings of Jefferson*, X, 171.

31. Jefferson to La Fayette, Dec. 26, 1820, Ford, ed., *Writings of Jefferson*, X, 179.

32. Jefferson to Adams, May 17, 1818, Ford, ed., *Writings of Jefferson*, X, 107.

33. "As the first step towards a cure, the government itself must be regenerated. Its will must be made subordinate to, or rather the same with, the will of the community." James Madison, "Universal Peace," February 2, 1792, Gaillard Hunt, ed., *The Writings of James Madison* (New York, 1900–1910), VI, 89.

34. Jefferson to Thomas Leiper, June 12, 1815, Paul Leicester Ford, ed., *The Works of Thomas Jefferson* (New York, 1904–5), XI, 477–78.

Bibliography

Primary Sources

Contemporary Books and Pamphlets

Barbé-Marbois, François. *The History of Louisiana*. Edited by E. Wilson Lyon. Baton Rouge and London, 1977 [1830].

Gentz, Friedrich. *Fragments on the Balance of Power*. London, 1806.

————. *On the State of Europe Before and After the French Revolution.* . . . London, 1802.

Gibbon, Edward. *The Decline and Fall of the Roman Empire*. Edited by J. B. Bury. 7 vols. London, 1930 [1776–88].

Hamilton, Alexander, John Jay, and James Madison, *The Federalist Papers*. Edited by Jacob Cooke. Middletown, Conn., 1961 [1787–88].

Hume, David. *Essays: Moral, Political, and Literary*. Edited by Eugene F. Miller. Indianapolis, Ind., 1985 [1777].

Jefferson, Thomas. *Notes on the State of Virginia*. Edited by William Peden. New York, 1954 [1782].

Magruder, Allen Bowie. *Political, Commercial and Moral Reflections on the Late Cession of Louisiana to the United States*. Lexington, Ky., 1803.

Montesquieu, Charles Louis de. *The Spirit of the Laws*. Edited by Franz Neumann. New York, 1949 [1748].

Ramsay, David. *An Oration, on the Cessation of Louisiana, to the United States*. Charleston, S.C., 1804.

Rousseau, Jean Jacques. "Abstract of the Abbe de Saint-Pierre's Project for Perpetual Peace," *The Theory of International Relations: Se-*

lected Texts from Gentili to Treitschke. Introduced and edited by
M. G. Forsyth, H. M. A. Keens-Soper, and P. Savigear. New York,
1970 [1761].

Smith, Adam. *An Inquiry into the Nature and Causes of the Wealth of
Nations.* New York, 1937 [1776].

Stephen, James. *War in Disguise; or, The Frauds of the Neutral Flags,*
2nd edition. London, 1805.

Vattel, Emmerich de. *The Law of Nations or the Principles of Natural
Law applied to the Conduct and to the Affairs of Nations and of
Sovereigns.* Edited by Charles G. Fenwick. 3 vols. Washington, D.C.
1916 [1758].

[Webster, Daniel]. *Considerations on the Embargo Laws.* Boston, 1808.

Correspondence, Papers, and Memoirs of Individuals

*The Adams–Jefferson Letters: The Complete Correspondence Between
Thomas Jefferson and Abigail and John Adams.* Edited by Lester
J. Cappon. 2 vols. Chapel Hill, N.C. 1959.

The Works of John Adams, Second President of the United States. . . .
Edited by Charles Francis Adams. 10 vols. Boston, 1854.

Memoirs of John Quincy Adams. Edited by Charles Francis Adams. 12
vols. Philadelphia, 1874–77.

The Writings of John Quincy Adams. Edited by Worthington Chauncey
Ford. 7 vols. New York, 1913–17.

The Works of Fisher Ames. Edited by W. B. Allen. 2 vols. Indianapolis,
Ind., 1983 [1854].

The Works of the Right Honourable Edmund Burke. 16 vols. London,
1826.

Selected Writings of Albert Gallatin. Edited by E. James Ferguson. New
York, 1967.

The Writings of Albert Gallatin. Edited by Henry Adams. 3 vols. New
York, 1960 [Philadelphia, 1879].

The Papers of Alexander Hamilton. Edited by Harold C. Syrett. 26 vols.
New York, 1961–79.

The Complete ANAS of Thomas Jefferson. Edited by Franklin B. Sawvel.
New York, 1903.

*Correspondence Between Thomas Jefferson and Pierre Samuel du Pont
de Nemours, 1798–1817.* Edited by Dumas Malone. Boston and
New York, 1930.

The Correspondence of Jefferson and Du Pont de Nemours. Edited by
Gilbert Chinard. Washington, D.C., 1939.

The Papers of Thomas Jefferson. Edited by Julian P. Boyd. 22 vols. to
date. Princeton, N.J., 1950–   .

Thomas Jefferson: Writings. Edited by Merrill D. Peterson. New York, 1984.

The Works of Thomas Jefferson. Edited by Paul Leicester Ford. 12 vols. New York, 1904–5.

The Writings of Thomas Jefferson. Edited by Paul Leicester Ford. 10 vols. New York, 1892–99.

The Writings of Thomas Jefferson. Edited by Andrew A. Lipscomb and Albert Bergh. 20 vols. Washington, D.C., 1904–5.

The Life and Correspondence of Rufus King, Comprising his Letters, Private and Official, His Public Documents and His Speeches. Edited by Charles R. King. 6 vols. New York, 1894–1900.

The Original Letters of Robert R. Livingston, 1801–1803. Edited by Edward Alexander Parsons. New Orleans, 1953.

Letters and Other Writings of James Madison. 4 vols. New York, 1884 [Philadelphia, 1865].

The Mind of the Founder: Sources of the Political Thought of James Madison. Edited with introduction and commentary by Marvin Meyers. Rev. ed. Hanover and London, 1981.

The Papers of James Madison. Edited by William T. Hutchinson, William M. E. Rachel, Charles F. Hobson, and Robert A. Rutland. 15 vols. to date. Chicago and Charlottesville, Va., 1962– .

The Papers of James Madison: Presidential Series. Edited by Robert A. Rutland, Thomas A. Mason, Robert J. Brugger, Susannah H. Jones, Jeanne K. Sissan, and Fredrika J. Teute. 1 vol. to date. Charlottesville, Va., 1984– .

The Writings of James Madison. Edited by Gaillard Hunt. 9 vols. New York, 1900–1910.

The Writings of James Monroe. Edited by Stanislaus Murray Hamilton. 7 vols. New York, 1898–1903.

Correspondance de Napoleon Ier. Edited by Albert Du Casse. 32 vols. Paris, 1858–70.

The Writings of Thomas Paine. Edited by Moncure Daniel Conway. 4 vols. New York, 1967.

William Plumer's Memorandum of Proceedings in the United States Senate, 1803–1807. Edited by Everett Somerville Brown. New York, 1923.

The Writings of George Washington, 1745–1799. Edited by John C. Fitzpatrick. 39 vols. Washington, D.C., 1931–44.

Official Documents and Other Sources

Affaires Etrangères, Correspondence Politique, Etats-Unis. Transcripts from French Archives, Library of Congress.

American State Papers: Foreign Relations. Edited by Walter Lowrie and Matthew St. Clair Clarke. 6 vols. Washington, D.C., 1832–61.

Annals of Congress. The Debates and Proceedings in the Congress of the United States; with an Appendix, Containing Important State Papers and Public Documents, 1789–1824. 42 vols. Washington, D.C., 1834–56.

Cobbett's Annual Register. London, 1803, 1804.

A Compilation of the Messages and Papers of the Presidents, 1789–1897. Edited by James D. Richardson. 10 vols. Washington, D.C., 1896–99.

The Constitutions and Other Select Documents Illustrative of the History of France, 1789–1907. Edited by F. M. Anderson. New York, 1908 [1904].

Consular Dispatches. Department of State Archives in National Archives.

The Controversy Over Neutral Rights Between the United States and France, 1797–1800: A Collection of American State Papers and Judicial Decisions. Edited by James Brown Scott. New York, 1917.

"Correspondence of the French Ministers to the United States, 1791–1797." Edited by Frederick J. Turner. *Annual Report of the American Historical Association for the Year 1903.* 2 vols. Washington, D.C., 1904.

The Debates in the Several State Conventions on the Adoption of the Federal Constitution. Edited by Jonathan Elliot. 4 vols. Washington, D.C., 1836.

"Despatches from the United States Consulate in New Orleans, 1801–1803, II." *American Historical Review,* 33 (January 1928).

"Despatches from U.S. Ministers to Spain, 1792–1906." National Archives, No. 31, Roll 9.

A Digest of International Law. Edited by John Bassett Moore. 8 vols. Washington, D.C., 1906.

Diplomatic Correspondence of the United States: Canadian Relations, 1784–1860. 4 vols. Edited by William R. Manning. Washington, D.C., 1940.

Documents Relating to New England Federalism, 1800–1815. Edited by Henry Adams. Boston, 1877.

Documents Relating to the Purchase and Exploration of Louisiana. American Philosophical Society. Boston and New York, 1904.

French State Papers. Henry Adams Transcripts in Library of Congress. (unpublished)

"Hamilton on the Louisiana Purchase: A Newly Identified Editorial from the *New-York Evening Post.*" Edited by Douglas Adair. *William and Mary Quarterly,* 3rd Series, 12 (1955).

"Instructions to the British Ministers to the United States, 1791–1812."

Edited by Bernard Mayo. *Annual Report of the American Historical Association, 1936.* Vol. 3. Washington, D.C., 1941.

Journals of the Continental Congress. Edited by Worthington C. Ford, Gaillard Hunt, John C. Fitzpatrick, and Roscoe R. Hill. 34 vols. Washington, D.C., 1904–37.

Louisiana Under the Rule of Spain, France, and the United States, 1785–1807. Edited by James Alexander Robertson. 2 vols. Cleveland, 1911.

Military Book, 1800–1803. National Archives.

National Intelligencer. Library of Congress. (microfilm)

Naval Documents Related to the United States Wars with the Barbary Powers. Edited by Captain Dudley W. Knox. 6 vols. Washington, D.C., 1939–44.

The Records of the Federal Convention of 1787. Edited by Max Farrand. 4 vols. New Haven, Conn., 1937.

State Papers and Correspondence Bearing Upon the Purchase of the Territory of Louisiana. Washington, D.C., 1903.

The Territorial Papers of the United States. Edited by Clarence Edwin Carter. 26 vols. Washington, D.C., 1934–60.

Treaties, Conventions, International Acts, Protocols and Agreements Between the United States of America and Other Powers, 1776–1909. Edited by William M. Malloy, 2 vols. Washington, D.C., 1910.

Secondary Sources

Adams, Henry. *History of the United States of America during the Administrations of Jefferson and Madison.* 9 vols. New York, 1903 [1889–91].

———. *The Life of Albert Gallatin.* New York, 1943 [1879].

Adams, Mary P. "Jefferson's Reaction to the Treaty of San Ildefonso," *Journal of Southern History,* 21 (May 1955).

Ammon, Harry. *The Genet Mission.* New York, 1973.

Appleby, Joyce. *Capitalism and Á New Social Order: The Republican Vision of the 1790s.* New York, 1984.

———. "What Is Still American in the Political Philosophy of Thomas Jefferson?," *William and Mary Quarterly,* 3rd ser., 39 (April 1982).

Ashworth, John. "The Jeffersonians: Classical Republicans or Liberal Capitalists?," *Journal of American Studies,* 18 (1984).

Banning, Lance. *The Jeffersonian Persuasion: Evolution of a Party Ideology.* Ithaca, N.Y., 1978.

Barnett, Correlli. *Bonaparte.* New York, 1978.

Bemis, Samuel Flagg. *A Diplomatic History of the United States.* New York, 1936.

342 BIBLIOGRAPHY

———. *Jay's Treaty: A Study in Commerce and Diplomacy.* New York, 1923.

———. *John Quincy Adams and the Foundations of American Foreign Policy.* New York, 1949.

———. *Pinckney's Treaty: A Study of America's Advantage from Europe's Distress, 1783–1800.* New York, 1926.

———, ed. *The American Secretaries of State and Their Diplomacy, 1776–1925,* 10 vols. New York, 1958 [1928].

Bowman, Albert H. "Jefferson, Hamilton, and American Foreign Policy," *Political Science Quarterly,* 71, No. 1 (March 1956).

———. *The Struggle for Neutrality: Franco-American Diplomacy During the Federalist Era.* Knoxville, Tenn., 1974.

Boyd, Julian P. *Number 7, Alexander Hamilton's Secret Attempts to Control American Foreign Policy.* Princeton, N.J., 1964.

———. "Thomas Jefferson's 'Empire of Liberty,'" *Virginia Quarterly Review,* 24 (1948).

Brant, Irving. *James Madison: Secretary of State, 1800–1809.* Indianapolis, Ind., and New York, 1953.

Brown, Everett S. *The Constitutional History of the United States, 1803–1812.* Berkeley, Calif., 1920.

Bruun, Geoffrey. *Europe and the French Imperium, 1799–1814.* New York, 1938.

Buel, Richard, Jr. *Securing the Revolution: Ideology in American Politics, 1789–1815.* Ithaca, N.Y., and London, 1972.

Bundy, William P., ed. *Two Hundred Years of American Foreign Policy.* New York, 1977.

Burt, A. L. *The United States, Great Britain and British North America: From the Revolution to the Establishment of Peace After the War of 1812.* New Haven, Conn., 1940.

Butterfield, Herbert. *Napoleon.* London, 1939.

Carr, E. H. *The Twenty Years' Crisis, 1919–1939,* 2nd ed. New York, 1946.

Charles, Joseph. *The Origins of the American Party System.* New York, 1956.

Chinard, Gilbert. *Thomas Jefferson: The Apostle of Americanism,* 2nd ed. Ann Arbor, Mich., 1957 [1st ed., 1929].

Christie, Ian R. *Wars and Revolutions: Britain, 1760–1815.* Cambridge, Mass., 1982.

Combs, Jerald A. *The Jay Treaty: Political Battleground of the Founding Fathers.* Berkeley, Calif., 1970.

Coquelle, P. *Napoleon and England, 1803–1813: A Study from Unprinted Documents.* Translated by Gordon D. Knox. London, 1904.

Cox, Isaac Joslin. *The West Florida Controversy, 1798–1813: A Study in American Diplomacy.* Baltimore, 1918.

Crackel, Theodore J. *Mr. Jefferson's Army: Political and Social Reform of the Military Establishment, 1801–1809*. New York, 1987.

Crawley, C. W. *The New Cambridge Modern History, IX: War and Peace in an Age of Upheaval, 1793–1830*. Cambridge, 1965.

Cress, Lawrence Delbert. *Citizens in Arms: The Army and the Militia in American Society to the War of 1812*. Chapel Hill, N.C., 1982.

Cunnnigham, Noble E., Jr. *In Pursuit of Reason: The Life of Thomas Jefferson*. Baton Rouge and London, 1987.

————. *The Process of Government Under Jefferson*. Princeton, N.J., 1978.

Dangerfield, George. *Chancellor Robert R. Livingston of New York, 1746–1813*. New York, 1960.

Darling, A. B. *Our Rising Empire: 1763–1803*. New Haven, Conn., 1940.

Davis, John W. "The Permanent Bases of American Foreign Policy," *Foreign Affairs*, 10, No. 1 (October 1931).

DeConde, Alexander. *The Quasi-War: The Politics and Diplomacy of The Undeclared War with France, 1797–1801*. New York, 1966.

————. *This Affair of Louisiana*. New York, 1976.

————. ed. *Encyclopedia of American Foreign Policy*. 3 vols. New York, 1978.

Degler, Carl N. "The American Past: An Unsuspected Obstacle in Foreign Affairs," *American Scholar*, 32 (Spring 1963).

Egan Clifford L. *Neither Peace Nor War: Franco-American Relations, 1803–1812*. Baton Rouge and London, 1983.

————. "The United States, France, and West Florida, 1803–1807," *Florida Historical Quarterly*, 47 (January 1969).

Ernst, Robert. *Rufus King: American Federalist*. Chapel Hill, N.C., 1968.

Fortescue, John. *History of the British Army*. 13 vols. London, 1910–30.

Geggus, David P. *Slaves, War and Revolution: The British Occupation of Saint Domingue, 1793–1798*. Oxford, 1982.

Geyl, Pieter. *Napoleon: For and Against*. New Haven, Conn., 1949.

Gilbert, Felix. *History: Choice and Commitment*. Cambridge, 1977.

————. *To the Farewell Address: Ideas of Early American Foreign Policy*. Princeton, N.J., 1961.

Glover, Michael. *The Napoleonic Wars: An Illustrated History, 1792–1815*. New York, 1978.

Goebel, Julius. *The Recognition Policy of the United States*. New York, 1915.

Graebner, Norman A. *Ideas and Diplomacy: Readings in the Intellectual Tradition of American Foreign Policy*. New York, 1964.

————, ed. *Traditions and Values: American Diplomacy, 1790–1865*. Lanham, Md., 1985.

Harlow, Vincent T. *The Founding of the Second British Empire: Discovery and Revolution*. 2 vols. London, 1952–64.

Haskins, George Lee, and Herbert A. Johnson. *History of the Supreme*

Court of the United States, II: Foundations of Power: John Marshall, 1801–15. New York, 1981.

Heaton, Herbert. "Non-Importation, 1806–1812," *Journal of Economic History,* 1 (1941).

Heckscher, Eli F. *The Continental System: An Economic Interpretation.* Oxford, 1922.

Hirschmann, Albert O. *The Passions and the Interests: Political Arguments for Capitalism Before Its Triumph.* Princeton, N.J., 1977.

Horsman, Reginald. "American Indian Policy in the Old Northwest, 1783–1812," *William and Mary Quarterly,* 18 (January 1961).

———. *The Causes of the War of 1812.* Philadelphia, 1962.

Hunt, Michael H. *Ideology and U.S. Foreign Policy.* New Haven, Conn., 1987.

Huntington, Samuel P. *American Politics: The Promise of Disharmony.* Cambridge, Mass., 1980.

James, C. L. R. *The Black Jacobins: Toussaint L'Ouverture and the San Domingo Revolution.* New York, 1963.

Johnstone, Robert M., Jr. *Jefferson and the Presidency: Leadership in the Young Republic.* Ithaca, N.Y., 1978.

Kaplan, Lawrence. *Entangling Alliances with None: American Foreign Policy in the Age of Jefferson.* Kent, Ohio, 1987.

———. *Jefferson and France.* New Haven, Conn., 1967.

Ketcham, Ralph. *James Madison: A Biography.* New York, 1971.

———. *Presidents Above Party: The First American Presidency, 1789–1829.* Chapel Hill, N.C., 1984.

Knorr, Klaus. *The Power of Nations: The Political Economy of International Relations.* New York, 1975.

Kurtz, Stephen G. *The Presidency of John Adams: The Collapse of Federalism, 1795–1800.* Philadelphia, 1957.

Lang, Daniel G. *Foreign Policy in the Early Republic: The Law of Nations and the Balance of Power.* Baton Rouge, 1985.

Langford, Paul. *Modern British Foreign Policy: The Eighteenth Century, 1688–1815.* New York, 1976.

Lecky, W. E. H. *The French Revolution: Chapters from the Author's History of England During the Eighteenth Century.* New York, 1904.

Lefebvre, Georges. *The French Revolution.* 2 vols. New York, 1962.

———. *Napoleon.* 2 vols. New York, 1969.

Lerner, Ralph. *The Thinking Revolutionary: Principle and Practice in the New Republic.* Ithaca, N.Y., 1987.

Levy, Leonard. *Jefferson and Civil Liberties: The Darker Side.* Cambridge, Mass., 1963.

Lint, Gregg L. "The American Revolution and the Law of Nations, 1776–1789," *Diplomatic History,* 1 (Winter 1977).

Loewenheim, Francis L. *The Historian and the Diplomat: The Role of*

History and Historians in American Foreign Policy. New York, 1967.

Logan, Rayford W. *The Diplomatic Relations of the United States with Haiti, 1776–1891.* Chapel Hill, N.C., 1941.

Lokke, Carl L. *France and the Colonial Question: A Study of Contemporary French Opinion, 1763–1801.* New York, 1932.

———. "Jefferson and the Leclerc Expedition," *American Historical Review,* 33 (October 1927).

———. "The Leclerc Instructions," *Journal of Negro History,* 10 (January 1925).

———. "Secret Negotiations to Maintain the Peace of Amiens," *American Historical Review,* 49 (October 1943).

Lycan, Gilbert L. *Alexander Hamilton and American Foreign Policy: A Design for Greatness.* Norman, Okla., 1970.

Lyon, E. Wilson. *Louisiana in French Diplomacy, 1759–1804.* Norman, Okla., 1934.

Mahan, Alfred Thayer. *Sea Power in Its Relations to the War of 1812.* 2 vols. Boston, 1905.

Malone, Dumas. *Jefferson and His Time.* 6 vols. Boston, 1948–81.

Mannix, Richard. "Gallatin, Jefferson, and the Embargo of 1808," *Diplomatic History,* 3, No. 2 (Spring 1979).

Marcus, G. J. *A Naval History of England, II: The Age of Nelson.* Sheffield, 1971.

Marks, Frederick W., III. *Independence on Trial: Foreign Affairs and the Making of the Constitution.* Baton Rouge, 1973.

McCoy, Drew R. *The Elusive Republic: Political Economy in Jeffersonian America.* New York, 1980.

McDonald Forrest. *Alexander Hamilton.* New York, 1979.

———. *The Presidency of Thomas Jefferson.* Lawrence, Kans., 1976.

Meinecke, Friedrich. *Machiavellism: The Doctrine of Raison D'Etat and Its Place in Modern History.* Translated by Douglas Scott. New Haven, Conn., 1957.

Merk, Frederick. *Manifest Destiny and Mission in American History: A Reinterpretation.* New York, 1963.

Miller, John C. *Alexander Hamilton: Portrait in Paradox.* New York, 1959.

———. *The Federalist Era, 1789–1801.* New York, 1960.

Miller, John Chester. *The Wolf by the Ears: Thomas Jefferson and Slavery.* New York, 1977.

Morgenthau, Hans J. *In Defense of the National Interest: A Critical Examination of American Foreign Policy.* New York, 1951.

———. *The Purpose of American Politics.* New York, 1960.

Ott, Thomas O. *The Haitain Revolution, 1764–1804.* Knoxville, Tenn., 1973.

Perkins, Bradford. *The First Rapprochement: England and the United States, 1795–1805*. Berkeley and Los Angeles, 1967.
———. *Prologue to War, 1805–1812: England and the United States.* Berkeley, Calif., 1961.
Peterson, Merrill D. *Adams and Jefferson: A Revolutionary Dialogue.* Athens, Ga., 1976.
———. *Thomas Jefferson and the New Nation: A Biography.* Oxford, 1970.
———, ed. *Thomas Jefferson: A Profile.* New York, 1967.
Phillips, W. Allison, and Arthur H. Reede. *Neutrality: Its History, Economics and Law, II: The Napoleonic Period.* New York, 1936.
Potter, David. *The Impending Crisis: 1848–1861.* Completed and edited by Don E. Fehrenbacher. New York, 1976.
Risjord, Norman K. *The Old Republicans: Southern Conservatism in the Age of Jefferson.* New York and London, 1965.
Ritcheson, Charles R. *Aftermath of Revolution: British Policy toward the United States, 1783–1795.* New York, 1971.
Roloff, Gustav. *Die Kolonialpolitik Napoleons I.* Munich, 1899.
Rose, J. H. *The Revolutionary and Napoleonic Era, 1789–1815.* Cambridge, 1895.
Rosen, Stephen Peter. "Alexander Hamilton and the Domestic Uses of International Law," *Diplomatic History,* 5, No. 3 (summer 1981).
Rutland, Robert A. *James Madison: The Founding Father.* New York, 1987.
Savelle, Max. "Colonial Origins of American Diplomatic Principles," *Pacific Historical Review,* 3, No. 3 (September 1934).
Sears, Louis Martin. *Jefferson and the Embargo.* New York, 1966 [1927].
Setser, Vernon G. *The Commercial Reciprocity Policy of the United States, 1774–1829.* New York, 1969. [1937].
Sheehan, Bernard W. *Seeds of Extinction: Jeffersonian Philanthropy and the American Indian.* Chapel Hill, N.C., 1973.
Skeen, C. Edward. *John Armstrong, Jr., 1758–1843: A Biography.* Syracuse, N.Y., 1981.
Sofaer, Abraham D. *War, Foreign Affairs and Constitutional Power: The Origins.* Cambridge, Mass., 1976.
Sorel, Albert. *L'Europe et la Révolution Français,* 7th ed. 8 vols. Paris, 1908.
Spivak, Bernard. *Jefferson's English Crisis: Commerce, Embargo, and the Republican Revolution.* Charlottesville, Va., 1979.
Sprout, Harold, and Margaret Sprout. *The Rise of American Naval Power, 1776–1918.* Princeton, N.J., 1944.
Stagg, J. C. A. *Mr. Madison's War: Politics, Diplomacy and Warfare in the Early American Republic, 1783–1830.* Princeton, N.J., 1983.
Steele, Anthony. "Impressment in the Monroe–Pinkney Negotiation, 1806–1807," *American Historical Review,* 57, No. 2 (January 1952).

Stourzh, Gerald. *Alexander Hamilton and the Idea of Republican Government*. Stanford, Calif., 1970.
———. *Benjamin Franklin and American Foreign Policy*. Chicago, 1954.
Stuart, Reginald C. *The Half-way Pacifist: Thomas Jefferson's View of War*. Toronto, 1978.
———. *United States Expansionism and British North America, 1775–1871*. Chapel Hill, N.C., and London, 1988.
———. *War and American Thought from the Revolution to the Monroe Doctrine*. Kent, Ohio, 1982.
Symonds, Craig L. *Navalists and Antinavalists: The Naval Policy Debate in the United States, 1785–1827*. Newark, N.J., 1980.
Tansill, Charles Callan. *The United States and Santo Domingo, 1798–1873*. Baltimore, 1938.
Treudley, Mary W. "The United States and Santo Domingo, 1789–1866," *Journal of Race Development*, 7 (July 1916).
Tucker, Robert W., and David C. Hendrickson, *The Fall of the First British Empire: Origins of the War of American Independence*. Baltimore, 1982.
Turner, Frederick J. "The Diplomatic Contest for the Mississippi Valley," *Atlantic Monthly*, 93 (June 1904).
———. *The Frontier in American History*. New York, 1920.
Varg, Paul. *Foreign Policies of the Founding Fathers*. East Lansing, Mich., 1963.
Vincent, R. J. *Nonintervention and International Order*. Princeton, N.J., 1974.
Watts, Steven. *The Republic Reborn: War and the Making of Liberal America, 1790–1820*. Baltimore and London, 1987.
Weinberg, Albert K. *Manifest Destiny: A Study of Nationalist Expansionism in American History*. Baltimore, 1935.
Weymouth, Lally, ed. *Thomas Jefferson: The Man . . . His World . . . His Influence*. New York, 1973.
Whitaker, Arthur P. *The Mississippi Question 1795–1803: A Study in Trade, Politics, and Diplomacy*. Gloucester, Mass., 1962 [New York, 1934].
———. "New Light on the Treaty of San Lorenzo: An Essay in Historical Criticism," *Mississippi Valley Historical Review*, 15 (March 1929).
———. *The Spanish–American Frontier*. Boston, 1927.
———. *The United States and the Independence of Latin America, 1800–1830*. New York, 1964 [1941].
Willis, Edmund P., ed. *Fame and the Founding Fathers*. Bethlehem, Pa., 1967.
Wright, J. Leitch, Jr. *Britain and the American Frontier, 1783–1815*. Athens, Ga., 1975.
Zimmerman, James Fulton. *Impressment of American Seamen*. New York, 1925.

Index